PEM FUEL CELLS

PEM FUEL CELLS

THERMAL AND WATER MANAGEMENT FUNDAMENTALS

YUN WANG, KEN S. CHEN, AND SUNG CHAN CHO

MP MOMENTUM PRESS

MOMENTUM PRESS, LLC, NEW YORK

First published by Momentum Press®, LLC
222 East 46th Street, New York, NY 10017
www.momentumpress.net

ISBN-13: 978-1-60650-245-7 (paperback)
ISBN-13: 978-1-60650-247-1 (e-book)

DOI: 10.5643/9781606502471

Cover design by Jonathan Pennell
Cover image: Basic equipment of scholarly created classical and modern methods of contemporary results of employing those methods
Interior design by Exeter Premedia Services Private Ltd.
Chennai, India

10 9 8 7 6 5 4 3 2 1

Printed in the United States of America

CONTENTS

PREFACE

Polymer electrolyte membrane (PEM) fuel cells, which convert the chemical energy stored in hydrogen fuel directly and efficiently to electrical energy with water as the only by-product, have the potential to reduce our energy usage, pollutant emissions, and dependency on fossil fuels. Tremendous efforts have been made so far, particularly during the last couple of decades or so, on advancing the PEM fuel cell technology and fundamental research. In addition to the large number of research and review paper publications, several classic books have been published and are available in the market, which are primarily for introductory level readers. There are, however, very few books that address the graduate-level or advanced aspects of PEM fuel cells and are based on the first principles or conservation laws, dimensionless analysis, time constant evaluation, and numerical simulation by solving partial differential equations. There are abundant knowledge regarding flow, heat transfer, and mass transport in general engineering, which has been successfully extended to the water and thermal management of PEM fuel cells. This book contributes to this aspect of PEM fuel cell technology; that is, it focuses on the fundamental understanding of phenomena or processes involved in PEM fuel cells.

More specifically, this book volume aims at delineating the fundamentals of water and thermal management for PEM fuel cells. In other words, the objective of this book is to provide an introduction to the essential knowledge required to perform effective water and thermal management and strategy optimization for PEM fuel cells; the book also presents and discusses the current status of the fundamental research in the field. In addition to providing an introduction in Chapter 1, the book outlines two basic subjects related to PEM fuel cells in Chapter 2, namely thermodynamics and electrochemical reaction kinetics. Chapter 3 of the book presents the fundamentals of heat and mass transfer pertaining to PEM fuel cells. Chapters 4-7 describe and discuss water transport and management within a PEM fuel cell. In these four chapters or sections, water transport within the polymer electrolyte membrane is first discussed and followed by vapor−phase water removal and management, liquid water removal from the electrodes and the effects of two-phase flow, and solid water dynamics and removal, particularly the specialized case of starting a fuel cell at sub-freezing temperatures or cold start and the various processes or phenomena related to ice formation. Chapter 8 focuses on thermal or heat transport and management. Lastly, Chapter 9 discusses the coupled thermal and water management or phase change effects.

YW (primary author) would like to acknowledge his indebtedness to a few special people; without their respective guidance, encouragement, and assistance, this book could never have been written. First, Professor Chao-Yang Wang led YW into this field of research and guided him toward his PhD degree and his career as a faculty at the UC Irvine. A portion of the contents in this book are based on YW's PhD thesis, his research papers co-authored with Professor Chao-Yang Wang, and his project experience at the Electrochemical Engine Center (ECEC) at the Pennsylvania State university. Professor Chao-Yang Wang has been a constant source of help and inspiration to YW. Second, Professor Derek Dunn-Rankin invited YW to contribute a book volume for a series he serves as an editor. Though YW previously had ideas of writing a book on PEM fuel cells, the tremendous unforeseen challenges and long-term commitment stopped him many times, till Professor Dunn-Rankin's invitation. Lastly, YW would like to thank Ms. Millicent Treloar and Mr. Joel Stein for their constant understanding and granting for extensions. Ms. Treloar read the manuscript and made many helpful suggestions. KSC (second author) would like to acknowledge Sandia National Laboratories and the US Department of Energy's Fuel Cell Technologies Program for funding PEM fuel cell projects for which he served as PI – it is these projects that enabled KSC to develop his expertise on PEM fuel cells. KSC would also like to note that the collaboration between YW and him on PEM fuel cell research for the last few years laid the ground work for writing this book. Lastly, we would like to acknowledge S. C. Cho (third author) for his assistance and contribution in writing this book.

Yun Wang
Ken S. Chen

Key Words/Terms

PEM fuel cells, energy, fundamental, water management, thermal management, two-phase flow, polymer electrolyte membrane, ice formation, subfreezing operation, heat transfer, phase change, voltage loss, liquid water removal, coupled thermal and water management, numerical simulation, CFD, multiphase mixture (M^2) formulation, analysis

LIST OF FIGURES

LIST OF TABLES

Nomenclature

A   electrode area, m^2; the Tafel slope; phase A

a   water activity; effective catalyst area per unit volume, m^2/m^3

a_o catalyst surface area per unit volume, m^2/m^3

aCL   anode catalyst layer

B   a constant depending on operation state and reactants; phase B

b   the concentration of a quantity such as momentum, heat, and species concentration

C   the double-layer capacity, $\mu F/cm^2$; molar concentration of a species, mol/m^3

C_D the drag coefficient

CR   compression ratio

C^k molar concentration of species k, mol/m^3

$C_{i,0}$ reference molar concentration of species i, mol/m^3

$C_n H_m O_p$ a generic carbonaceous fuel

C_α^k molar concentration of species k in the α phase, mol/m^3

C_{Ox} surface concentrations of the oxidized form of the reactant

C_{Re} surface concentrations of the reduced form of the reactant

cCL   cathode catalyst layer

c_{H^+} the molar concentration of protons

c_p specific heat, J/kg K

D   species diffusivity, m^2/s

$D^e{}_{12}$ the effective mutual diffusion coefficient of hydronium ion (H_3O^+)

Da the Damkohler number

$D_{H^+}^{eff}$ the effective diffusivity of proton, m^2/s

$D_{H^+}^m$ diffusivity for proton interaction with the membrane matrix

$D_{H^+}^w$ diffusivity for proton interaction with water

$D_{H^+}^{0,w}$ the proton diffusivity at infinite dilution in water

D_i^K the Knudsen diffusion coefficient of species i

D_i^M the molecular diffusion coefficient of species i

D_c^l capillary "diffusion" coefficient

d	molecular diameter, m; the average pore size, m
$d*$	the finite distance at which the inter-particle potential is zero
E	potential
E_a	activation energy
E_{eq}	equilibrium potential
ECSA	electrochemical surface area
E'	thermodynamic potential assuming all the enthalpy change of reactions is converted to electric energy
EW	equivalent weight of dry membrane, kg/mol
F	Faraday's constant, 96,487 C/equivalent
$\vec{F}$	the flux of b
f_v	the volume fraction (f_v) of the aqueous phase (i.e., water) in the polymer
f_{v0}	a percolation threshold of the volume fraction (f_v) of the aqueous phase (i.e., water) in the polymer
f_x	heat, species, or momentum flux in the x direction
G	Gibbs free energy
G_v	volumetric formation or generation rate
G_s	surface formation or generation rate
$\vec{G}_k$	Diffusive flux of species k
$\vec{G}_{w,perm}$	Hydraulic permeation flux
g	Gibbs free energy per mole of species, gravity
h	the Planck's constant; heat transfer coefficient; enthalpy
h_m	mass transfer coefficient
h_{sl}	the latent heat during the phase change between liquid water and ice
h_{sg}	the latent heat during the phase change between water vapor and ice
$\hbar$	dimensionless nonuniformity parameter of the oxygen reduction reaction within the cathode catalyst layer
I	current density, A/cm²; the intensity of the transmitted neutrons
I_o	the intensity of the incident neutrons beam
IEC	the ion exchange capacity of the PEM (usually expressed in units of milli-equivalents of acid per gram of dry polymer)
i	superficial current density, A/cm²
i_e	Protonic current flux
i_L	limiting current density, A/cm²
$\vec{J}_i$	molar flux of species i relative to the mass-average velocity
j	transfer current density, A/cm³
j_f	transfer current density of forward process, A/cm³
j_b	transfer current density of backward process, A/cm³

j_0	transfer current density at the catalyst layer–membrane interface, A/cm^3
j_δ	transfer current density at the catalyst layer–GDL interface, A/cm^3
$\vec{j}_l$	mass flux of liquid phase, kg/m^{2}s
K	permeability, m^2
Kn	the Knudsen number defined as λ/l_D
k	thermal conductivity, W/m K
k_B	the Boltzmann's constant
k_r	relative permeability
k_{rl}	the relative permeability for the liquid phase
k_{rg}	the relative permeability for the gas phase
L	length, m
M	molecular weight, kg/mol
m_w	the mass of water absorbed
$m_{m,0}$	the mass of the dry membrane
mf_l^k	mass fraction of species k in the liquid phase
$\dot{m}$	mass flow flux, kg/s m^2
N	the atomic density of the material through which neutrons pass
N_{AV}	Avogadro's number, $6.02214129(27)\times10^{23}$ mol^{-1}
$\vec{N}_i$	molar flux of species i
n	the direction normal to the surface; number of electrons involved in the electrode reaction
$\vec{n}$	the direction normal to the surface
n_d	electro-osmotic coefficient, H$_2$O/H$^+$
n_k	exponent in the power expression of the relative permeability
P	pressure, Pa
q	the critical exponent
q_{irrev}	irreversible heat
q_{rev}	reversible heat
$\dot{q}$	heat generation rate
$\dot{q}_x$	heat flux component in the x direction
R	universal gas constant, 8.134 J/mol K
$R_{contact}$	contact resistance
R_{e-}	electronic resistance
R_Ω	ohmic resistance
R_{Vi}	volumetric reaction rate of species i
r	compression ratio; the distance between the particles, m
r_{pore}	the average pore radius, m
$r_{123}^{\sigma m}$	ratio of $\mathbf{S}_m^{eff}$ in the sublayer 1, 2, and 3

r_{123}^{aj0} ratio of $aj_{0,T}^{c,\text{ref}}$ in the sublayer 1, 2, and 3

S source term

Sh the Sherwood number

s stoichiometry coefficient in the electrochemical reaction; liquid saturation

s_e effective saturation

s_{ir} residual liquid saturation

t time, s

T temperature, K

T_h the thickness of the neutron beam attenuating medium

U_G the superficial gas-phase velocity, m/s

U_L the superficial liquid-phase velocity, m/s

U_o equilibrium potential, V; reference velocity, m/s

$\vec{u}$ velocity vector, m/s

u_i species mobility

u_{mol} the molecular velocity magnitude

V molar volume of the hydrated membrane

V_{cell} cell potential, V

V_m volume of the membrane

$\bar{V}_m$ molar volume of the dry membrane

$\bar{V}_w$ molar volumes of water

$\vec{v}_i$ the diffusive velocity of species i averaged over a differential volumetric element

W work, Joule

X the Lockhart–Martinelli parameter

x the oxygen-to-fuel molar ratio

x_g the quality of gas

x_i the gas-phase mole fraction of species i

x_l the qualities of liquid

$\bar{x}^*$ the position of transition from the single- to two-phase regions

Y_o the position of transition from the single- to two-phase regions

Z' the real part of the impedance response

z_i the charge number of species i

Greek

1ϕ single phase

2ϕ two phases

α transfer coefficient; net water flux per proton flux; ratio of molar volumes of water and the dry membrane

α_m mass accommodation coefficient which is usually around 0.01–0.07

α_T	thermal diffusivity
ρ	density, kg/m^3
μ_w	water attenuation coefficient
v	kinematic viscosity, m^2/s
θ_a	advancing contact angle, °
θ_c	contact angle, °
θ_r	receding contact angle, °
θ_s	static contact angle, °
ϕ_2	phase potential, V; the absolute humidity
ϕ_G^2	the two-phase flow multiplier
Φ	electrolyte potential, V
$\Phi^{(m)}$	electrolyte potential, V
$\Phi_I^{(m)}$	electrolyte potential in the sublayer I, V
$\Phi_{II}^{(m)}$	electrolyte potential in the sublayer II, V
$\Phi_\delta^{(m)}$	electrolyte potential at the catalyst layer–GDL interface, V
$\Phi^{(s)}$	electronic potential, V
ψ	the inter-molecular potential energy; the steam-to-carbon ratio
κ	proton conductivity, S/m
κ_{30}	proton conductivity at 30° C; Ω^{-1} cm^{-1} or S/cm
ξ	stoichiometric flow ratio
λ	membrane water content; the mean free path
λ_k	mobility of phase k
$l_{molecule}$	the mean free path
ε	porosity; the depth of the potential well; emissivity
ε_i	the volume fraction of solute i
ε_0	uncompressed porosity of a GDL
η	surface overpotential, V; efficiency
τ	shear stress, N/m^2; time constant, s
$\tau_{m,D}$	time constant of water diffusion in the membrane, s
$\tau_{m,H}$	time constant of membrane hydration, s
γ	the specific heat ratio
γ_c	correction factor for species convection
γ_T	correction factor for thermal convection
δ	thickness, m; the ratio of the effective diffusivity of H$_3$O$^+$ interaction with water to that with membrane
δ_{land}	the land width
σ	electronic conductivity, S/m; surface tension, N/m; the cross-section area of the neutron beam attenuating medium; the Stefan–Boltzmann constant (= 5.67×10^{-8} W/m^2 K^4)

Superscripts and Subscripts

A	phase A
a	anode
act	activation polarization
ad	adsorbed
B	phase B
BP	bipolar plate
c	cathode; capillary
CL	catalyst layer
D	diffusion
dl	double layer
e	electrolyte; effective
eff	effective value
fg	phase change
fr	friction
g	gas phase
GDL	gas diffusion layer
H	hydration
h	hydraulic
in	inlet
ir	irreducible or residual
k	species; liquid or gas phase
l	liquid
m	membrane phase; mass; mean
o	gas channel inlet value; reference value
r	radiation
ref	reference value
S	surface; source term
s	solid; surface
sat	saturate value
trans	transport polarization
w	water
v	volumetric
*	dimensionless

CHAPTER 1

INTRODUCTION

This chapter provides an overview of current challenges on energy and associated environmental issues as well as their relevance to nonrenewable energy resources. PEM (polymer electrolyte membrane) fuel cell is a promising technology that can effectively address these issues, offering advantages of high efficiency and zero emissions when using hydrogen gas as a fuel. The basic operation principle of PEM fuel cells is introduced in comparison with both conventional and emerging energy-conversion or power-generation methods, along with their present status and the importance of water and thermal managements on achieving high fuel cell performance.

1.1 ENERGY CHALLENGES

Due to the rapid growth of society and pace of industrialization, particularly in emerging economic powers, such as China and India, world energy consumption has been steadily rising in past decades. World energy consumption has been projected to increase by 49 percent or 1.4 percent per year, from 2007 to 2035; see Figure 1-1 [1]. That is an increase from 495 quadrillion BTU in 2007 to 739 quadrillion BTU in 2035. The carbon-based energy sources, including coals, crude oils, and natural gases, account for the major portion of world energy consumption. In 2008, the US transportation sector alone accounted for 70% of all petroleum consumption, 57% of which came from non-US sources [2]. Those sources are projected to account for an increasing proportion of the world electricity generation; see Figure 1-2.

Carbon-based energy sources are of significant concern for two reasons. First, they are rapidly depleting resources. Based on the published estimates [3], at the present rate of world energy consumption,

- petroleum will run out within the next 40–50 years;
- the world's natural gas reserves are forecasted to last around 60 more years;
- coal reserves are anticipated to be depleted within 200 years;
- the world's reserves of uranium are projected to last about 50 years.

The second significant concern about carbon-based energy sources is that they cause pollution and CO_2 buildup in the atmosphere. These are widely regarded as the two most

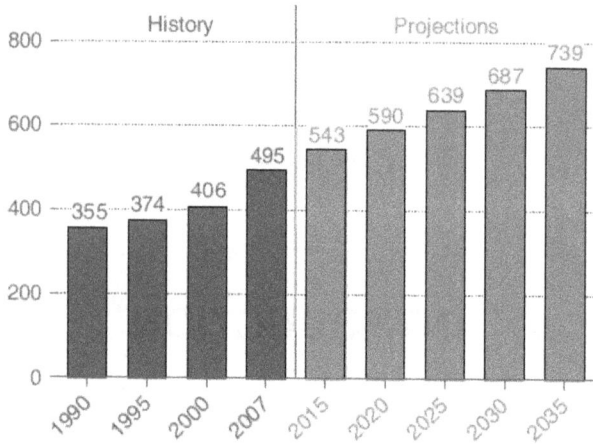

Figure 1-1. World marketed energy consumption, 1990–2035 (unit: quadrillion BTU) [1].

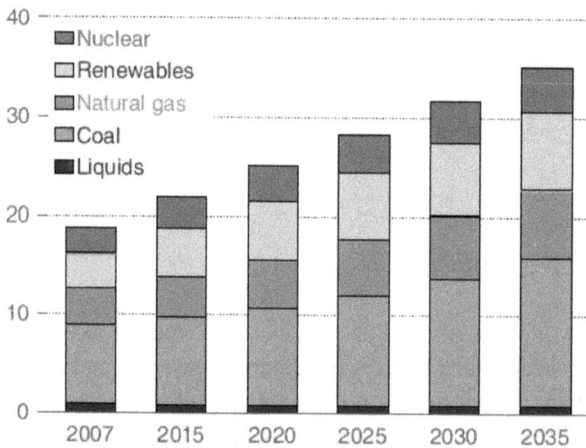

Figure 1-2. World net electricity generation by fuel, 2007–2035 (unit: trillion kilowatt-hours) [1].

important environmental hazards faced by the world today [1]. Pollution has long been recognized as causing serious health problems, and contributions of CO_2 to global warming and climate change are not disputed in serious science circles. Carbon dioxide emitted from combustion is the likely cause of global warming and climate change, endangering the Earth's ecosystem. Carbon dioxide in the atmosphere is now at its highest level in 650,000 years [1] and increasing by approximately 2 ppm per year. Figure 1-3 displays atmospheric CO_2 concentrations measured at Mauna Loa Observatory in past decades, indicating an increasing trend. To date, Mauna Loa constitutes the longest record of direct measurements of CO_2 in the atmosphere, initiated by C. David Keeling in 1958 at a facility of the National Oceanic and Atmospheric Administration [5]. Among the energy sector, in the United States the transportation sector contributes most to the nation's greenhouse gas emissions, accounting for ~ 30% of all CO_2 emissions [6].

Atmospheric CO_2 at Mauna Loa Observatory

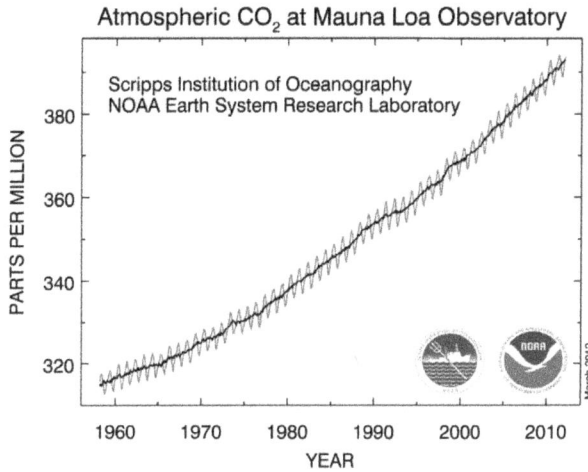

Figure 1-3. Atmospheric CO_2 concentrations measured at Mauna Loa Observatory. The black curve represents the seasonally corrected data [4].

1.2   FUEL CELLS AND THEIR ROLES IN ADDRESSING THE ENERGY CHALLENGES

Fuel cell technology has received much attention in recent years owing to their high efficiencies and low emissions. Fuel cells, classified according to their electrolyte, are electrochemical devices that directly convert chemical energy stored in fuels to electrical energy. Its practical efficiency can reach as high as 60% in direct electrical energy conversion.

Fuel cells are electrochemical devices that convert energy through electrochemical processes instead of through a heat-engine cycle. Their efficiency, therefore, is not limited by the Carnot cycle. The overall efficiency of fuel cells can be over 80% in co-generation of electrical and thermal energies. Depending on the types of fuel cells, the thermal energy can be high grade in SOFCs (solid oxide fuel cells) or MCFCs (molten carbonate fuel cells) or low grade in PEM fuel cells. Both grades of thermal energy can be reused for residential applications such as hot bath water.

When using hydrogen gas as a fuel, no carbon dioxide is produced during energy conversion with fuel cells. Hydrogen gas is currently the major candidate as the fuel for low-temperature fuel cells. Hydrogen is the most abundant element in the universe, accounting for about 75% mass of the universe. On Earth, hydrogen is the 10th most abundant element. Hydrogen gas is renewable and can be produced through a variety of methods, such as the steam reformation of methane, ammonia dissociation, water electrolysis, bacterial fermentation, thermal decomposition, and solar thermochemical splitting of water. Hydrogen has been produced and used for industrial purposes for over 100 years. Presently, methane reformation is the major technology used in industrial hydrogen production. Hydrogen gas is widely used in ammonia production, food and beverage industry, pharmaceutical industry, and space propulsion.

Fuel cells can be divided into the following five major categories:

1. Polymer electrolyte membrane (PEM) fuel cells or PEMFCs
2. Solid oxide fuel cells (SOFCs)

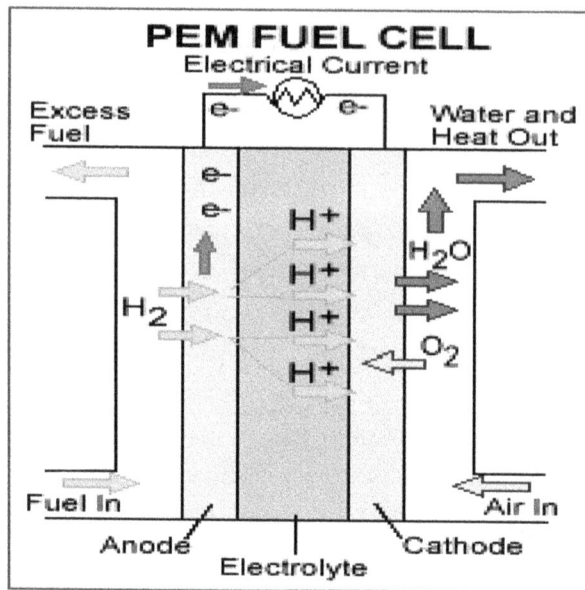

Figure 1-4. Schematic of PEM fuel cell operation [7].

3. Alkaline fuel cells (AFCs)
4. Phosphoric acid fuel cells (PAFCs)
5. Molten carbonate fuel cells (MCFCs)

PEM fuel cells, the focus of this book, are constructed by employing polymer electrolyte membranes (notably Nafion®) as a proton conductor and platinum (Pt) or Pt-based alloys as a catalyst. Figure 1-4 shows a schematic of PEM fuel cell operation. They work at relatively low operating temperatures (<120°C), produce high power density, and can be scaled-up easily. PEM fuel cells are primarily considered for

• transportation propulsion
• small-scale stationary power
• portable power

The power requirements for electric passenger cars, utility vehicles, and buses range from 20 to 250 kW. The stationary power generated by PEM fuel cells has a wide range, 1–50 MW. Some small-scale stationary generation, a remote telecommunication application for example, requires a power amount ranging from 100 W to 1 kW. The portable power is usually in the range of 5–50 W. Figure 1-5 displays the portion of PEM fuel cell units installed around the world in each category in 2008. Up to 2010, over 200 fuel cell vehicles, more than 20 fuel cell buses, and about 60 fueling stations are operating in the United States. Approximately 75,000 fuel cells for stationary power, auxiliary power, and specialty vehicles are in use worldwide, among them about 24,000 systems were manufactured in 2009, an approximate 40% increase over 2008 [9].

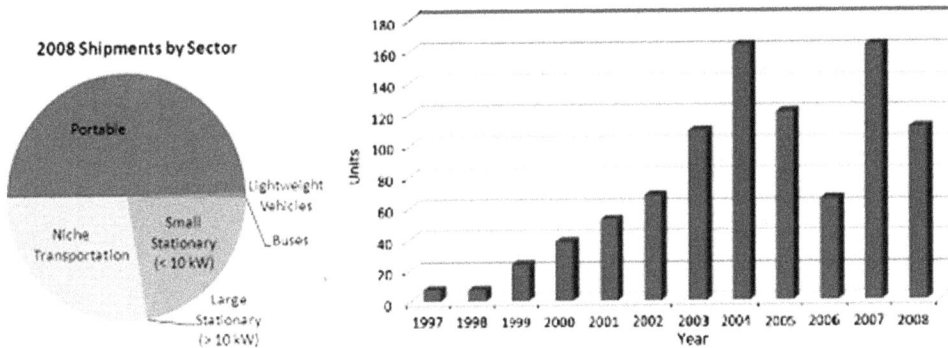

Figure 1-5. PEM fuel cell units installed for each application in 2008 (left) and number of fuel cell automobiles manufactured each year (right) [8].

1.3 PEM FUEL CELLS

The phenomena occurring during PEM fuel cell operation are complex. Specifically, they involve heat transfer, species and charge transport, multiphase flow, and electrochemical reactions. Fundamental knowledge of these multiphysics phenomena and their relevance to material properties are critically important to overcome the two major barriers, durability and cost. A PEM fuel cell consists of the following major components:

- Catalyst layers (CLs) and membrane, which are components that make up the MEA (membrane electrode assembly)
- Gas diffusion layers (GDLs) and microporous layers (MPLs)
- Bipolar plates (BPs), gas flow channels (GFCs), and/or cooling units

PEM fuel cells differ from

- Solid oxide fuel cells (SOFCs) or molten carbonate fuel cells (MCFCs) in that the electrolyte membrane used in PEM fuel cells is based on polymer membranes, usually Nafion-based series, which exhibits good ionic conductivity at low temperature.
- Batteries in that PEM fuel cells have gas flow channels (GFCs) for reactant supply and product removal; therefore, their energy capacity is determined by their fuel tank capacity.
- Internal combustion engines (ICE) in that PEM fuel cells convert energy through electrochemical reactions.

1.3.1 PEM FUEL CELL OPERATION

In PEM fuel cells, ambient air is the common oxidizer. Early experiments also considered using pure oxygen, which, however, is not practical for common applications. PEM fuel cells can directly utilize hydrogen gas, methanol, alcohol, and glucose as their fuels. Hydrogen gas is widely used for PEM fuel cells due to its relatively rapid reactivity (thus low voltage loss). During operation, hydrogen gas and air are fed, respectively, into the anode and cathode GFCs. While flowing down the gas channels, hydrogen and oxygen diffuse to the respective catalyst sites via the diffusion media, with the following electrochemical reactions occurring on the active reaction surface:

Hydrogen oxide reaction (HOR): $H_2 \rightarrow 2H^+ + 2e^-$ (anode)

Oxygen reduction reaction (ORR): $O_2 + 4e^- + 4H^+ \rightarrow 2H_2O$ (cathode)

Specifically, as schematically shown in Figure 1-6, the following multiphysics, highly coupled and nonlinear transport and electrochemical phenomena take place during PEM fuel cell operation:

1. Hydrogen gas and air are forced (by pumping) to flow down the anode and cathode gas flow channels (GFCs), respectively.
2. H_2 and O_2 flow through the respective porous GDLs/MPLs and diffuse into the respective catalyst layers (CLs).
3. H_2 is oxidized in the anode CL, forming protons and electrons.
4. Protons migrate and water is transported through the membrane.
5. Electrons are conducted via carbon support to the anode current collector, and then to the cathode current collector via an external circuit.
6. O_2 is reduced with protons and electrons in the cathode CL to form water.
7. Product water is transported out of the cathode CL, through cathode GDL/MPL, and eventually out of the cathode GFC.
8. Heat is generated due to inefficiencies, mainly in the cathode CL due to the sluggish ORR, and is conducted out of the cell via GDL/MPL and bipolar plates (BPs).

The transport phenomena are three-dimensional because the flows of H_2 fuel and O_2 oxidant in the respective anode and cathode GFCs are normal to proton transport through the membrane and gas transport through the respective GDLs/MPLs and CLs. Under relatively high inlet humidity, liquid water emerges inside the PEM fuel cell.

Differing from heat engines, PEM fuel cell operation is based on the electrochemical reactions which directly convert chemical energy to electricity; consequently, its efficiency is not limited by the Carnot cycle. For some heat engines, to achieve high efficiency the cycle needs to reach a high temperature, causing pollutant emissions such as NOx. Differing from ICEs, PEM fuel cells exhibit a higher efficiency and no moving parts are involved during the energy-conversion operation. PEM fuel cells are considered to replace internal combustion engines (ICEs) for automotive applications. Differing from batteries, fuel cell reactants are not stored inside fuel cells, but in external fuel tanks; therefore, charging/recharging is similar to that for ICEs. Furthermore, comparing with batteries which are essentially energy-storage devices, fuel cells are energy convertors.

Differing from solid oxide fuel cells (SOFCs), PEM fuel cells work at relatively low temperatures, usually less than 120°C; in contrast, SOFCs operate around 1000°C. Consequently, PEM fuel cells require catalysts to improve their electrochemical kinetics due to low temperature

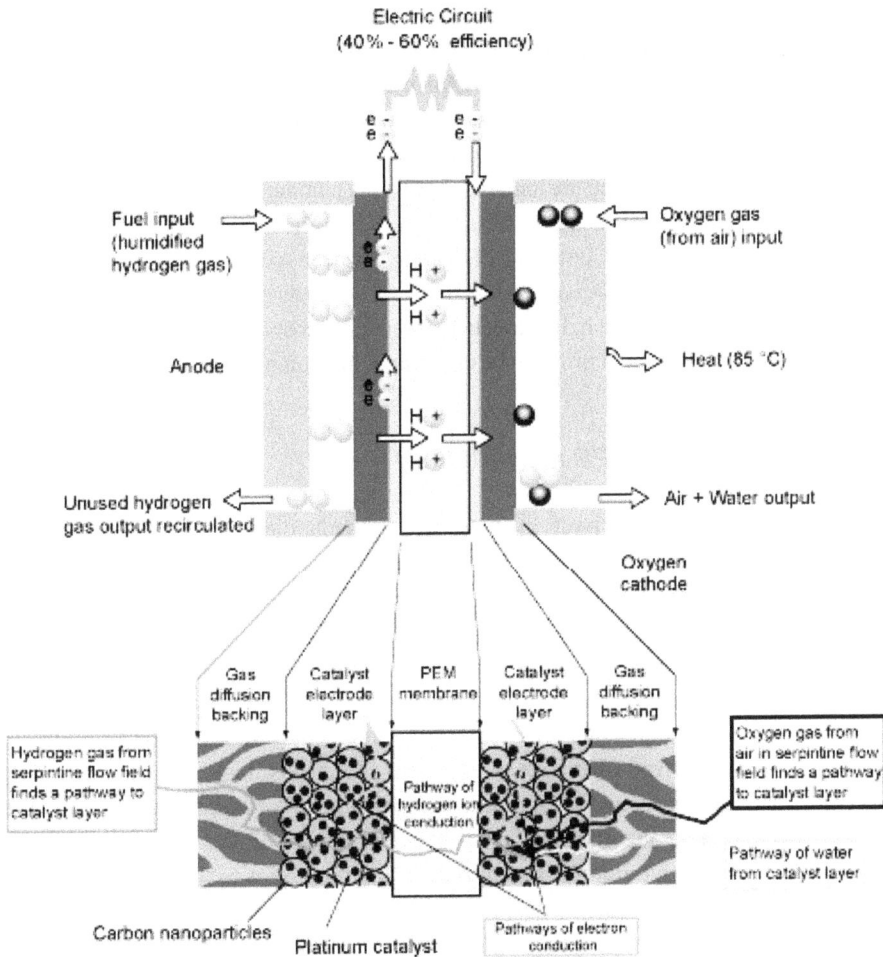

Figure 1-6. Phenomena occurring in a PEM fuel cell: two-dimensional sectional view [10].

operation. High temperature enables reforming hydrogen-carbon-based fuels; thus, SOFCs can directly work with a wider range of fuels without an auxiliary fuel processor.

1.3.2 CURRENT STATUS OF PEM FUEL CELLS

The very first fuel cell was invented by Sir William Robert Grove (an English lawyer turned scientist) in 1839, though no practical use was found for another century. General Electric (GE) Company began developing fuel cells in the 1950s and was awarded the contract for the Gemini space mission in 1962. The 1 kW Gemini fuel cell system had a platinum loading of 35 mg Pt/cm^2 and provided 37 m A/cm^2 at 0.78 V. In the 1960s, improvements were made by incorporating Teflon in the catalyst layer directly adjacent to the electrolyte, as was done with GE fuel cell at the time. Considerable improvements were made from the early 1970s onward with the adoption of the fully fluorinated Nafion® membrane. However, research and development in PEM fuel cells didn't receive much attention and funding from the federal government, in particular, the US Department of Energy (DOE), and industry until a couple of decades ago or so when breakthrough methods for reducing the amount of platinum required for PEM fuel cells were developed and subsequently improved by Los Alamos National Laboratory (LANL) and others. Notably, Raistrick (1986) of LANL came up with a catalyst-ink technique for fabricating the electrodes. This breakthrough method made it possible to increase the utilization of active catalyst and at the same time to reduce the amount of precious platinum metal needed. Consequently, PEM fuel cell performance has been significantly improved. The output voltage can reach around 0.7 V at a current density of 1.0 A/cm^2. The Pt loading has been reduced to around 0.3 mg Pt/cm^2. Though many technical and associated fundamental breakthroughs have been achieved during the last couple of decades, many challenges such as reducing cost and improving durability while maintaining performance remain prior to the commercialization of PEM fuel cells [8].

1.3.3 THERMAL AND WATER MANAGEMENT

Water management is a central issue in PEM fuel cell technology. Briefly, the goal of water management is to balance two opposite or conflicting effects. One effect is to hydrate the electrolyte. The ionic conductivity of the electrolyte depends on the membrane hydration level, and dryness increases membrane ionic resistance. The other effect is to avoid severe "flooding" (which refers to the presence of excessive liquid water) in fuel cells, which increases mass transport limitation. The Nafion-based membranes are still the best candidate as electrolyte for PEM fuel cells. In this type of membrane, the water molecules attach to the sulfonic groups, facilitating protons to transport between the sulfonic sites. Thus, high water content increases the proton conductivity of membranes, reducing the ohmic voltage loss. Two methods are frequently adopted to hydrate membranes. One is to externally humidify inlet reactant gases. This strategy requires external humidifiers and thus adds auxiliary units to the fuel cell system. The other method is to utilize the water produced by the PEM fuel cell to humidify reactant gases internally, which requires a sophisticated flowfield design to accomplish the optimal internal humidification. As for liquid water, its presence hinders reactant transport to the catalyst sites, increasing the concentration polarization. Since liquid water formation is determined by the water vapor saturation pressure (which is a strong function of temperature), the temperature field and its coupling with water condensation and/or evaporation are critical to the study of two-phase transport and the ensuing cathode "flooding" in a PEM fuel cell. Figure 1-7 shows the membrane temperature in a commercial fuel cell with an active area of 200 cm^2. The high variation of temperature, from inlet to outlet and from land to channel, is evident.

In addition to performance, water and thermal management plays a critical role in fuel cell durability:

- Liquid water can dissolve materials such as sealing materials leading to contamination of membranes and catalyst layers.
- Excess water floods electrodes, hindering reactant transport to reaction sites and consequently leading to reactant starvation and material degradation.
- Dry operation subjects the electrolyte membrane to cracking or pinhole formation.
- Hot spot formation causes localized dryout and consequently membrane pinhole formation.
- Thermal cycles degrade fuel cell materials.
- Freezing/thawing cycles reduce the total area of active or catalytic surfaces, and hence electrode performance and durability.

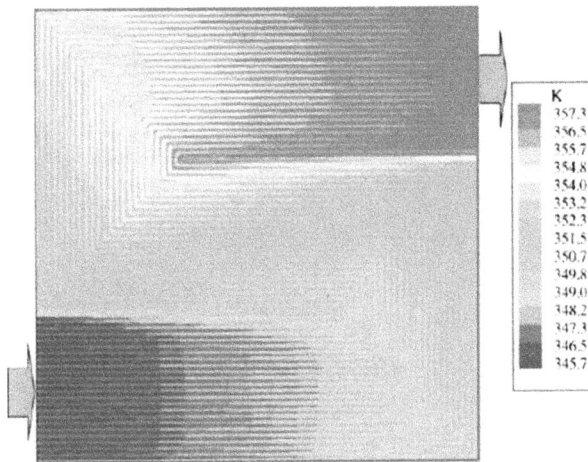

Figure 1-7. Temperature contours in the electrolyte membrane of a commercial PEM fuel cell [10].

REFERENCES

1. U.S. Department of Energy (DOE), International Energy Outlook 2010, U.S. Energy Information Administration (EIA), Office of Integrated Analysis and Forecasting, (2010). DOE/EIA-0484.

2. U.S. Department of Energy (DOE), http://www.eia.doe.gov, Energy Information Agency (EIA), (2012).

3. Press, R.J., Santhanam, K.S.V., Miri, M.J., Bailey, A.V. and Takacs, G.A., Introduction to Hydrogen Technology, Hoboken, NJ: John Wiley, (2009). doi: http://dx.doi.org/10.1111/j.2153-3490.1976.tb00701.x

4. Earth System Research Laboratory (ESRL), Trends in Atmospheric Carbon Dioxide, http://www.esrl.noaa.gov/gmd/ccgg/trends/, (2012).

5. Keeling, C.D., Bacastor, R.B., Bainbridge, A.E., Ekdahl, C.A., Guenther, R.R. and Waterman, L.S., Atmospheric Carbon Dioxide Variations at Mauna Loa Observatory, Hawaii, Tellus, 28 (1976) 538–551.

6. Environmental Protection Agency (EPA), Inventory of Greenhouse Gas Emissions and Sinks: 1990–2007, (2009). Report No: EPA 430-R-09-004.

7. U. S. Depart of Energy (DOE) – Energy Efficiency Renewable Energy (EERE), FCT Fuel Cells: Types of Fuel Cells, https://www1.eere.energy.gov/hydrogenandfuelcells/fuelcells/fc_types.html, (2012).

8. Wang, Y., Chen, K.S., Mishler, J., Cho, S.C. and Adroher, X.C., A review of polymer electrolyte membrane fuel cells: technology, applications, and needs on fundamental research, *Applied Energy*, 88 (2011) 981–1007. doi: http://dx.doi.org/10.1016/j.apenergy.2010.09.030

9. Papageorgopoulos, D., DOE Fuel Cell Technology Program Overview and Introduction to the 2010 Fuel Cell Pre-Solicitation Workshop, DOE Fuel Cell Pre-Solicitation Workshop, (2010).

10. National Institute of Standards and Technology (NIST), PEM Fuel Cells, http://www.physics.nist.gov/MajResFac/NIF/pemFuelCells.html, (2012).

11. Wang, Y. and Wang, C.Y., Ultra large-scale simulation of polymer electrolyte fuel cells, *J. Power Source*, 153(1) (2006) 130–5. doi: http://dx.doi.org/10.1016/j.jpowsour.2005.03.207

CHAPTER 2

BASICS OF PEM FUEL CELLS

This chapter provides an overview of the basic principles underlying the operation of PEM fuel cells, namely thermodynamics and electrochemistry. The first section presents explanations of the major thermodynamics concepts relevant to PEM fuel cells, such as internal energy, enthalpy, entropy, and Gibbs free energy, as well as the fundamental laws of thermodynamics. The second section introduces the electrochemical theory pertinent to PEM fuel cells, such as the rates and routes of hydrogen oxidation reaction (HOR) and oxygen reduction reaction (ORR). The third section discusses the voltage loss (which is the major concern on PEM fuel cell efficiency) and describes the major voltage loss mechanisms and a simple model of cell output voltage. In addition, two appendices are provided for this chapter: Appendix II.A lists the thermodynamic properties of air, hydrogen gas, and water vapor, and Appendix II.B gives an example of calculating the enthalpy, entropy, and Gibbs free energy for a substance and the PEM fuel cell reaction.

2.1 THERMODYNAMICS

Thermodynamics is a branch of physical science, which deals with energy and its conversion among heat (thermal energy), work (or mechanical energy), electricity (or electrical energy), and other forms of energy. Since PEM fuel cells are energy converters, thermodynamics is useful in analyzing the fuel-cell theoretical conversion efficiency or the upper bound of electrical energy that can be withdrawn. In this section, we first introduce important thermodynamic concepts that are closely related to PEM fuel cell operation as well as water and thermal management. Secondly, we will discuss the fundamentals related to the reversible thermodynamic potential and its determining parameters.

In addition to the fundamental laws of thermodynamics, several basic concepts are essential to the thermodynamics analysis of PEM fuel cells: internal energy, enthalpy, entropy, Gibbs free energy, chemical potential, humidity, and phases.

2.1.1 INTERNAL ENERGY AND THE FIRST LAW OF THERMODYNAMICS

Internal energy is the total intrinsic energy contained in a system. It is the energy necessary to create the system, but excludes that of making room for the system by displacing its surroundings and establishing the volume and pressure. Fuel cells convert the energy stored in a fuel such as hydrogen to electric power. Therefore, internal energy is essential to understanding how much energy is available in a system. This energy can be explained through molecule movement and interaction: the internal energy consists of two important parts: one is the kinetic energy of constituent particles due to molecule movement and the other is the potential energy due to the interparticle or intermolecular force such as repulsion and attraction. The change in internal energy can be expressed as follows:

$$dU = TdS - pdV. \tag{Eq. 2-1}$$

In a simplest system consisted of ideal gases, the particles are on average far away from each other. Under standard conditions, the intermolecular distance in the air is about 10 times of the molecular dimension. As the intermolecular force decays rapidly with distance, the interaction becomes weak when the constituent particles or molecules are far apart from each other, and the portion of potential energy is negligible when calculating internal energy change. In that case, the internal energy can be evaluated through adding the kinetic energies of all the constituent particles. Under the typical operating conditions of PEM fuel cells, both hydrogen gas and air are ideal gases. Consequently, the internal energy is only a function of temperature. The internal energies of hydrogen fuel, oxygen oxidant, water vapor product, and air in a wide range of temperatures are listed in Appendix II.A.

The first law of thermodynamics describes how energy is conserved: the internal energy of a system can be changed through heat transfer or work exchange with surroundings, expressed mathematically:

$$dU = dQ - W. \tag{Eq. 2-2}$$

Equation 2-2 holds for a closed system, where only heat and work are exchanged with the surroundings, that is, no matter crosses the system boundaries. For an open system, the internal energy is changed by what is brought in or out by the exchanged matter. An operating PEM fuel cell is an open system with the hydrogen fuel and oxygen reactant being fed in and water product being moved out. The internal energy changes due to the fuel and oxidant being fed-in, the outlet mass flows, the heat transfer, and electrical work exchanged with the system's surroundings.

Under constant volume, W becomes zero. Thus, the internal energy can be expressed as

$$dU = C_v(T)dT, \tag{Eq. 2-3}$$

where C_v is the specific heat at constant volume, defined as the energy required to raise the temperature of a unit mass by one degree as the volume is maintained constant. The internal energy change is then obtained by integrating Eq. 2-3 (from State 1 to State 2) to reach Eq. 2-4:

$$\Delta U = \int_1^2 C_v(T)dT \approx C_{v,avg}(T_2 - T_1). \tag{Eq. 2-4}$$

Inside a PEM fuel cell, however, the gas reactants undergo a fairly isobaric process; therefore, the specific heat at constant pressure C_p is frequently used. For incompressible substances such as liquid water and solid components, the two specific heats are equal to each other. For compressible gases, C_p is closely related to enthalpy change, to be discussed in the next subsection. The typical values of C_v for air, hydrogen, and water vapor at room temperature are listed in Appendix II.A.

2.1.2 ENTHALPY CHANGE

Enthalpy $(H = U + pV)$ is the sum of the internal energy and the amount of energy required to make room for the system by displacing its surroundings and establishing its volume and pressure, expressed in the differential form as

$$dH = dU + pdV + Vdp. \qquad \text{(Eq. 2-5)}$$

Given the first law of thermodynamics, under constant pressure the work W is done only through expansion; that is, pdV. This situation usually holds true in PEM fuel cell operation, that is, the enthalpy change is equal to the amount of heat exchange:

$$dH = dQ. \qquad \text{(Eq. 2-6)}$$

For ideal gases, the enthalpy is directly related to the specific heat at constant pressure, that is

$$dH = C_p(T)dT. \qquad \text{(Eq. 2-7)}$$

Using the equation of state for ideal gases, one can relate the specific heat at constant pressure with that at constant volume as follows:

$$c_p = c_v + R, \qquad \text{(Eq. 2-8)}$$

where R is the gas constant. c_p and c_v are the heat capacities expressed on a per mole basis at constant pressure and constant volume, respectively. Both c_p and c_v are temperature dependent. Appendix II.B lists the coefficients for a third-degree polynomial expression for c_p. The enthalpy change can be calculated by integrating the above differential form, Eq. 2-8, to obtain

$$\Delta H = \int_1^2 C_p(T)dT \approx C_{p,avg}(T_2 - T_1). \qquad \text{(Eq. 2-9)}$$

In a reaction, the steady-flow energy balance requires that the heat transfer during this process equal to the difference between the enthalpy of products and that of reactants, that is,

$$Q = H_{product} - H_{reactant}. \qquad \text{(Eq. 2-10)}$$

This heat is solely due to the changes in the chemical composition of the system if the products and reactants are at the same state, which can be treated as a reaction property. The enthalpy of reaction is then defined to quantify this heat in a complete reaction. For combustion, it is often referred to as the enthalpy of combustion and denoted as h_c.

As an example, the enthalpy of combustion of H_2 as described in Eq. 2-11 at 25°C and 1 atm can be evaluated using enthalpy-of-formation data from Table 2-1, as shown in Eq. 2-12:

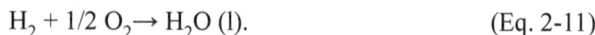

$$H_2 + 1/2\ O_2 \rightarrow H_2O\ (l). \qquad \text{(Eq. 2-11)}$$

Both the reactants and products are at the reference state of 25°C and 1 atm. The enthalpy of formation for H_2 and O_2 is zero. Thus,

$$h_c = h_{product} - h_{reactant} = h_{f,\ H_2O} = -2.85 \times 10^5\ \text{kJ/kmol.} \qquad \text{(Eq. 2-12)}$$

In PEM fuel cells, the chemical energy stored in hydrogen fuel is converted to electricity, similar to the conversion in combustion, though in a completely different process. In combustion, the concept of the heating value of fuels is widely used to quantify the amount of heat that can be released from the combustion of a given amount of a fuel. It is measured in unit of released energy per unit of the substance fuel, usually mass, such as kJ/kg. Three heating values are frequently used: they are the higher heating value (HHV), lower heating value (LHV), and gross heating value (GHV).

The HHV is determined by bringing all the products of combustion back to the original temperature prior to combustion, typically the room temperature of 25°C, and in particular to the product in its condensed state. That is, the HHV is the heat when assuming that the water produced by combustion is in the liquid state. The HHV takes into account the latent heat of vaporization of water product and is useful in calculating heating values for fuels where condensation of reaction products is practical.

Differing from the HHV, the LHV subtracts the amount of the latent heat of water vaporization, h_{fg}. Table 2-2 compares the heating values for several typical fuels, showing that hydrogen gas has the highest LHV per mass among the fuels listed. Heat values can be evaluated using the enthalpy change as follows:

$$\text{LHV} = h_c \text{ when produced water is in vapor phase} = \text{HHV} - h_{fg}. \quad \text{(Eq. 2-13)}$$

The gross heating value (GHV) accounts for exhaust water as vapor, including liquid water in the fuel prior to combustion. This value is important for fuels such as woods or coals, which contain certain amount of water prior to burning. The GHV quantifies the thermal energy stored in a fuel or energy that can be released from a fuel.

2.1.3 ENTROPY CHANGE AND THE SECOND LAW OF THERMODYNAMICS

Entropy is a thermodynamic property that is a measure of the energy unavailable for useful work in a thermodynamic process. It can be viewed as a degree of disorder in a system and its change represents the level of disorganization or randomness generated in a process. In the microscopic view, it can be expressed in terms of the number of thermodynamically possible microscopic states of a system, P, through the Boltzmann relation as follows:

Table 2-1. Enthalpy of formation (h_f), Gibbs function of formation (g_f), and absolute entropy (s_f) of hydrogen, oxygen, water, and other substances at 25°C, 1 atm [1, 2]

Substance	Formula	h_f (kJ/kmol)	g_f (kJ/kmol)	s_f (kJ/kmolK)
Hydrogen	$H_2(g)$	0	0	130.68
Oxygen	$O_2(g)$	0	0	205.04
Nitrogen	$N_2(g)$	0	0	191.61
Carbon	$C(s)$	0	0	5.74
Carbon monoxide	$CO(g)$	−110,530	−137,150	197.65
Carbon dioxide	$CO_2(g)$	−393,520	−394,360	312.80
Water vapor	$H_2O(g)$	−241,820	−228,590	188.83
Water	$H_2O(l)$	−285,830	−237,180	69.92
Methane	$CH_4(g)$	−74,850	−50,790	186.16
Hydrogen	$H(g)$	+218,000	+203,290	114.72
Oxygen	$O(g)$	+249,190	+231,770	161.06
Nitrogen	$N(g)$	+472,650	+455,510	153.30
Ammonia	$NH_3(g)$	−45,940	−16,400	192.77
Sodium hydroxide	$NaOH(l)$	−416,880	−419.2	75.91
Nitrogen dioxide	$NO_2(g)$	+33,100	51,300	240.04
Nitrogen monoxide	$NO(g)$	+90,290	87,600	210.76
Sodium chloride	$NaCl(l)$	−385,920	−393,170	95.06
Ethylene	$C_2H_4(g)$	+52,470	+68,430	219.32
Hydrobromic acid	$HBr(g)$	−36,440	−53,400	198.70
Ethanol	$C_2H_5OH(l)$	−276,000	−174,800	159.86
Hydrogen chloride	$HCl(g)$	−92,310	−95,300	186.90
Hydrogen sulfide	$H_2S(g)$	−20,500	−33,400	205.77
Sulfur dioxide	$SO_2(g)$	−296,840	−300,130	248.21
Benzene	$C_6H_6(l)$	49,000	129,700	173.26

$$S = k \log P, \qquad \text{(Eq. 2-14)}$$

where k is the Boltzmann constant, equal to 1.3806×10^{-23} J/K. Entropy increases when the molecular uncertainty or randomness of a system increases. In the macroscopic view, entropy is mathematically defined as

Table 2-2. Heating values for several typical fuels [2]

Fuel	HHV (MJ/kg)	LHV (MJ/kg)
Hydrogen	141.80	121.00
Methane	55.50	50.00
Ethane	51.90	47.80
Propane	50.35	46.35
Butane	49.50	45.75
Gasoline	47.30	44.40
Kerosene	44.80	43.00
Diesel	44.80	43.40
Ethane	51.90	47.80
Methanol	22.70	19.93
Ethanol	29.70	28.87
Acetylene	49.90	48.24
Benzene	41.80	40.17
Carbon	32.80	32.80

$$dQ_{rev} = TdS. \tag{Eq. 2-15}$$

The concept of entropy is often linked to the second law of thermodynamics, which states that the entropy of a closed system always increases or remains constant as expressed as follows:

$$dS \geq \frac{dQ}{T}. \tag{Eq. 2-16}$$

In Eq. 2-16, the terms are equal only for a reversible process. With irreversibility, the entropy will increase.

In PEM fuel cell operation, energy conversion is realized through the electrochemical reactions. Only a portion of the usable energy, that is, the Gibbs free energy, can be possibly converted to electrical power due to irreversibility arising from the ohmic resistance, the overpotential at interfaces, or other factors. The entropy change in a reversible process is used to evaluate the reversible heating by fuel cells. Given that the Gibbs free energy represents the maximum usable work that can be extracted, the waste heat production in the reversible process can be calculated as follows:

$$dQ_{rev} = TdS = dH - dG. \tag{Eq. 2-17}$$

Appendix II.B documents how to calculate the enthalpy, entropy, and Gibbs free energy of a substance and a reaction.

2.1.4 GIBBS FREE ENERGY AND THERMODYNAMIC VOLTAGE

The Gibbs free energy, G, is the energy accounting for the internal energy and part of the energy required to make room for it by displacing its surrounding and establishing its volume and pressure, but excluding that involves heat transfer. The Gibbs free energy represents the useful energy that can be extracted from a system; mathematically, it can be expressed as follows:

$$G = U + pV - TS = H - TS. \qquad \text{(Eq. 2-18)}$$

The differential form of the Gibbs free energy is thus given by

$$dG = dU + Vdp + pdV - TdS - SdT. \qquad \text{(Eq. 2-19)}$$

The internal energy change consists of heat exchanged with and work done by the environment:

$$dU = dQ - W. \qquad \text{(Eq. 2-20)}$$

Note that in a reversible process, $dQ = (TdS)_{\text{rev}}$; therefore, dU takes the following form:

$$dU = (TdS)_{\text{rev}} - W, \qquad \text{(Eq. 2-21)}$$

where W is the work done by the system in a reverse process. Substituting the above into the expression for dG yields

$$dG = (TdS)_{\text{rev}} + Vdp + pdV - TdS - SdT - W. \qquad \text{(Eq. 2-22)}$$

Under constant temperature ($dT = 0$) and constant pressure ($dp = 0$), which are good approximations for fuel cell operation under the steady state, Eq. 2-22 is reduced to

$$dG = (TdS)_{\text{rev}} - TdS + pdV - W. \qquad \text{(Eq. 2-23)}$$

Note that in real-world fuel cell operation, $(TdS)_{\text{rev}} \leq TdS$ and pdV represents the mechanical work. The first two items on the right-hand side of Eq. 2-23 will be negative (irreversible) or zero (reversible). W represents all the work exchanged with environment, including the mechanical work PdV and electric work W_{electr}. Thus, $dG = (TdS)_{\text{rev}} - TdS + W_{\text{electr}}$. For a reversible process, $-dG = W_{\text{electr}}$; with irreversibility, $-dG > W_{\text{electr}}$. Therefore, dG represents the maximum energy that can be converted to electric energy. The change in the Gibbs free energy of a reaction can be obtained through the enthalpy and entropy changes, as explained in Appendix II.A. Note that the change is dependent on temperature. Table 2-3 summarizes the free energy change of the H_2 and O_2 reaction at various temperatures and water states. The theoretical maximum efficiency can be defined through the ratio of the Gibbs free energy to the enthalpy changes:

$$\eta = \frac{\Delta G}{\Delta H}. \qquad \text{(Eq. 2-24)}$$

Table 2-3. The Gibbs free energy change of the H_2 and O_2 reaction at various temperatures and water states $(H_2 + \frac{1}{2}O_2 \leftrightarrow H_2O)$

Temperature (K)	Free energy change (Δg)	State of water (H_2O)
280	−240.127	
300	−236.836	
320	−233.598	Liquid
340	−230.360	
360	−225.783	
380	−224.857	
400	−223.914	
500	−219.063	Vapor
600	−214.008	
800	−203.489	
1000	−192.570	

Both the Gibbs free energy and enthalpy are functions of temperature; therefore, the theoretical efficiency depends on temperature. In comparison, the efficiency of heat engines is limited by the Carnot cycle:

$$\eta_{Carnot} = 1 - \frac{T_L}{T_H} \qquad \text{(Eq. 2-25)}$$

A practical example of the Carnot cycle is steam turbines, which are based on the Rankine cycle. As for the Otto cycle-based gasoline engines, the theoretical efficiency is determined by its compression ratio r as given by

$$\eta_{Otto} = 1 - \frac{1}{r^{\gamma-1}}, \qquad \text{(Eq. 2-26)}$$

where $\gamma (\equiv \frac{C_p}{C_v})$ is the specific heat ratio. Figure 2-1 plots the theoretical efficiencies of fuel cells and heat engines as a function of temperature.

In a reversible fashion, the relationship between the electrical work and potential can be written as

$$E \cdot C = W_{electr} = -dG \qquad \text{(Eq. 2-27)}$$

where C ($\equiv znF$) is the coulombs of charges transferred with z being the mole of charges (electrons) transferred per mole reactant, and F being the Faraday constant ($=96,487$ C/mol). Thus, the maximum potential (also called the reversible potential) is calculated by

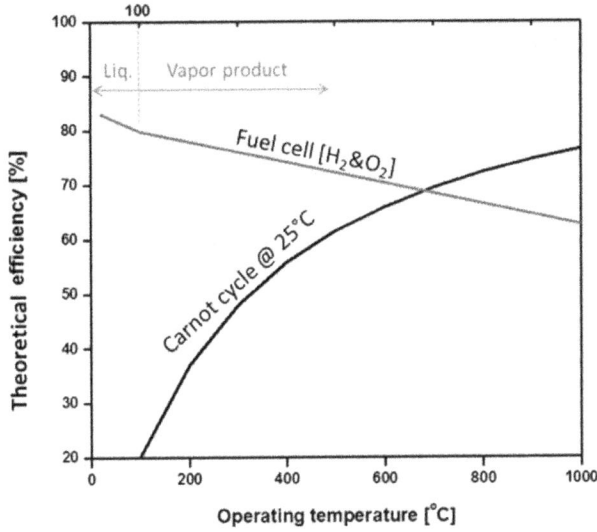

Figure 2-1. Theoretical efficiencies of fuel cells and Carnot cycle as a function of temperature (redrawn after [3] with modification).

$$E = -\frac{dG}{znF} = -\frac{dg}{zF},$$ (Eq. 2-28)

where dg is the Gibbs free energy change per mole of the reactant. Table 2-3 lists the Gibbs free energy at various temperatures for the H_2 and O_2 reactions. For the electrochemical reactions that produces liquid water at standard conditions, the Gibbs free energy change is -237 kJ/mol, yielding E of 1.23 V.

Because of temperature dependence of the Gibbs free energy, E is a function of temperature as expressed below:

$$E = \frac{1}{zF}\left(-dg_o - \frac{\partial(dg)}{\partial T}\Big|_o (T - T_o)\right)_{P=\text{const}} = E_o + \frac{ds_o}{zF}(T - T_o).$$ (Eq. 2-29)

The above uses the thermodynamic relationship of

$$dG = Vdp - SdT \text{ or } \left(\frac{\partial(dG)}{\partial T}\right)_{P=\text{const}} = -S.$$

2.1.5 CHEMICAL POTENTIAL AND NERNST EQUATION

In multicomponent systems, the chemical potential of each species is part of the internal energy:

$$dU = TdS - pdV + \sum_{i=1}^{N} \mu_i dn_i,$$ (Eq. 2-30)

where N is the number of species, μ_i is the chemical potential of species i, and n_i is the amount of species i. The corresponding change on Gibbs free energy can be expressed by

$$dG = dG_0 + \sum_{i=1}^{N} \mu_i dn_i,$$

(Eq. 2-31)

where dG_0 is the free energy change excluding the chemical potential. The chemical potential is determined by the species activity a_i as follows:

$$\mu_i = \mu_{i,0} + RT \ln a_i.$$

(Eq. 2-32)

The species activity a_i equals to p_i / p_o (for ideal gases), C_i / C_o (for ideal solutions), or unity (for pure substance). Using the following reaction in Eq. 2-33 as an example,

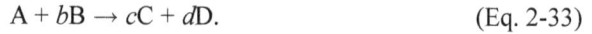

$$A + bB \rightarrow cC + dD.$$

(Eq. 2-33)

The change in the Gibbs free energy is written as follows:

$$dg = dg_0 - RT \ln \left(\frac{a_A a_B^b}{a_C^c a_D^d} \right).$$

(Eq. 2-34)

The above is known as the Nernst equation. For $2H_2 + O_2 \rightarrow 2H_2O(v)$, the reversible potential is written as

$$E = E^0 + \frac{RT}{2F} \ln \left(\frac{a_{H_2} a_{O_2}^{1/2}}{a_{H_2O}} \right).$$

(Eq. 2-35)

2.1.6 RELATIVE HUMIDITY AND PHASE CHANGE

PEM fuel cells operate at relatively low temperatures, ranging from -40 to $120°C$. Phase change is frequently encountered, and the condensed phase such as liquid water significantly influences fuel cell performance. The occurrence of water phase-change is determined by the gas-phase humidity. Below we will briefly introduce the concepts of relative humidity and phase change in the general context of thermodynamics.

(1) Mixtures of Ideal Gases and Relative Humidity (RH)

The anode and cathode reactant streams in PEM fuel cells are essentially mixtures, containing H_2 and water in the anode, and O_2, N_2, water, and impurities in the cathode. Dalton law states that the total pressure of a mixture consisting of nonreactive gases is equal to the sum of the partial pressures of individual gases. For example, in a binary mixture of A and B ideal gases, the total pressure p is equal to the right-hand side of Eq. 2-36:

$$p = p_A + p_B$$

where the partial pressure $p_i = c_i p$

$$\text{and the molar fraction } c_i = \frac{n_i}{n_{\text{total}}} \qquad \text{(Eq. 2-36)}$$

n_{total} is the total number of moles. Likewise, the mass fraction $y_i = \dfrac{m_i}{m_{\text{total}}}$ can be defined for parameters or variables on the mass basis.

The vapor content in reactant streams is important for the water management of PEM fuel cells. Dry reactants dehydrate the membrane and thus decrease its ionic conductivity. Humidity is a term quantifying the amount of water vapor in a gas mixture. Humidity can be defined in terms of the vapor content of this mixture, called the absolute humidity, as follows:

$$\phi = \frac{p_w}{p_g} \qquad \text{(Eq. 2-37)}$$

The humidity ratio for an airwater vapor mixture is frequently used to represent the ratio of the mass of water vapor to the mass of dry air. In everyday usage, the humidity ratio is commonly referred to as relative humidity (RH), expressed as a percent in weather forecasts and on household humidistats. RH measures the absolute humidity relative to that of the saturated mixture. It is most commonly defined as the partial pressure of water vapor in an air–water mixture, given as a percentage of the saturated vapor pressure:

$$RH = \frac{\phi}{\phi_{\text{sat}}} = \frac{p_w}{p_{w,\text{sat}}} \qquad \text{(Eq. 2-38)}$$

where the saturated vapor condition is determined by temperature, as given by the following correlation [4]:

$$\log_{10} p_{w,\text{sat}} = -2.1794 + 0.02953\,(T - 273.15) - 9.1837 \times 10^{-5}$$

$$(T - 273.15)^2 + 1.4454 \times 10^7\,(T - 273.15)^3 \qquad \text{(Eq. 2-39)}$$

The *RH* is often used instead of absolute humidity in situations where the rate of water evaporation is important, as it takes into account the saturated vapor pressure. *RH* is important for two-phase operation of PEM fuel cells. When *RH* reaches unity, liquid water emerges upon further water addition, leading to two-phase transport.

The correlation as described in Eq. 2-39 presents a good approximation for temperature over 0°C. At subfreezing temperature, the saturated vapor pressure $p_{w,\text{sat}}(T)$ can be calculated through a six-degree polynomial function proposed by Rasmussen [5], which is accurate (relative error <0.0007) from −50 to 50°C as follows:

$$p_{w,\text{sat}} = \sum_{n=0}^{6} C_n T^n \qquad \text{(Eq. 2-40)}$$

Coefficient C_n	Water	Ice
C_0	6.1070422	6.1090668
C_1	4.4411566×10^{-1}	5.0249291×10^{-1}
C_2	1.4320982×10^{-2}	1.8684567×10^{-2}
C_3	2.6513961×10^{-4}	4.0559217×10^{-4}
C_4	3.0099985×10^{-6}	5.4323745×10^{-6}
C_5	2.0088796×10^{-8}	4.2255374×10^{-8}
C_6	$6.1926232 \times 10^{-11}$	$1.4687829 \times 10^{-10}$

Under subfreezing conditions, ice will be produced in fuel cell electrodes upon further water addition by the ORR when *RH* reaches unity.

(2) Phase Change

In a single phase of a thermodynamic system, the state of matter is the same everywhere, that is, physical properties are uniform. A phase change is the transformation of a thermodynamic system from one phase or state of matter to another. During phase transition of a given medium as a result of external heating, certain properties of the medium change, often discontinuously. For example, ice becomes liquid water upon heating to the melting point and the continuing addition of heat results in an abrupt change in volume. In the microscopic view, the added energy is absorbed by furthering the distance among constituent particles, that is, increasing the molecular potential. Figure 2-2 shows a phase-change diagram of water.

The amount of energy absorbed or released during phase change is called the latent heat. Fuel cell operation involves several types of phase changes in water, including the liquid–vapor, liquid–ice, and vapor–ice phase changes. When the RH reaches unity, condensation or sublimation occurs upon further water addition in the vapor phase if no supercooled state exists. The latent heat can be calculated through the enthalpy change during phase change as follows:

$$h_{fg} = h_g - h_f$$
$$h_{sg} = h_g - h_s.$$

(Eq. 2-41)

Both liquid water and ice have small diffusivity to oxygen and hydrogen gas as opposed to gas phases. In general, gaseous species diffusion is hindered by the presence of other species due to collision. Therefore, diffusivity is closely related to the mean free path, as elucidated by the lattice model in Chapter 3. Diffusion in liquid media can be evaluated by using a hydrodynamic model. The hydrodynamic approach is developed by assuming that the resistance of solute molecule movement is caused by the viscous force, which is similar to particle movement in a viscous fluid. In a dilute liquid, the approach yields the following well-known Stokes–Einstein equation:

$$D^0 = \frac{k_B T}{6\pi r \mu},$$

(Eq. 2-42)

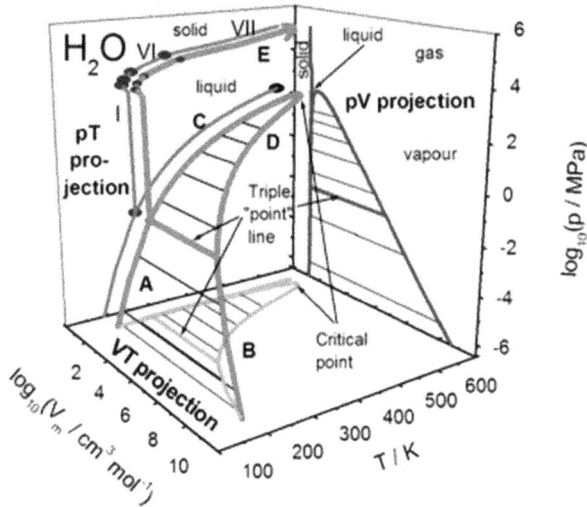

Figure 2-2. Phase diagram of ordinary water substance [6] (Courtesy of American Chemical Society).

where k_B is the Boltzmann's constant, r is the radius of the species molecule, and μ is the liquid viscosity. In a nonideal solution (such as a solution that is not infinitely dilute), modification can be made by introducing the volume fraction of solute i, ε_i,

$$D^{0*} = D^0(1+1.45\varepsilon_i).$$

(Eq. 2-43)

Ghai et al. (1973) reviewed models for species diffusion in liquids [7]. In solids such as crystals, diffusion can occur through substitution diffusion, in which the solute particle such as atoms or molecules moves only along paths connecting lattice sites, or through interstitial diffusion, in which the paths take other ways. Figure 2-3 shows the approximate ranges of binary diffusivity in different phases.

2.2 ELECTROCHEMICAL REACTION KINETICS

Electrochemistry is a branch of chemistry that focuses on the studies of chemical reactions occurring at the electrode–electrolyte interface in a solution and involving charged chemical species or ions. Different from the thermodynamic view that is concerned with whether or not a chemical reaction proceeds spontaneously, reaction kinetics focuses on how fast or the rate a chemical reaction occurs. Factors that influence reaction kinetics include catalyst materials, electrode structure, temperature, and reactant/product species concentrations. In contrast to common chemical reactions that involve neutral species, an electrochemical reaction results in charge transfer between the electrode and electrolyte. The surface overpotential is a primary driving force for the electrochemical reaction and thus a key factor on controlling its rate. A detailed description and discussion of modern theories regarding the electrode–electrolyte interface is provided by Bard and Faulkner [9], and Newman and

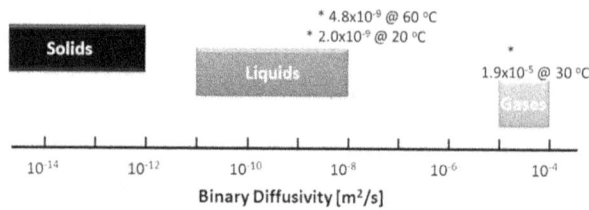

Figure 2-3. Ranges of species diffusivities in different phases (redrawn after [8] with modification). The diffusion coefficients refer to oxygen diffusivity in liquid water and air, respectively.

Thomas-Alyea [10]. Due to space limitation, this section is not intended to describing and discussing detailed electrochemical phenomena or exploring reaction paths or steps. Rather, a brief discussion on several theories and approximation methods is presented, pertaining to the water and thermal management in PEM fuel cells. Readers who are interested in a detailed discussion are referred to consult with texts by Bard and Faulkner, and Newman and Thomas-Alyea.

2.2.1 ELECTROCHEMICAL KINETICS

The Butler–Volmer Equation

The very first fuel cell was invented in 1839 by Sir William Robert Grove (an English lawyer turned scientist) (see Fig. 2-4). He designed a primitive fuel cell, called then a gas voltaic battery, and proved his hypothesis of reverse water electrolysis.

Both forward and backward reactions take place on the surface of a general electrode. The forward and backward reactions on a hydrogen anode surface are written as follows:

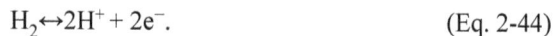

$$H_2 \leftrightarrow 2H^+ + 2e^-. \tag{Eq. 2-44}$$

At equilibrium, both of the processes in Eq. 2-44 occur at equal rates but in opposite directions; therefore, the currents generated in these two reactions balance each other, resulting in zero net charge production. In a single-step electrode reaction, the current density is determined by the reactant concentration at the reaction surface. At nonequilibrium, the net reaction rate is the difference between the rates of the forward and backward reactions as shown below:

$$j = j_f - j_b, \tag{Eq. 2-45}$$

where j_f and j_b are the charge production rates or current density of forward and backward processes, respectively, in an electrochemical reaction. They can be expressed by

$$j_f = nFk_f C_{Ox}$$
$$j_b = nFk_b C_{Re}, \tag{Eq. 2-46}$$

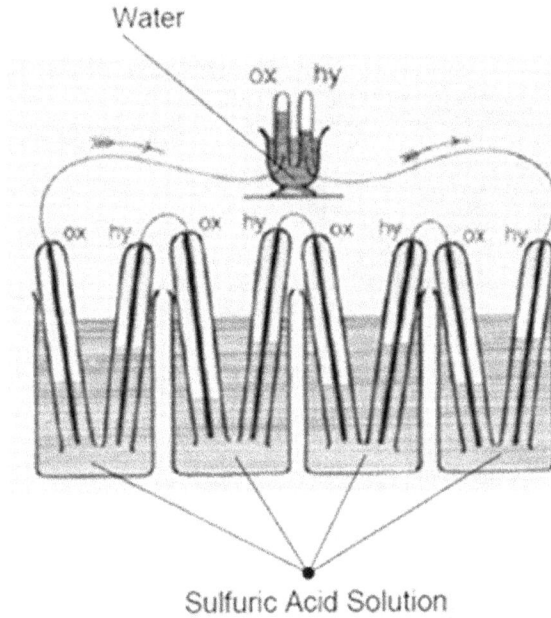

Figure 2-4. 1839 William Grove fuel cell [11].

where C_{ox} and C_{Re} stand for the surface concentrations of the oxidized and reduced forms of the reactant, respectively. The rate constant k is a function of the Gibbs energy of activation, written by Eq. 2-47, according to the transition state theory [12]:

$$k = \frac{k_B T}{h} \exp\left(\frac{-\Delta G}{RT}\right),$$ (Eq. 2-47)

where k_B and h are Boltzmann's constant and Planck's constant, respectively. The Gibbs energy of activation for the reduction and oxidation reactions are illustrated in Figure 2-5, and when there exists electric potential, the activation energy will change, expressed as follows:

$$\Delta G_{Re} = \Delta G_0 + \alpha n F E \quad \text{and} \quad \Delta G_{Ox} = \Delta G_0 + \Delta G_{rxn} - (1-\alpha)n F E. \quad \text{(Eq. 2-48)}$$

In the two equations of Eq. 2-48, the first terms on the right represent the chemical component of the Gibbs free energy (see Fig. 2-5), and the last terms denote the electrical component. Here, n is the number of electrons transferred in the electrochemical reaction, and α is related to the symmetry of the reaction, representing the fraction of the applied potential E that promotes the reduction reaction, usually in a range between 0.2 and 0.5. Combining the above equations yields the well-known Butler–Volmer expression by using the surface overpotential η (see Eq. 2-49):

$$j = a i_o \left(e^{\alpha n F \eta / RT} - e^{-(1-\alpha)n F \eta / RT}\right),$$ (Eq. 2-49)

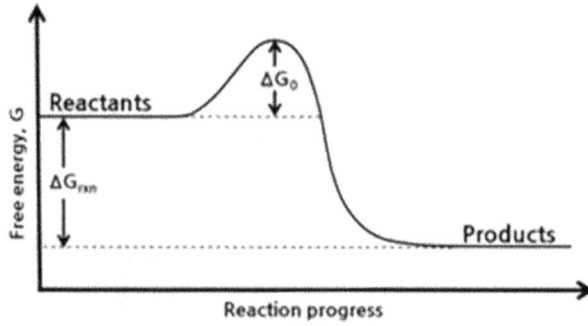

Figure 2-5. The activation energy of a reaction.

where i_o is the exchange current density; physically, it represents the forward/backward reaction rate at equilibrium as expressed by

$$i_o = nFk_f C_{Ox} = nFk_b C_{Re}.$$ (Eq. 2-50)

The surface overpotential η is defined by

$$\eta = E - E_{eq}.$$ (Eq. 2-51)

In PEM fuel cell modeling, the following form of the Butler–Volmer equation, Eq. 2-52, is frequently adopted (see Fig. 2-6 for an illustration of the Butler–Volmer equation):

$$j = a\, i_0 \left\{ \exp\left(\frac{\alpha_a F}{RT} \eta \right) - \exp\left(-\frac{\alpha_c F}{RT} \eta \right) \right\},$$ (Eq. 2-52)

where a is the effective surface area determined by factors such as electrode roughness, and the surface overpotential is then expressed as follows:

$$\eta = \Phi^s - \Phi^m - U_o.$$ (Eq. 2-53)

The equilibrium potential U_o is zero in the anode, while in the cathode it is a function of temperature, described by

$$U_o = 1.23 - 0.0009 \times (T - 298).$$ (Eq. 2-54)

2.2.2 ELECTROCHEMICAL MECHANISMS IN PEM FUEL CELLS

Electrocatalysis of the hydrogen oxidation reaction (HOR) plays a vital role in the development of the PEM fuel cell anode. Electrocatalysis, in general, refers to a specific type of heterogeneous catalysis whereby reaction occurs on the catalyst surface. The intrinsic kinetic rate or exchange current density strongly depends on the catalyst material and interface morphology. For example, the exchange current density can vary from about 10^{-4} A/cm^2 at a platinum

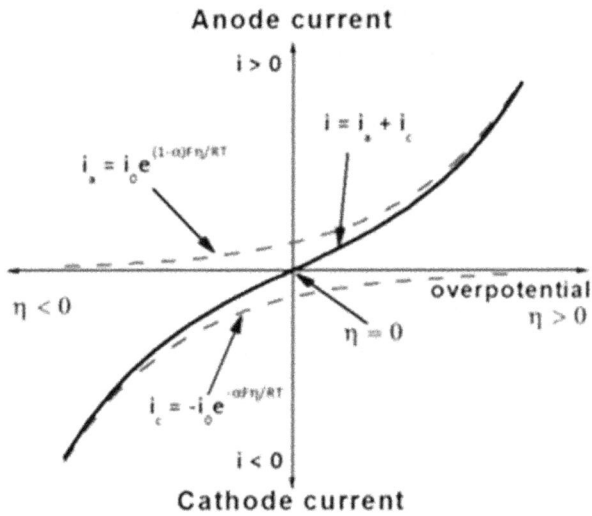

Figure 2-6. The Butler–Volmer equation and approximations.

electrode to 10^{-13} A/cm^2 at a Pb electrode for the HOR as presented in Table 2-4. Platinum-group metals, including Pt, Ru, Pd, Ir, Os, and Rh, are traditionally known as catalysts for the HOR. These metal catalysts show high catalytic activities, exhibiting rapid H adsorption, which is a rate-limiting step for the HOR. The HOR on platinum electrodes in acid electrolyte can take different pathways, such as the Tafel–Volmer and Heyrovsky–Volmer mechanisms, both of which involve hydrogen adsorption on the Pt surface (Pt–H_{ad}) and a fast charge-transfer step [13], as follows:

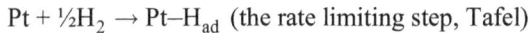

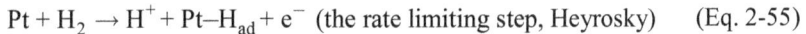

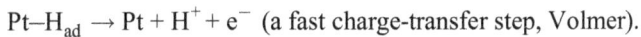

$$Pt + \tfrac{1}{2}H_2 \rightarrow Pt\text{–}H_{ad} \text{ (the rate limiting step, Tafel)}$$

$$Pt + H_2 \rightarrow H^+ + Pt\text{–}H_{ad} + e^- \text{ (the rate limiting step, Heyrosky)} \quad \text{(Eq. 2-55)}$$

$$Pt\text{–}H_{ad} \rightarrow Pt + H^+ + e^- \text{ (a fast charge-transfer step, Volmer)}.$$

In the mechanisms described above, it has been recognized that H_{ad} is the reactive intermediate in the HOR, and the HOR kinetics is mainly determined by the Pt–H_{ad} interaction on the surface.

The mechanism of the oxygen reduction reaction (ORR) is a widely studied reaction mechanism in electrode processes. Under the operating conditions of a PEM fuel cell, the ORR is sluggish; consequently, catalyst such as Pt is usually used to lower the activation energy level so as to improve the cathode kinetics. The ORR on Pt is a multielectron process, involving a number of elementary steps and various intermediates. A simplified path of the reaction mechanism is shown in Figure 2-7; the only two products are H_2O_2 and H_2O. Figure 2-7 indicates possible pathways for ORR on the Pt surface: $O_{2,ad}$ can be reduced either directly or electrochemically to water (a reduction reaction involving four electrons) at the rate constant k_1, or to adsorbed hydrogen peroxide ($H_2O_{2,ad}$) at the rate v k_2 (a reduction reaction involving two electrons). $H_2O_{2,ad}$ can be further converted to water (k_3), chemically decomposed on the electrode surface (k_4), and/or desorbed from the electrode surface (k_5).

Table 2-4. i_0 of the hydrogen electrode for various metals in an acid electrolyte [14–16]

Metal	i_0 (A cm^{-2})
Pb	2.5×10^{-13}
Zn	3×10^{-11}
Ni	6×10^{-6}
Pt	10^{-3} to 10^{-4}
Ir	10^{-4}
Pd	4×10^{-3}
Ta	10^{-7}
Sn	10^{-8}
Al	10^{-10}
Cd	10^{-12}
Hg	10^{-12}
Au	10^{-6}
Fe	10^{-6}
Mo	10^{-7}
Ag	10^{-5} to 10^{-7}

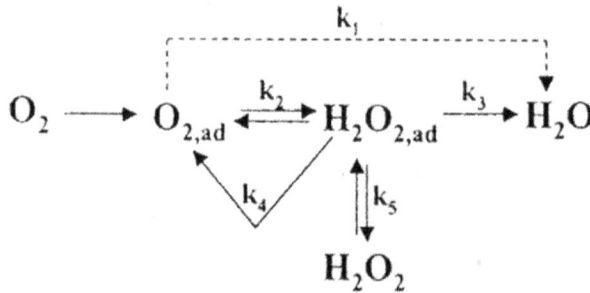

$$O_2 \longrightarrow O_{2,ad} \overset{k_2}{\underset{}{\rightleftharpoons}} H_2O_{2,ad} \overset{k_3}{\longrightarrow} H_2O$$

$$\overset{k_1}{\cdots\cdots\cdots\cdots\cdots}$$

$$k_4 \qquad k_5$$

$$H_2O_2$$

Figure 2-7. Possible pathways of the ORR on Pt [17] (Courtesy of Elsevier).

The ORR pathway can be explained through the electronic structure and density functional theory. The dissociative and associative mechanisms were proposed for low and high current densities, respectively [13], as shown in Eqs. 2-56 and 2-57.

(1) Dissociative Pathway

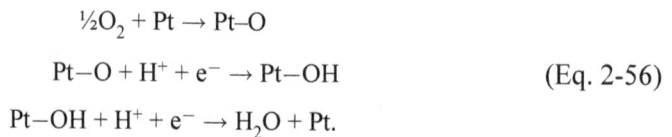

$$\tfrac{1}{2}O_2 + Pt \rightarrow Pt-O$$

$$Pt-O + H^+ + e^- \rightarrow Pt-OH \qquad \text{(Eq. 2-56)}$$

$$Pt-OH + H^+ + e^- \rightarrow H_2O + Pt.$$

No H_2O_2 species is involved in this pathway. At the Pt surface, O_2 adsorption breaks the O–O bond, forming adsorbed oxygen Pt–O. Pt–O further gains electrons in the latter two steps to produce water and release the site. This mechanism can be understood by considering the detailed steps of the direct four-electron pathway (k_1), as shown in Eq. 2-57:

(2) Associative Pathway

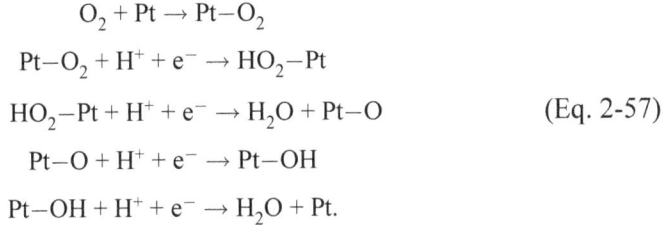

$$O_2 + Pt \rightarrow Pt\text{–}O_2$$

$$Pt\text{–}O_2 + H^+ + e^- \rightarrow HO_2\text{–}Pt$$

$$HO_2\text{–}Pt + H^+ + e^- \rightarrow H_2O + Pt\text{–}O \qquad \text{(Eq. 2-57)}$$

$$Pt\text{–}O + H^+ + e^- \rightarrow Pt\text{–}OH$$

$$Pt\text{–}OH + H^+ + e^- \rightarrow H_2O + Pt.$$

Though no H_2O_2 is explicitly involved in the mechanism shown in Eq. 2-57, as is the case with the dissociative pathway, the O–O bond in adsorbed oxygen (Pt–O_2) may not be broken properly in the subsequent steps, leading to the formation of H_2O_2. H_2O_2 could either be further reduced to H_2O or remain unchanged as a final product, similar to that outlined in Figure 2-7.

2.2.3 LINEAR APPROXIMATION AND TAFEL EQUATION

A detailed understanding of the reaction mechanisms is essential to catalyst selection and electrochemical process control. In the context of thermal and water management, the Butler–Volmer equation usually suffices. This is in light of the fact that the thermal and water management requires knowledge of heat, species, electrochemical reaction kinetics, and their coupling; therefore, a certain degree of simplification on electrochemical detail is beneficial. For PEM fuel cells, the HOR on Pt is approximately 5–7 orders of magnitude faster than the ORR. It has one of the fastest known specific rate constants in aqueous solutions. Because the HOR kinetics is fast, a small anode overpotential occurs. Using the Taylor series expansion of exponential functions with the higher orders of the overpotential truncated, the Butler–Volmer equation is reduced to a linear kinetic equation as follows:

$$j_a = a\, i_{0,a} \left(\frac{C_{H_2}}{C_{H_2}^{ref}} \right)^{1/2} \left(\frac{\alpha_a + \alpha_c}{RT} \cdot F \cdot \eta \right). \qquad \text{(Eq. 2-58)}$$

Note that in the above, the overpotential is in a linear relationship with the current density, which enables efficient numerical treatment in computer simulation implementation. For the oxygen reduction reaction (ORR), its sluggish kinetics results in a large cathode overpotential. Thus, the first term on the right-hand side of Eq.2-52 is small as opposed to the second term (note the overpotential is negative), yielding the Tafel equation as follows:

$$j_c = -a\, i_{0,c} \left(\frac{C_{O_2}}{C_{O_2}^{ref}} \right) \exp \left(-\frac{\alpha_c F}{RT} \cdot \eta \right). \qquad \text{(Eq. 2-59)}$$

2.3 VOLTAGE LOSS MECHANISMS AND A SIMPLIFIED MODEL

One of the most important concerns in PEM fuel cells is their output voltage loss. Output voltage, along with output current, directly determines PEM fuel cell's power output. Because there exists a theoretical maximum voltage that a fuel cell can achieve from the hydrogen fuel, voltage loss is a key factor determining the output voltage and efficiency (the ratio of the output voltage to the theoretical one). In this section, we highlight the major voltage loss mechanisms and further introduce a simple model to predict the voltage loss, along with the parameters affecting voltage losses. An introduction to voltage loss mechanisms is also well explained in Ref. [3].

2.3.1 OPEN CIRCUIT VOLTAGE (OCV)

Open circuit voltage (OCV) is the electrical potential difference between the two electrodes when no external load is applied, that is, the external circuit is open or disconnected. Ideally, the OCV is equal to the thermodynamic potential as calculated by Eq. 2-35. In operation, fuel or internal current may transport through the membrane (also called "crossover"), lowering the voltage observed at the open-circuit state.

Though electrolyte membrane materials are selected to prevent electron or reactant transport, small amounts of electrons and species are able to cross over. For example, the diffusivity of H_2 in the Nafion membrane is on the order of 10^{-12} m^2/s; see Figure 2-3. In degraded membranes with defects such as thinning or pin-hole formation, the cross-over of reactants and electrons can be significant, yielding an appreciable reduction in the OCV. Given a crossover current of $i*$, the resulting OCV can be estimated as follows:

$$OCV = E - \frac{RT}{2\alpha F}\ln i*. \qquad \text{(Eq. 2-60)}$$

The cross-over process is irreversible, contributing to the ohmic waste heat generation. In PEM fuel cell operation, when power is withdrawn, output voltage becomes lower, caused by the activation loss, ohmic loss, and mass transport loss. The portion of energy relevant to voltage loss is released in the form of waste heat.

2.3.2 ACTIVATION LOSS

The activation loss is determined by the electrochemical reaction kinetics on the electrode surface. In order to drive a reaction that transfers electrons to or from electrodes, surface overpotential is required, reducing the output voltage. The activation loss can be evaluated using the Butler–Volmer equation as follows:

$$i = i_0 \exp\left(\frac{2\alpha F\Delta V_{act}}{RT}\right) \text{ or } \Delta V_{act} = \frac{RT}{2\alpha F}\ln\left(\frac{i}{i_0}\right). \qquad \text{(Eq. 2-61)}$$

Providing that the activation loss occurs on both anode and cathode sides, the overall activation loss can be expressed as follows:

$$\Delta V_{act} = \frac{RT}{2\alpha_a F} \ln\left(\frac{i}{i_{0,a}}\right) + \frac{RT}{2\alpha_c F} \ln\left(\frac{i}{i_{0,c}}\right).$$

(Eq. 2-62)

For pure hydrogen gas, the activation loss in the anode is usually much smaller than that in the cathode, due to the HOR electrochemical kinetics being much faster. When other fuels such as methanol or glucose are used, the anode activation loss can be significant. The activation voltage loss is determined by several factors, such as the catalyst material, electrode roughness, and temperature. Pt-based materials are still the best candidate as the catalyst for PEM fuel cells. The catalyst material contributes to a major portion of fuel cell cost. Thus, catalyst loading reduction (for Pt-based catalyst) and the search for alternative cost-effective catalysts remain as key areas of research in fuel cell electrode development. In practice, the reaction surface is fabricated to be highly tortuous to increase the active reaction area per unit volume; the specific reaction area can be increased by 100–1000 times, considerably improving the electrochemical kinetics. In addition, i_0 is sensitive to temperature. Raising operating temperature improves the reaction kinetics and thus yields a smaller activation voltage loss.

2.3.3 OHMIC LOSS

When protons or electrons are conducted through a membrane or carbon structure, respectively, the ohmic resistances of these components contribute to voltage loss, which is proportional to their current flow as follows:

$$\Delta V_\Omega = R_\Omega I,$$

(Eq. 2-63)

where I is the current density (in units of A/m^2) and R_Ω (in units of Φ m^2) is the overall resistance that consists of the membrane ionic resistance, ionic/electric resistance in catalyst layers, and contact resistance. R_Ω can be estimated from

$$R_\Omega = \frac{\delta_m}{\kappa_m} + \frac{\delta_{aCL}}{2\kappa_{aCL}\varepsilon_m^{\tau k}} + \frac{\delta_{cCL}}{2\kappa_{cCL}\varepsilon_m^{\tau k}} + R_{e-} + R_{contact},$$

(Eq. 2-64)

where K is the ionic conductivity, δ is the thickness of membrane or catalyst layer (CL), R_{e-} is the electronic resistance, and $R_{contact}$ is the contact resistance. The first three terms on the right-hand side of Eq. 2-64 represent the ionic resistances in the membrane and anode and cathode catalyst layers, respectively. The ionic conductivity of Nafion membranes is determined by their hydration and temperature, to be discussed in Chapter 4. Thus, water and thermal management is critical to the ohmic voltage loss. The latter two terms on the right-hand side of Eq. 2-64 represent the electronic resistances in GDLs, catalyst layers, and bipolar plates, and contact resistance between components, respectively. Obviously, thin membranes and good contacts between components are beneficial to the ohmic loss reduction.

2.3.4 TRANSPORT VOLTAGE LOSS

This voltage loss mechanism can be directly seen from the dependence of the Gibbs free energy on reactant concentration (i.e., the Nernst equation) as indicated in Eq. 2-35. A simple model can be developed to evaluate the transport voltage loss, as follows:

$$\Delta V_{\text{trans}} = -B \ln(1 - \frac{i}{i_L}), \qquad \text{(Eq. 2-65)}$$

where i_L is the limiting current density and B is a parameter depending on the operation state and reactants. The physical meaning of the limiting current can be elucidated using oxygen transport as an example. The limiting current density due to oxygen transport in GDLs is given below:

$$i_{L,\text{GDL}} = \frac{4FC_{O_2}}{\delta_{\text{GDL}}} D, \qquad \text{(Eq. 2-66)}$$

where C_{O_2} is the oxygen concentration at the outer GDL surface. Ambient air (which contains around 21% oxygen by volume) is usually used as the fuel cell oxidizer. As the cathode air flow travels downstream, oxygen is depleted by the ORR, increasing the transport voltage loss. To reduce the transport loss, materials with large diffusivity are preferred. In addition, liquid water emerges in PEM fuel cells, hindering gaseous reactant supply. Thus, liquid water management is essential for reducing the transport voltage loss.

2.3.5 CURRENT–VOLTAGE (I–V) CURVE AND OPERATION EFFICIENCY

The final output voltage of a PEM fuel cell can be estimated by subtracting the three voltage losses from the OCV (assuming OCV = E) as follows:

$$V_{\text{cell}} = E - \frac{RT}{2\alpha F} \log i_0 - \Delta V_{\text{act}} - \Delta V_{\Omega} - \Delta V_{\text{trans}}. \qquad \text{(Eq. 2-67)}$$

Figure 2-8 shows a typical polarization profile estimated by Eq. 2-67. Given E is the maximum potential (that is, the thermodynamic potential), the operation efficiency η is defined as follows:

$$\eta = \frac{V_{\text{cell}}}{E}. \qquad \text{(Eq. 2-68)}$$

It must be pointed out that Eq. 2-68 is not representative of the conversion efficiency from fuel to electricity, because E only accounts for the Gibbs free energy change.

Figure 2-9 schematically shows the relative importance of different mechanisms of voltage loss: cathodic losses contribute a major portion of voltage loss, whereas anodic losses are usually negligible. The I–V curve in the region under low current density where activation dominates shows a semiexponential behavior. In the intermediate region, the curve is generally

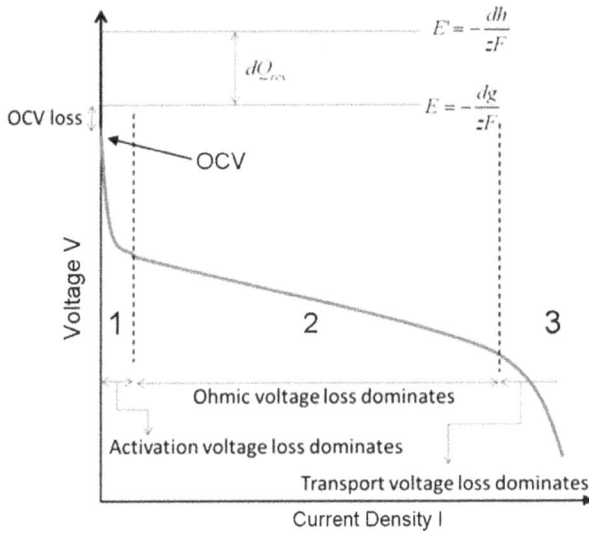

$$E = -\frac{dh}{zF}$$

$$E = -\frac{dg}{zF}$$

Figure 2-8. A typical polarization curve (also called the I–V curve) of a PEM fuel cell: (1) activation loss regime; (2) ohmic loss regime; (3) transport loss regime.

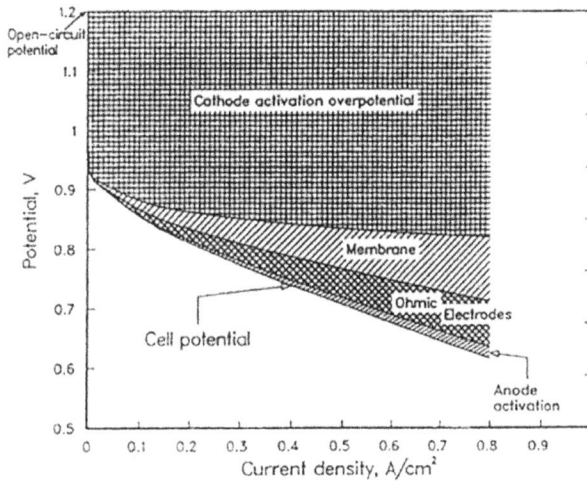

Figure 2-9. Breakdowns of voltage losses in a PEM fuel cell [18] (Courtesy of the Electrochemical Society).

linear, in which the ohmic losses in the MEA (membrane electrode assembly) dominate. The third region, dominated by reactant transport in the cathode, is not reported in Figure 2-9 but is shown in Figure 2-8, and in general it exhibits a fast drop with current density.

2.3.6 ROLE OF WATER AND THERMAL MANAGEMENT

The importance of water and thermal management in terms of the cell voltage losses can be delineated as follows:

1. Temperature is an important parameter determining the electrochemical kinetics; lower temperature operation usually leads to a larger activation loss.
2. The ionic conductivity of a Nafion membrane is determined by its hydration state and temperature; dryness (or insufficient hydration) or low temperature yields a high ionic resistance, increasing the ohmic voltage loss.
3. Excess water or low temperature causes water condensation and consequently liquid formation, which increases the reactant transport resistance and hence the transport voltage loss.
4. Thermal cycles increase the contact resistance among fuel cell components, increasing the ohmic voltage loss.
5. Liquid water in gas flow channels can cause flow maldistribution, leading to uneven distribution of reactants and increase in the transport voltage loss.

In addition to the voltage loss, thermal and water management is important to fuel cell durability. For example,

1. High temperature or hot spot formation degrades fuel cell materials and consequently lowers long-term performance.
2. Dry operation accelerates membrane degradation, such as formation of cracks or pinholes.
3. Excess water results in flooding and consequently reactant starvation, which causes electrolysis of carbon support in the catalyst layer.
4. Freezing/thawing operation degrades the reaction interface and thus reduces the active catalyst surface.

2.4 CHAPTER SUMMARY

This chapter provides an overview of two basics or essential subjects pertaining to PEM fuel cells, namely thermodynamics and electrochemistry, which are the first principles underlying the operation of PEM fuel cells:

- Internal energy represents the energy available in a matter whereas enthalpy includes both internal energy and that making room for the matter. The Gibbs free energy determines the maximum or thermodynamic potential; and the Nernst equation describes the dependence of thermodynamic potential on the reactant partial pressure.
- Reaction kinetics describes the routes and rates of the hydrogen oxidation reaction (HOR) and oxygen reduction reaction (ORR). The Butler–Volmer equation is customarily adopted to describe the reaction rates, and the HOR and ORR are usually approximated using the linear and Tafel kinetics, respectively.
- The output voltage of PEM fuel cells is mainly determined by the activation, ohmic, and transport voltage losses – each mechanism of voltage loss dominates different regimes of PEM fuel cell operation. Cathodic losses contribute a major portion of voltage loss, whereas anodic losses are usually much smaller, and effective thermal and water management results in the reduction of all three major losses.

REFERENCES

1. Cengel, Y.A., and Boles, M.A. *Thermodynamics: An Engineering Approach*, 6th Ed. McGraw-Hill, (2007).

2. National Institute of Standards and Technology (NIST), Chemistry WebBook, http://webbook.nist.gov /chemistry/, (2012).

3. Larminie, J., and Dicks, A. *Fuel Cell Systems Explained*, 2nd Ed. Wiley, (2003).

4. Springer, T.E., Zawodinski, T.A., and Gottesfeld, S. Polymer electrolyte fuel cell model. *J. Electrochem. Soc.* 138 (1991): 2334. doi: http://dx.doi.org/10.1149/1.2085971

5. Rasmussen, L.A. On the approximation of saturation vapor pressure. *J. Appl. Meteorol.* 17 (1978): 1564. doi: http://dx.doi.org/10.1175/1520-0450(1978)017<1564:OTAOSV>2.0.CO;2

6. Glasser, L. Water, water, everywhere: phase diagrams of ordinary water substance. *J. Chem. Educ.* 81 (2004): 414–18. doi: http://dx.doi.org/10.1021/ed081p414

7. Ghai, R.K., Ertl, H., and Dullien, F.A.L. Liquid diffusion of non-electrolytes, Part I. *AIChE J.* 19 (1973): 881. doi: http://dx.doi.org/10.1002/aic.690190502

8. Deen, W.M. *Analysis of Transport Phenomena*. Oxford University Press, (1998).

9. Bard, A.J., and Faulkner, L.R. *Electrochemical Methods: Fundamentals and Applications*, 2nd Ed. Wiley, (2000).

10. Newman, J., and Thomas-Alyea, K.E. *Electrochemical Systems*, 3rd Ed. Wiley-Interscience, (2004).

11. Grove, W.R. On a new voltaic combination. *Philos. Maga. J. Sci.* 13 (1838): 430.

12. Truhlar, D.G., Garrett, B.C., and Klippenstein, S.J. Current status of transition-state theory. *J. Phys. Chem.* 100 (1996): 12771–800. doi: http://dx.doi.org/10.1021/jp953748q

13. Zhang, J. *PEM Fuel Cell Electrocatalysts and Catalyst Layers: Fundamentals and Applications*. Springer, (2008). doi: http://dx.doi.org/10.1007/978-1-84800-936-3

14. Rayment, C., and Sherwin, S. *Introduction to Fuel Cell Technology*. University of Notre Dame, (2003).

15. Vielstich, W., Lamm, A., and Gasteiger, H.A. *Handbook of Fuel Cells*, Vol. 2. John Wiley and Sons, (2003).

16. Berger, C. (Ed.). *Handbook of Fuel Cell Technology*, Prentice-Hall, (1968).

17. Markovi, N.M., and Ross, P.N. Surface science studies of model fuel cell electrocatalysts. *Surf. Sci. Rep.* 45 (2002): 117–229. doi: http://dx.doi.org/10.1016/S0167-5729(01)00022-X

18. Bernardi, D.M., and Verbrugge, M.W. A mathematical model of the solid-polymer-electrolyte fuel cell. *J. Electrochem. Soc.* 139 (1992): 2477. doi: http://dx.doi.org/10.1149/1.2221251

CHAPTER 3

FUNDAMENTALS OF HEAT
AND MASS TRANSFER

3.1 INTRODUCTION

A PEM fuel cell is an energy system that converts chemical energy stored in reactant species to electrical energy or electricity with water as the sole by-product. Due to the inefficiency in the energy conversion, a portion of chemical energy is converted to waste heat instead of electricity. Reactant and heat transport plays a critical role in the operation of PEM fuel cells: reactants must be adequately supplied to the catalyst site at a proper rate. Product water and waste heat must be efficiently removed from catalyst layers, GDLs, and channels to maintain a high efficiency in power generation. Two major mechanisms are frequently encountered in species transport: the convective transport due to bulk motion and the molecular or diffusive transport that is induced by random motions of molecules. The general governing equations can be derived, based on the conservation law, by accounting for these two transport mechanisms. The constitutive equations are needed to relate the fluxes to local field variables (species concentrations, and temperature) and material properties. The most common constitutive equation for describing energy flux due to the temperature gradient is the Fourier law. For species fluxes driven by concentration gradients the constitutive equation is Fick's first law.

In the context of transport fundamentals, the framework of conservation equations and analysis is well established. In this chapter, the general conservation equations for transport and ordinary constitutive equations will be introduced. Readers who are interested in detailed discussions of the subject are referred to consult with texts by Bird et al. [1], William [2], and Kay et al. [3]. In addition, three appendices are provided for this chapter: Appendix III.A presents mass, momentum, and energy conservation equations in the Cartesian, cylindrical, and spherical coordinates; Appendix III.B lists several important mathematical basics and relations in general engineering; and Appendix III.C presents Henry's constant for selected gases in water at moderate pressure.

3.2 CONSERVATION EQUATIONS

The governing equations of transport consist of two major parts: constitutive equations and conservation equations. The former relates molecular-motion-based fluxes to the field variables and material properties. The latter expresses the conservation laws in terms of rates of accumulation, transport, and generation/consumption. In other words, these governing equations can be derived from first principles and expressed in universal forms.

3.2.1 GENERAL FORMS

(1) Integral Form of Conservation Equation

Conservation equations over a finite volume can be derived as an integral form. Figure 3-1 shows a control volume with its volume V and surface S. In Figure 3-1, b denotes the concentration of a quantity. G_V represents the formation or generation rate, such as chemical reaction consumption/production or energy input from external sources. $\vec{u}$ is the bulk flow velocity.

For a fixed control volume, V and S stay constant. Let $\vec{F}$ represent the flux of b and $\vec{n}$ the unit normal vector on the surface with an outward direction (see Fig. 3-1). Then, the rate at which the quantity b enters the control volume across a differential element of surface is $-\vec{F} \cdot \vec{n} \, ds$ and the rate at which the quantity accumulates is bdV. Balancing the rate of accumulation with those of transport and generation gives

$$\left\{ \frac{d}{dt} \int_V b \, dV \right\} = \left\{ -\int_S \vec{F} \cdot \vec{n} \, ds \right\} + \left\{ \int_V G_V \, dV \right\}$$

$$\left\{ \text{accumulation rate} \right\} = \left\{ \text{net transport rate through boundary} \right\} + \left\{ \text{generation rate} \right\}.$$

(Eq. 3-1)

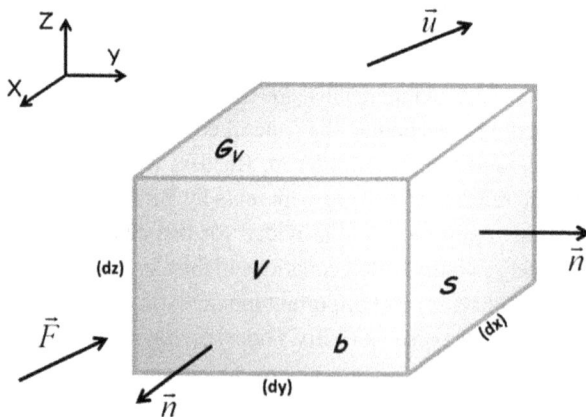

Figure 3-1. A control volume for deriving the conservation equation of a quantity b.

(2) Differential Form of Conservation Equation

From the integral form of the conservation equation, an expression for a point of continuum can be extracted by using the divergence theorem to obtain a differential form without dependence on the arbitrarily chosen control volume.

Through the divergence theorem, the surface integration in the integral form (Eq. 3-1) is converted to the volume integration:

$$\int_S \vec{F} \cdot \vec{n}\, ds = \int_V \nabla \cdot \vec{F}\, dV. \qquad \text{(Eq. 3-2)}$$

Then, Eq. 3-1 is modified to yield

$$\int_V \frac{\partial b}{\partial t}\, dV + \int_V \nabla \cdot \vec{F}\, dV - \int_V G_V\, dV = 0. \qquad \text{(Eq. 3-3)}$$

This equation can be written as

$$\int_V \left(\frac{\partial b}{\partial t} + \nabla \cdot \vec{F} - G_V \right) dV = 0. \qquad \text{(Eq. 3-4)}$$

Because the above is valid for any chosen control volume, the integrand must be zero everywhere in the domain, that is

$$\frac{\partial b}{\partial t} + \nabla \cdot \vec{F} - G_V = 0. \qquad \text{(Eq. 3-5)}$$

(3) Interfacial Condition

The integral form of the conservation equation is applied to obtain an interfacial condition. We fix the coordinate at the interface and extend the space normal to the surface by a small distance δ, which yields a control volume. The production rate of the quantity b is defined as G_S at the interface S_I, whereas G_V is the volumetric production rate in v; see Figure 3-2. Applying Eq. 3-1 to the control volume with the surface integral on each surface expressed separately yields

$$\int_V \frac{\partial b}{\partial t}\, dV = \int_{V_A} \frac{\partial b}{\partial t}\, dV + \int_{V_B} \frac{\partial b}{\partial t}\, dV = -\int_{S_A} \left(\vec{F} \cdot \vec{n} \right)_A dS - \int_{S_B} \left(\vec{F} \cdot \vec{n} \right)_B dS$$
$$- \int_{S_E} \left(\vec{F} \cdot \vec{n} \right)_E dS + \int_V G_V\, dV + \int_{S_I} G_S\, dS, \qquad \text{(Eq. 3-6)}$$

where G_S is the surfacte generation rate at the interface. As the thickness of the control volume approaches to zero (i.e., $\delta \to 0$), both the edge surface area and volume go to zero ($S_E \to 0$, $V \to 0$) and the outer surface of A and B approaches the interface ($S_E \to S_I$, $S_B \to S_I$). Eq. 3-6 is thus changed to

$$\int_{S_I} \left(\vec{F}_B - \vec{F}_A \right) \cdot \vec{n}\, dS = \int_{S_I} G_S\, dS \quad \text{or} \quad \int_{S_I} \left[\left(\vec{F}_B - \vec{F}_A \right) \cdot \vec{n} - G_S \right] dS = 0, \qquad \text{(Eq. 3-7)}$$

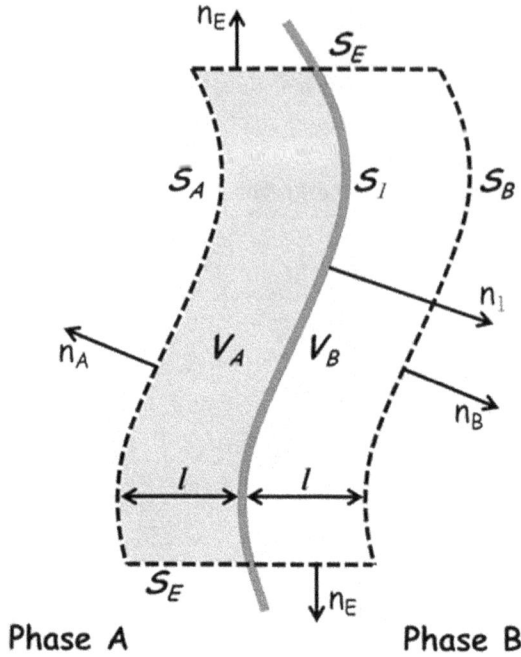

Figure 3-2. A control volume enclosing the interface of phases A and B.

Because the second equation on the above holds true for any chosen S_I, the integrand in the second integral must be zero, that is,

$$\left(\vec{F}_B - \vec{F}_A\right) \cdot \vec{n} = G_S.$$

(Eq. 3-8)

3.2.2 MASS AND MOMENTUM CONSERVATION

(1) CONTINUITY EQUATION

Applying the general form of the conservation equation, Eq. 3-5, to mass conservation yields the continuity equation. In this case, the quantity b becomes the mass density. There are no diffusive flux; thus, $\vec{F} = \rho \vec{u}$, and no sources or sinks for mass. Then, Eq. 3-5 becomes

$$\frac{\partial \rho}{\partial t} + \nabla \cdot \left(\rho \vec{u}\right) = 0.$$

(Eq. 3-9)

For constant density (i.e., ρ = constant), Eq. 3-9 further reduces to

$$\nabla \cdot \vec{u} = 0.$$

(Eq. 3-10)

This simplified equation is valid for incompressible fluid flows. Also, flows under low Mach numbers follow Eq. 3-10 since their compressible effects are negligible.

(2) Momentum Equation

For the momentum equation, the quantity b is set as the product of the density and velocity vectors (i.e., $b = \rho\vec{u}$). The flux F from the convection and shear stress is written as $\vec{F} = -\rho\vec{u}\vec{u} + \bar{\tau}$. If the pressure and gravitational force are included in the source $(\vec{G}_V = -\nabla p + \rho\vec{g})$, then Eq. 3-5 becomes

$$\frac{\partial \rho\vec{u}}{\partial t} + \nabla \cdot \left(\rho\vec{u}\vec{u}\right) = \rho\vec{g} - \nabla p + \nabla \cdot \bar{\tau}. \qquad \text{(Eq. 3-11)}$$

For incompressible Newtonian flow, the above equation reduces to

$$\rho\frac{\partial \vec{u}}{\partial t} + \nabla \cdot \left(\rho\vec{u}\vec{u}\right) = \rho\vec{g} - \nabla p + \mu\nabla^2\vec{u}. \qquad \text{(Eq. 3-12)}$$

3.2.3 ENERGY EQUATION

Heat transfer is important to PEM fuel cells because optimal operating temperature needs to be maintained to ensure high energy-conversion efficiency. In Eq. 3-5, we choose the quantity b as $\rho C_p T$ (i.e., $b = \rho C_p T$). The flux considered in the energy equation is the heat flux q, which combines the fluxes of conduction and convection. Following Fourier's law, the conductive flux equals to $-k\Delta T$, and the convective component is $\rho C_p \vec{u} T$. The total flux is given by

$$\vec{F} = -k\nabla T + \rho C_p \vec{u} T. \qquad \text{(Eq. 3-13)}$$

G_v is the heat generation rate $\dot{q}$ by chemical reaction, resistive heating, or fluid friction. Then, the energy conservation equation becomes

$$\rho C_p \frac{\partial T}{\partial t} + \nabla \cdot (\rho C_p \vec{u} T) = \nabla \cdot \left(k\nabla T\right) + \dot{q}. \qquad \text{(Eq. 3-14)}$$

For constant density (ρ = constant), the continuity equation $(\nabla \cdot \vec{u} = 0)$ can be incorporated to rearrange the above equation to

$$\rho C_p \left(\frac{\partial T}{\partial t} + \vec{u} \cdot \nabla T\right) = \nabla \cdot \left(k\nabla T\right) + \dot{q}. \qquad \text{(Eq. 3-15)}$$

The thermal conductivity k is usually a function of temperature. When the temperature range of interest is narrow, which is the case for PEM fuel cell operation, k can be taken to be constant; consequently, the coefficient k is placed outside of the derivative. The total derivative, also called the material derivative, is defined through combining the partial derivative and the convective flux (i.e., $\dfrac{D}{Dt} = \dfrac{\partial}{\partial t} + \vec{u} \cdot \nabla$), yielding

$$\rho C_p \frac{DT}{Dt} = k\nabla^2 T + \dot{q}. \tag{Eq. 3-16}$$

When the coordinate is fixed with the moving fluid (i.e., $\vec{u} = 0$), the material derivative changes back to the partial differential to yield the familiar heat conduction equation:

$$\rho C_p \frac{\partial T}{\partial t} = k\nabla^2 T + \dot{q} \tag{Eq. 3-17}$$

In the absence of heat generation (i.e., $\dot{q} = 0$), the above equation can be reduced to a simple one, using the thermal diffusivity $\alpha_T \left(= \dfrac{k}{\rho C_p} \right)$ as follows:

$$\frac{\partial T}{\partial t} = \frac{k}{\rho C_p} \nabla^2 T = \alpha_T \nabla^2 T. \tag{Eq. 3-18}$$

3.2.4 SPECIES TRANSPORT EQUATION

For the conservation of chemical species, one can set the quantity b as the molar concentration of species i (i.e., $b = C_i$), $\vec{F} = C_i\vec{u} + \vec{J}_i$ (where $\vec{J}_i$ is the molar flux of species i relative to the mass-average velocity), $G_V = R_{Vi}$ where R is the reaction rate, the species transport equation becomes

$$\frac{\partial C_i}{\partial t} + \nabla \cdot (\vec{u} C_i) = -\nabla \cdot \vec{J}_i + R_{Vi} \tag{Eq. 3-19}$$

For constant mixture density (ρ = constant), the above equation, along with the continuity equation ($\nabla \cdot \vec{u} = 0$), changes to

$$\frac{\partial C_i}{\partial t} + \vec{u} \cdot \nabla C_i = -\nabla \cdot \vec{J}_i + R_{Vi}. \tag{Eq. 3-20}$$

Following the same manner as the energy equation, the left-hand side of Eq. 3-20 is the material derivative:

$$\frac{DC_i}{Dt} = \left(\frac{\partial}{\partial t} + \vec{u} \cdot \nabla \right) C_i = -\nabla \cdot \vec{J}_i + R_{Vi}. \tag{Eq. 3-21}$$

The molar flux $\vec{J}_i$ for a binary mixture of A and B follows Fick's law:

$$\vec{J}_A = -D_{AB}\nabla C_A. \qquad \text{(Eq. 3-22)}$$

For species i that is present in a dilute gas mixture (pseudo-binary situation), Fick's law gives a good approximation:

$$\vec{J}_i = -D_i\nabla C_i. \qquad \text{(Eq. 3-23)}$$

For constant mixture density and species diffusivity, Eq. 3-21 simplifies to

$$\frac{DC_i}{Dt} = D_i\nabla^2 C_i + R_{Vi}. \qquad \text{(Eq. 3-24)}$$

Appendix III.A summarizes the mass, momentum, and energy conservation equations in the Cartesian, cylindrical, and spherical coordinates.

3.3   CONSTITUTIVE EQUATIONS

The basic constitutive equations relate the fluxes of heat, species, and momentum to the field variables and material properties. A simple molecular model is presented below to provide a qualitative estimate (on the order of magnitude) of the transport coefficients.

3.3.1   A LATTICE MODEL

Diffusion stems from the random molecular walks of constituent particles. Figure 3-3 shows a simple Lattice model with an assumption that the movement of a molecule occurs only along any of the six directions. For a particle, the probability of moving in one direction is 1/6, assuming that the chance for movement in each direction is the same. The flux in the x direction, denoted as f_x (which can be heat, species or momentum flux), is expressed as

$$f_x = \frac{1}{6}u_{\text{mol}}\left[b(x,y,z)-b(x+l,y,z)\right]. \qquad \text{(Eq. 3-25)}$$

where u_{mol} and b are, respectively, the molecular velocity magnitude and targeted quantity. Table 3-1 lists the quantity b and its corresponding flux for heat, species, and momentum transfer.

We consider the situation when the jumping length, l, is sufficiently small that the continuum assumption is satisfied; applying the Taylor series expansion for one jump yields

$$b(x+l,y,z)=b(x,y,z)+l\frac{\partial b}{\partial x}\bigg|_{(x,y,z)}+\frac{l^2}{2}\frac{\partial^2 b}{\partial x^2}+O(l^3). \qquad \text{(Eq. 3-26)}$$

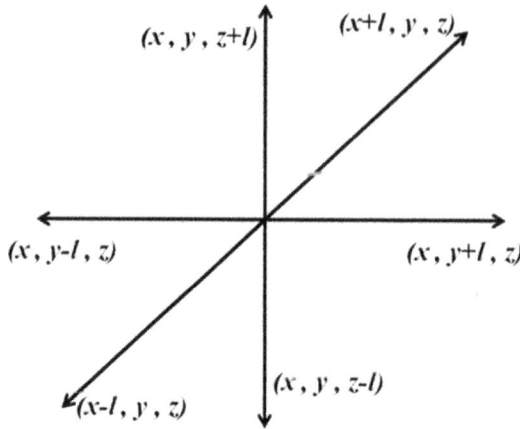

Figure 3-3. A lattice model for evaluting transport properties.

Table 3-1. Quantity b and its corresponding flux f_x

	b	f_x
Heat	$\rho C_p T$	q_x
Species i	C_i	J_{ix}
Momentum	ρu_x	τ_x

By retaining only the first two terms and assuming that the $O(l^3)$ term is small and negligible, substituting Eq. 3-26 into Eq. 3-25 leads to

$$f_x = -\frac{u_{\text{mol}} l}{6} \frac{\partial b}{\partial x}$$

(Eq. 3-27)

For gases, we can use the mean molecular velocity and mean free path, which can be obtained from the elementary kinetic theory of low-density gases, to approximate u_{mol} and l, respectively

$$u_{\text{mol}} = \left(\frac{8RT}{\pi M}\right)^{1/2},$$

(Eq. 3-28)

$$l = \frac{1}{\sqrt{2}\pi d^2 N_{AV} C} = \frac{RT}{\sqrt{2}\pi d^2 N_{AV} p},$$

(Eq. 3-29)

where N_{AV}, M, and d, respectively, denote Avogadro's number (6.02214129 × 10²³ per mole), molar weight, and molar diameter. For a mixture of A and B, the average molar weight and diameter are used:

$$M = 2\left(\frac{1}{M_A} + \frac{1}{M_B}\right)^{-1} \text{ and } d = \frac{d_A + d_B}{2}. \qquad \text{(Eq. 3-30)}$$

By incorporating the expressions of u_{mol} and l in Eqs. 3-28 and 3-29, we can express the thermal conductivity, molecular diffusivity, and momentum diffusivity (or viscosity) as follows:

$$\text{conductivity: } k = \rho C_p \frac{u_{mol}l}{6} = \frac{1}{3\pi^{3/2}} \frac{(MRT)^{1/2} C_p}{N_{AV} d^2} \qquad \text{(Eq. 3-31)}$$

$$\text{diffusivity: } D_{AB} = \frac{u_{mol}l}{6} = \frac{1}{3\pi^{3/2}} \frac{(RT)^{3/2}}{N_{AV} d^2 M^{1/2} p} \qquad \text{(Eq. 3-32)}$$

$$\text{viscosity: } \mu = \rho\frac{u_{mol}l}{6} = \frac{1}{3\pi^{3/2}} \frac{(MRT)^{1/2}}{N_{AV} d^2} \qquad \text{(Eq. 3-33)}$$

The above imply that $\nu = \alpha = D_{AB}$ or $Pr = Sc = 1$ where Pr denotes the Prandtl number $\left(Pr = \frac{\nu}{\alpha}\right)$ and Sc the Schmidt number $\left(Sc = \frac{\nu}{D_{AB}}\right)$. For gases at standard conditions ($T = 293K$, $P = 1$ atm) with $M = 30$g/mol, $d = 0.3$nm, $u_{mol} = 4.5 \times 10^2$ m/s and $l = 1.0 \times 10^{-7}$ m, D_{AB} is around 0.8×10^{-5} m^2/s. The Prandtl (Pr) and Schmidt (Sc) numbers are around unity, for example, $Pr \sim 0.7$–0.8 for air and many other gases. The thermal conductivity (k) and viscosity (μ) are dependent on T but independent of p. However, it is a simplified model, yielding inaccuracies, for example, k and μ are proportional to T instead of $T^{1/2}$; D_{AB} is proportional to $T^{7/4}$ instead of $T^{3/2}$. One possible cause for the inaccuracies is the exclusion of molecular attraction and repulsion. The above derivation is based on the simplest random walk of molecules using the average free path length and mean velocity developed in the kinetic theory. A major barrier to molecular movement in gases is the collision with other molecules. An example of intermolecular interaction model is the Lenard–Jones potential formula, expressed as follows:

$$\psi(r) = 4\varepsilon\left[\left(\frac{d^*}{r}\right)^{12} - \left(\frac{d^*}{r}\right)^6\right], \qquad \text{(Eq. 3-34)}$$

where $\psi(r)$ is the intermolecular potential energy, and ε, d^*, and r are, respectively, the depth of the potential well, the distance at which the interparticle potential is zero, and the distance between the particles. Readers who are interested in a detailed discussion of the Lattice model are referred to consult the reference text by Dean [2].

3.3.2 FOURIER'S LAW AND FICK'S LAW

Fourier's law of heat conduction states that the time rate of heat conduction through a material is proportional to the negative gradient of temperature, that is

$$\vec{q} = -k\nabla T. \tag{Eq. 3-35}$$

where k is a parameter called thermal conductivity; it is not a constant, instead varies with temperature. For most pure metals, k decreases with temperature, whereas for gases it increases. It is an important property of a material that controls the conductive heat flow in the medium. Thermal conductivity varies widely among various materials, ranging from around 0.01 W/m K to over 100 W/m K. Pure metals are highly thermally conductive (i.e., k is large), whereas gases or vapors exhibit low conductivities (see Appendix VIII.A).

Fick's first law of diffusion describes the relation between species diffusive flux and concentration gradient, originally developed for binary mixtures, but has been customarily used for dilute solutions; mathematically, it is expressed as

$$\vec{J}_i = -D_i\nabla C_i, \tag{Eq. 3-36}$$

where D_i is the diffusion coefficient of species i, which is a function of temperature. Gaseous diffusivity can be approximated using the Lattice model as described in the preceding section. The coefficient in solids at different temperatures is often found to follow the form of the Arrhenius constitutive relation (which describes exponential temperature dependence). Diffusion in a liquid can be analyzed using a hydrodynamic model—this approach is developed by assuming that the resistance of solute molecular movement arises from the viscous force, similar to particle movement in viscous fluids. In dilute liquids, the approach yields the well-known Stokes–Einstein equation:

$$D = \frac{k_B T}{6\pi r\mu}, \tag{Eq. 3-37}$$

where k_B is Boltzmann's constant, r is the radius of the oxygen molecule, and μ is the fluid viscosity. The diffusion coefficients in solid or liquid are usually much smaller than that in gases, as shown in Figure 2-3.

3.4 SCALING AND DIMENSIONLESS GROUPS

3.4.1 SCALING AND DIMENSIONLESS EQUATIONS

Scaling an equation yields a mathematical description that is independent of physical units. Scaling is also often used to estimate the orders of magnitude for the terms of interest. In the previous section, the conservation equations of momentum, energy, and species concentration are derived; they are summarized as follows:

$$\rho\frac{D\vec{u}}{Dt} = \rho g - \nabla p + \mu\nabla^2\vec{u}, \qquad\qquad \text{(Eq. 3-38)}$$

$$\rho c_p\frac{DT}{Dt} = k\nabla^2 T + \dot{q}, \qquad\qquad \text{(Eq. 3-39)}$$

$$\frac{DC_i}{Dt} = -\nabla\cdot\vec{J}_i + R_{Vi} = D_i\nabla^2 C_i + R_{Vi}. \qquad\qquad \text{(Eq. 3-40)}$$

For each field variable, a reference value can be chosen to define a dimensionless variable:

$$\vec{u}^* = \frac{\vec{u}}{U_0},\, x^* = \frac{x}{L},\, y^* = \frac{y}{L},\, z^* = \frac{z}{L},$$

$$C_i^* = \frac{C_i}{C_{i,0}},\, t^* = \frac{t}{L/U_0},\, T^* = \frac{T}{T_1 - T_0}, \qquad\qquad \text{(Eq. 3-41)}$$

$$p^* = \frac{p - p_0}{\rho_0 U_0^2}\ \text{where}\ \rho = \rho_0 = \text{const.}$$

With the dimensionless variables, the conservation equations are rearranged to yield nondimensional forms. For the momentum equation (Eq. 3-38), the following dimensionless form can be derived:

$$\frac{D\vec{u}^*}{Dt^*} = -\nabla^* p^* + \left(\frac{g_0 L}{U_0^2}\right)\vec{g}^* + \left(\frac{\mu}{LU_0\rho_0}\right)\nabla^{*2}\vec{u}^*, \qquad\qquad \text{(Eq. 3-42)}$$

or

$$\frac{D\vec{u}^*}{Dt^*} = -\nabla^* p^* + \frac{1}{Fr}\vec{g}^* + \frac{1}{Re}\nabla^{*2}\vec{u}^*, \qquad\qquad \text{(Eq. 3-43)}$$

where $Fr\left(\equiv\dfrac{U_0^2}{g_0 L}\right)$ is the Froude number, which is defined as the ratio of the inertial to gravitational forces, and $Re\left(\equiv\dfrac{LU_0\rho_0}{\mu}\right)$ denotes the Reynolds number, which is the ratio between the inertial and viscous forces.

Similarly, substituting the dimensionless variables into the energy equation, Eq. 3-39, yields

$$\rho C_p\frac{T_1 - T_0}{L/U_0}\frac{DT^*}{Dt^*} = k\frac{T_1 - T_0}{L^2}\nabla^{*2}T^* + \dot{q}. \qquad\qquad \text{(Eq. 3-44)}$$

Re-organizing the above equation leads to

$$\frac{DT^*}{Dt^*} = \frac{k}{\rho C_p U_0 L} \nabla^{*2} T^* + \frac{\dot{q}}{\rho C_p \frac{T_1 - T_0}{L / U_0}} = \frac{k}{\rho C_p \nu} \frac{\mu}{L \rho_0 U_0} \nabla^{*2} T^* + \dot{q}^*, \text{ (Eq. 3-45)}$$

or

$$\frac{DT^*}{Dt^*} = \frac{1}{Pr} \frac{1}{Re} \nabla^{*2} T^* + \dot{q}^*, \tag{Eq. 3-46}$$

where $Pr \left(\equiv \frac{\rho C_p \nu}{k} = \frac{\nu}{\alpha_T} \right)$ is the Prandtl number, denoting the importance of momentum diffusion relative to thermal diffusion. Lastly, the dimensionless species transport equation takes the following form:

$$\frac{C_{i,0}}{L / U_0} \frac{D^* C_i^*}{D^* t^*} = D_i \frac{C_{i,0}}{L^2} \nabla^{*2} C_i^* + R_{Vi}, \tag{Eq. 3-47}$$

Rearranging the above equation, one will reach

$$\frac{D^* C_i^*}{D^* t^*} = \frac{D_i}{U_0 L} \nabla^{*2} C_i^* + \frac{R_{Vi}}{C_{i,0} U_0} L, \tag{Eq. 3-48}$$

or

$$\frac{D^* C_i^*}{D^* t^*} = \frac{1}{Pe} \nabla^{*2} C_i^* + R_{Vi}^*, \tag{Eq. 3-49}$$

where $Pe \left(\equiv \frac{U_0 L}{D_i} \right)$ is the Peclet number for species i, representing the ratio between convection and diffusion effects. Dimensionless boundary conditions can be derived in a similar way. For the energy equation, the boundary condition of convection and conduction (or Newton's law of cooling) reads

$$-k_l \frac{\partial T_l}{\partial y} = h(T_w - T_\infty), \tag{Eq. 3-50}$$

where y is the direction outward normal to the boundary surface. Substituting Eq. 3-41 into Eq. 3-50 leads to

$$-\frac{\partial T_l^*}{\partial y^*} = \frac{Lh}{k_l} \left(T_w^* - T_\infty^* \right). \tag{Eq. 3-51}$$

The lumped coefficient on the right side is a dimensionless number, called the Nusselt number, $Nu \left(\equiv \frac{Lh}{k_l} \right)$, which evaluates the relative importance of convection to conduction within a fluid:

$$-\frac{\partial T_l^*}{\partial y^*} = Nu\left(T_w^* - T_\infty^*\right). \qquad \text{(Eq. 3-52)}$$

Similarly in the chemical species equation, the dimensionless form of the convection-diffusion boundary condition can be derived as follows:

$$-D_i \frac{\partial C_i}{\partial y} = h_i \left(C_{i,0} - C_{i,\infty}\right), \qquad \text{(Eq. 3-53)}$$

$$-\frac{\partial C_i^*}{\partial y^*} = \frac{h_i L}{D_i}\left(C_{i,0}^* - C_{i,\infty}^*\right). \qquad \text{(Eq. 3-54)}$$

The Sherwood number, $Sh \left(\equiv \dfrac{h_i L}{D_i}\right)$, represents the ratio of convection and diffusion effects in

a fluid. This boundary condition is then rewritten as

$$-\frac{\partial C_i^*}{\partial y^*} = Sh\left(C_{i,0}^* - C_{i,\infty}^*\right). \qquad \text{(Eq. 3-55)}$$

As for the energy equation in a solid object, the convection boundary condition reads

$$-k_s \frac{\partial T_s}{\partial y} = h(T_s - T_\infty). \qquad \text{(Eq. 3-56)}$$

Similarly, its dimensionless form is derived as

$$-\frac{\partial T_s^*}{\partial y^*} = \frac{Lh}{k_s}\left(T_s^* - T_\infty^*\right) = Bi\left(T_s^* - T_\infty^*\right), \qquad \text{(Eq. 3-57)}$$

which is similar to Eq. 3-52, but with a different dimensionless number, called the Biot number, Bi $\left(\equiv \dfrac{Lh}{k_s}\right)$, which compares the thermal resistance in a solid to the flow resistance in a fluid. In the occasion of a sufficiently small Biot number ($Bi << 1$), the temperature variation within the solid is small and negligible in analysis; as a result, the lumped-capacity system assumption is applicable.

Likewise, a Biot number (Bi_j) for species transport is defined as the ratio of the diffusion resistance in a solid to that in a fluid. The boundary condition of the species equation thus reads

$$-D_{is} \frac{\partial C_{is}}{\partial y} = h_i \left(C_{if,0} - C_{if,\infty}\right) = \frac{h_i}{k_i}\left(C_{is,0} - k_i C_{if,\infty}\right), \qquad \text{(Eq. 3-58)}$$

Table 3-2. List of dimensionless groups

Dimensionless number	Numerical expression	Physical meaning
Reynolds number	$Re = \dfrac{\rho u L}{\mu}$	$\dfrac{\text{Inertial force}}{\text{Viscous force}}$
Froude number	$Fr = \dfrac{u^2}{gL}$	$\dfrac{\text{Inertial force}}{\text{Gravitational force}}$
Prandtl number	$Pr = \dfrac{C_p \mu}{k} = \dfrac{\nu}{\alpha_T}$	$\dfrac{\text{Momentum diffusion}}{\text{Thermal diffusion}}$
Peclet number	$Pe = \dfrac{uL}{\alpha_T}$	$\dfrac{\text{Heat convection}}{\text{Thermal diffusion}}$
Peclet number for species i	$Pe_i = \dfrac{uL}{D_i}$	$\dfrac{\text{Species convection}}{\text{Species diffusion}}$
Schmidt number	$Sc = \dfrac{\nu}{D_i}$	$\dfrac{\text{Momentum diffusion}}{\text{Mass diffusion}}$
Lewis number	$Le = \dfrac{\alpha_T}{D_i}$	$\dfrac{\text{Thermal diffusion}}{\text{Mass diffusion}}$
Sherwood number	$Sh = \dfrac{h_i L}{D_i}$	$\dfrac{\text{Convection in a fluid}}{\text{Diffusion in a fluid}}$
Biot number	$Bi = \dfrac{hL}{k_s}$	$\dfrac{\text{Solid thermal resistance}}{\text{Fluid thermal resistance}}$
Biot number for species i	$Bi_i = \dfrac{h_i L}{k_i D_{i,s}}$	$\dfrac{\text{Solid resistance of species transfer}}{\text{Fluid resistance of species transfer}}$
Nusselt number	$Nu = \dfrac{hL}{k_l}$	$\dfrac{\text{Convection in a fluid}}{\text{Conduction in a fluid}}$
Stanton number	$St = \dfrac{h}{\rho u C_p} = \dfrac{Nu}{Re\,Pr}$	$\dfrac{\text{Convection into a fluid}}{\text{Heat capacity of a fluid}}$
Damkohler number	$Da = \dfrac{k_{sn} C_{i0}^{n-1} L}{D_i}$	$\dfrac{\text{Reaction rate}}{\text{Diffusion rate}}$

$$-\frac{\partial C_{is}^{*}}{\partial y^{*}} = \frac{h_i L}{k_i D_{is}}\left(C_{is,0}^{*} - k_i C_{if,\infty}^{*}\right) = Bi_i \left(C_{is,0}^{*} - k_i C_{if,\infty}^{*}\right), \qquad \text{(Eq. 3-59)}$$

where the subscript f represents fluid and k_i is Henry's constant for species i. Appendix III.C lists Henry's constant for selected gases in water at moderate pressure.

3.4.2 DIMENSIONLESS GROUPS

In the preceding section, a number of dimensionless groups or numbers appear in the dimensionless conservation equations and boundary conditions. In general heat and mass transfer, these dimensionless groups play vital roles in fundamental understanding, theoretical analysis, and engineering design. The most frequently used dimensionless numbers, along with their physical meanings, are summarized in Table 3-2.

3.5 CHAPTER SUMMARY

This chapter presents the fundamental governing equations of mass, heat, and species transport, the basic constitutive correlations, and the major dimensionless groups that are relevant to PEM fuel cell analysis. Key points are summarized below:

- The general conservation equation is derived, along with the conditions at the interfaces between components. The mass, heat, and species transport equations are obtained from the general formulation of conservation laws, showing the similarities among the three types of transports.
- Two basic constitutive correlations, which relate the fluxes to the field variables and material properties, are presented. Fourier's and Fick's laws are explained, and a lattice model is introduced to provide a qualitative analysis of two different diffusive processes from the prespective of molecular random movement.
- The dimensionless forms of the governing equations are derived for mass, momentum, heat, and species transport. Major dimensionless groups, pertinent to the heat and mass transport in PEM fuel cells, are defined and tabulated.

REFERENCES

1. Bird, R.B., Stewart, W.E., and Lightfoot, E.N. *Transport Phenomena*, 2nd Ed. Wiley, (2001).
2. Deen, W.M. *Analysis of Transport Phenomena*. Oxford University Press, (1998).
3. Kays, W.M., Crawford, M.E., and Weigand, B. *Convective Heat and Mass Transfer*, 4th Ed. McGraw-Hill, (2004).

CHAPTER 4

WATER AND ITS TRANSPORT IN THE POLYMER ELECTROLYTE MEMBRANE

A polymer electrolyte membrane is essential to PEM fuel cell operation: the membrane facilitates the transport of hydrogen ions, that is, protons, from the anode to the cathode so that they are available for the ORR (oxygen reduction reaction) that takes place in the cathode. Electrons needed for the ORR are conducted from the anode to the cathode via an external wire rather than the membrane; in fact, the membrane is designed to resist electron transport so as to avoid creating a "short circuit." Furthermore, the membrane must not allow reactant gases to crossover to avoid the direct contact with each other. The ohmic voltage loss is directly affected by the membrane ionic conductivity and membrane thickness. In this chapter, we present an overview of membrane materials, ionic conductivity correlations, water transport phenomena (or physics), ionic conductivity in catalyst layer, and local reaction rate in multilayered catalyst layers. Readers who are interested in more detailed discussions on the fundamentals (e.g., chemistry, structure, and properties) of perfluorinated ionomer membranes or perfluorosulfonic acid-based membranes such as Nafion (which is currently the workhorse membrane for PEM fuel cells) are referred to consult with the review papers by, respectively, Heitner-Wirguin [1] and Mauritz and Moore [2].

In addition, two appendices are provided for this chapter: Appendix IV.A summarizes PEM fuel cell performance and membrane properties for various membrane materials, including acid-doped PBI (phosphoric acid-doped polybenzimidazole) membranes, inorganic-organic composite membranes, and modified SPFA membranes, and Appendix IV.B provides a brief introduction to ion transport in electrolyte solutions.

4.1 INTRODUCTION TO THE POLYMER ELECTROLYTE MEMBRANE

In PEM fuel cells, the membrane refers to the thin layer of solid polymer electrolyte (~10–100 μm, e.g., 18 μm for Gore 18 and 175 μm for Nafion® 117), which conducts protons from the anode to cathode; see Figure 1-6. Desirable membrane materials are those that simultaneously

- exhibit high ionic conductivity,
- prevent electron transport, and
- inhibit the cross-over of hydrogen and oxygen reactant gases.

In addition, they must be chemically stable in an environment with HO- and HOO radicals, thermally stable throughout the operating temperature and mechanically robust. Membranes currently in use are mostly based on perfluorosulfonic acid, the most prominent of which, Nafion, was first developed by DuPont Company in the 1960s. Nafion® has a backbone structure of polytetrafluoroethylene (or PTFE, known by the trade name of Teflon), which provides the membrane with physical strength. The sulfonic acid functional groups in Nafion® provide charge sites for proton transport. Besides Nafion, other perfluorinated polymeric materials such as Neosepta-F™ (Tokuyama), Gore-Select™ (W.L. Gore and Associates, Inc.), Flemion™ (Asahi Glass Company), and Asiplex™ (Asahi Chemical Industry) are also adopted as the membrane for PEM fuel cell applications. In addition, membrane materials that can withstand high temperatures (100°C–200°C) are preferred for PEM fuel cells that operate at high temperatures, which have the advantages of better catalyst tolerance to CO poisoning, more efficient cooling strategies, and most importantly better water management.

The Nafion-based membranes are costly due to their complex fabrication process. Research on cost-effective high-performance electrolyte materials has been active from the very beginning of fuel cell development. Solvay Solexis has been developing Hyflon® ion ionomers, also known as short side chain (SSC) ionomer (which was originally developed by Dow Chemicals Company but abandoned later on), that can exhibit better performance than Nafion® under certain conditions. However, severe degradation was observed for this membrane material. A PBI membrane is a promising material for high temperature applications due to its high proton conductivity at temperatures up to 200°C and low methanol or ethanol permeability. However, there are concerns on low proton conductivity at low temperatures (important for cold start) and low solubility of oxygen. Hydrocarbon-based membranes have been attempted by Sandia National Laboratories and PolyFuel for fuel-cell applications. Appendix IV.A summarizes PEM fuel cell performance and membrane properties for several membrane materials, including acid-doped PBI membranes, inorganic–organic composite membranes, and modified SPFA membranes. As membrane materials are critical to PEM fuel cell performance, their properties determine the major research direction of PEM fuel cells: for example, Nafion needs to be hydrated in order to be a good ionic conductor. Accordingly, water management has attracted a lot of attention. Nevertheless, Nafion is the best candidate for PEM fuel cells to date. Currently, research on membrane and water management is very active and ongoing. Fundamental understanding of the proton transport process provides a foundation of knowledge for electrolyte material design and development. The desirable properties of an electrolyte-membrane material include high ionic conductivity, high electronic resistance, low fuel crossover, low manufacturing cost, good chemical stability, and sufficient mechanical strength. Among them, transport property plays the most important role in evaluating electrolyte-membrane materials. Specifically, ion- and water-transport properties are the most important ones in electrolyte membranes for PEM fuel cells.

4.2   ION TRANSPORT AND IONIC CONDUCTIVITY

4.2.1   PROTON TRANSPORT

Ionic conductivity is an essential feature of electrolyte for fuel cell applications. Because of this feature, the protons produced in the anode are able to transport through the electrolyte to the cathode to react with oxygen. Fuel cells are classified according to the electrolyte material used. At low temperatures (e.g., 80°C), the Nafion or Nafian-based materials exhibit good ionic conductivity and can be fabricated into thin membranes to reduce overall resistance. Gierke and Hsu [3] described the polymeric membrane characteristics using a cluster model in terms of an inverted micellar structure in which the ion-exchange sites are separated from the fluorocarbon backbone, forming spherical clusters (pores) that are connected by short narrow channels; see Figure 4-1. The cluster sizes grow with increasing water content. The main driving force for proton transport is the gradient of electrical potential in the electrolyte. That is, protons transport across the membrane mainly due to the presence of the electrolyte potential gradient. Water in the membrane is essential for proton transport: one mechanism is called the "vehicular" diffusion. By forming hydronium ions (H_3O^+), protons can "ride along" or transport with the movement of water so this phenomenon is generally called the *vehicular diffusion*. This vehicular-diffusion mechanism largely depends on the water diffusivity in membranes. Another mechanism is through the "hopping" process that takes place when sufficient water is present so that the side chains of sulfonic groups are connected; and consequently, protons can move directly from one site to another. In addition, protons can be transported via self-diffusion (driven by the concentration gradient of protons), but it is usually slow in comparison with migration (which is driven by the electrolyte potential gradient); therefore, proton transport via self-diffusion is neglected in most situations.

4.2.2   IONIC CONDUCTIVITY CORRELATIONS

One key technical challenge in evaluating fuel cell performance is the ability to accurately compute the potential drop across the electrolyte membrane. Ohm's law states that the voltage (or electrical potential) drop is proportional to the electrical current with the proportionality being

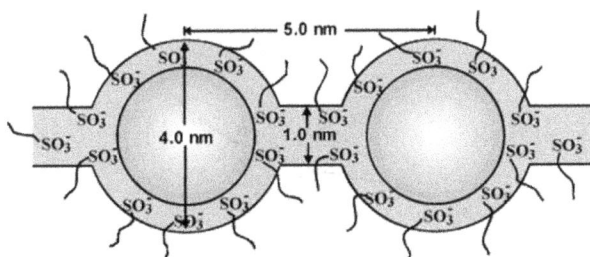

Figure 4-1. Cluster-network model for the morphology of hydrated Nafion [2, 3] (Courtesy of American Chemical Society).

the electrical resistance. Adopting Ohm's law to describe proton transport through the membrane, electrolyte potential drop is proportional to ionic current with the proportionality being the inverse of ionic conductivity. Thus, the accuracy in computing the potential drop across the membrane (or the ionic current through the membrane) depends on the ability to properly estimate the conductivity of the polymer electrolyte.

(1) The Empirical Correlation of Springer et al.

One of the most prevalently used proton-conductivity models is the empirical correlation given in Eq. 4-1, which was developed by Springer *et al.* [6] for the Nafion 117 membrane (for $\lambda > 1$):

$$\kappa_{30} = 0.005193\,\lambda - 0.00326, \qquad \text{(Eq. 4-1)}$$

where λ is the membrane's water content, defined as the number of moles of water per mole of sulfonic acid groups or sites attached to the membrane (namely, SO_3H), and is a function of water activity, as shown in Figure 4-2. κ_{30} is proton conductivity at 30°C in units of $\Omega^{-1}\,cm^{-1}$ or $S\,cm^{-1}$. Springer *et al.*'s empirical correlation, Eq. 4-1, was based on the experimental data of Zawodzinski *et al.* [4, 7]. To account for the temperature dependence of proton conductivity, Springer *et al.* [6] then used a standard Arrhenius relationship and an activation energy of 2.52 kcal mol^{-1} (using data from Zawodzinski *et al.* [4, 7]) to obtain Eq. 4-2:

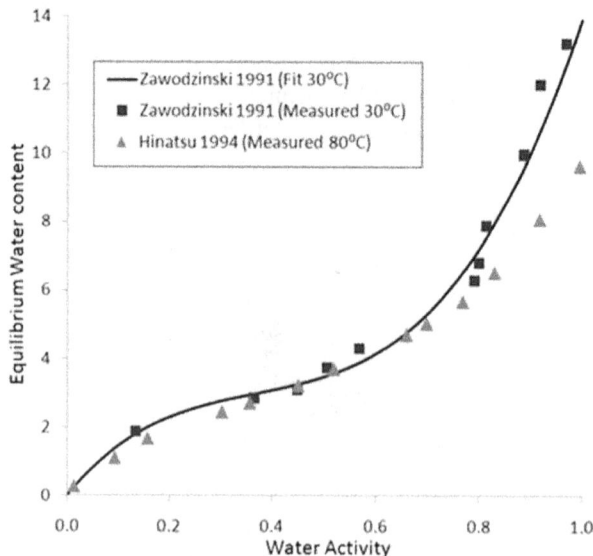

Figure 4-2. Equilibrium water uptake relation or water content as a function of water activity) for Nafion® membranes at 30°C [4, 5].

$$\kappa = \exp\left[1268\left(\frac{1}{303} - \frac{1}{273+T}\right)\right]\kappa_{30}. \qquad \text{(Eq. 4-2)}$$

A limitation of the above correlation is subfreezing temperature, in which a portion of the water freezes and hence has a negligible contribution to the ionic conductivity. Wang *et al.* [8] experimentally measured the Nafion membrane conductivity and developed a correlation as a function of temperature and water content, as shown in the following equation (also see Fig. 4-3):

$$\kappa = (0.01862\lambda - 0.02854)\exp\left[4029\left(\frac{1}{303} - \frac{1}{T}\right)\right] \text{ or}$$

$$\kappa = (0.004320\lambda - 0.006620)\exp\left[4029\left(\frac{1}{273} - \frac{1}{T}\right)\right] = \kappa_0(\lambda)\exp\left[4029\left(\frac{1}{273} - \frac{1}{T}\right)\right] \text{ for } \lambda \leq 7.22$$

$$\kappa = \kappa(\lambda = 7.22) \text{ for } \lambda > 7.22. \qquad \text{(Eq. 4-3)}$$

Figure 4-3 indicates that at $-30°C$ the ionic conductivity extended from Springer *et al.*'s correlation is almost an order-of-magnitude larger than the experimental data. One reason for the observed lower conductivity is that a portion of water in the membrane freezes, and consequently it has a small contribution to the membrane ionic conductivity.

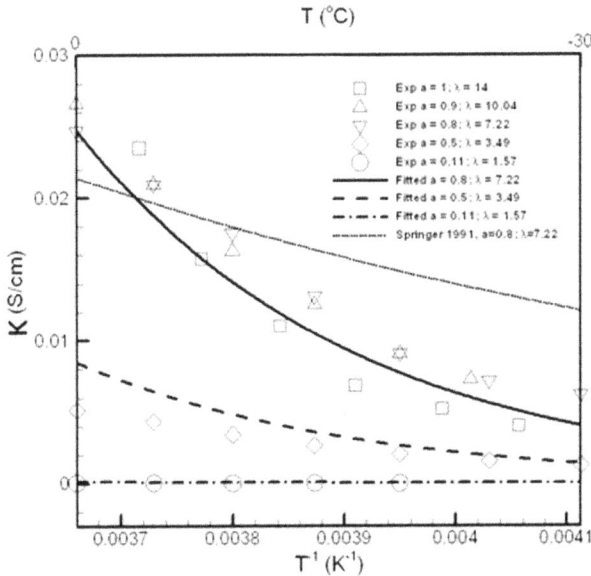

Figure 4-3. Ionic conductivity of Nafion® 117 membrane at subfreezing conditions [8].

(2) Derivation of a Constitutive Model

Appendix IV.B provides a brief introduction to ion transport in electrolyte solutions and derives the relevant governing equations. In order to develop a more general correlation, one can derive from the relationship shown in Eq. 4-4, that is, current density is generally related to the species molar flux using Faraday's law (cf., Newman and Thomas-Alyea [9]):

$$\vec{i} = F\sum_{i=1}^{n}[z_i(\vec{J}_i + c_i\vec{v})],$$
(Eq. 4-4)

where F is the Faraday's constant ($\equiv 96487$ C mol^{-1}), z_i, c_i, and $\vec{J}_i$ are, respectively, the charge number, molar concentration, and molar migration/diffusive flux of species i, and $\vec{v}$ is mixture-fluid velocity due to convection. Due to electro-neutrality (i.e., $\sum z_i c_i = 0$), the second term at the right-hand side of Eq. 4-4 (i.e., the convective term) drops out. Also, in most PEM fuel cell operations, migration flux dominates over diffusive flux, and this leads to the following equation:

$$\vec{i} = F\sum_{i=1}^{n}z_i\vec{J}_i = -F^2\sum_{i=1}^{n}z_i^{2}u_i c_i\nabla\Phi,$$
(Eq. 4-5)

where u_i is the species mobility and Φ is the electrolyte potential. Using the Nernst–Einstein equation to relate species mobility to diffusivity and providing that proton is the only mobile and charged species in the membrane, we obtain Eq. 4-6 by setting $z_{H^+} = 1$:

$$\vec{i} = -\left(\frac{F^2 D_{H^+}^{\text{eff}} c_{H^+}}{RT}\right)\nabla\Phi = -\kappa\nabla\Phi,$$
(Eq. 4-6)

where κ is the proton conductivity given by

$$\kappa = \frac{F^2 D_{H^+}^{\text{eff}} c_{H^+}}{RT},$$
(Eq. 4-7)

with R being the universal gas constant, T temperature, $D_{H^+}^{\text{eff}}$ the effective diffusivity, and c_{H^+} the molar concentration of protons. Next, c_{H^+} and $D_{H^+}^{\text{eff}}$ as functions of water content (λ) are unknown and need to be expressed. It turns out that there are two different ways of deriving the same constitutive model, depending on how c_{H^+} is determined. A simpler way will be presented first and then the same constitutive model using an approach similar to that taken by Thampan et al. [10] will be derived.

To determine c_{H^+}, the assumptions can be made that the membrane swells freely and that the molar volume of the hydrated membrane, V, changes based on the assumption of constant additive molar volumes of water and the dry membrane (Weber and Newman [11]), as expressed in the following equation:

$$V = \bar{V}_m + \lambda\bar{V}_w.$$
(Eq. 4-8)

Then, the molar concentration of protons is simply given by

$$c_{\text{H}^+} = \frac{1}{\bar{V}_m + \lambda \bar{V}_w} = \frac{1}{\bar{V}_m (1 + \alpha \lambda)}, \tag{Eq. 4-9}$$

where $\alpha \equiv \bar{V}_w / \bar{V}_m$. Now, the effective diffusivity of protons can be related to the continuum diffusion coefficient, D_{H^+}, by employing a Bruggeman correction exponent of q and adopting a percolation threshold of f_{v0} (Weber and Newman [11]), as follows:

$$D_{\text{H}^+}^{\text{eff}} = D_{\text{H}^+} (f_v - f_{v0})^q, \tag{Eq. 4-10}$$

where f_{v0} is the threshold value for the volume fraction (f_v) of the aqueous phase (i.e., water) in the polymer, and q is the critical exponent. f_{v0} can be expressed explicitly in terms of water content as follows:

$$f_v = \frac{\lambda \bar{V}_w}{\bar{V}_m + \lambda \bar{V}_w} = \frac{\alpha \lambda}{1 + \alpha \lambda}, \tag{Eq. 4-11}$$

where $\bar{V}_w$ is the molar volume of water and $\bar{V}_m$ is the molar volume of dry membrane. Equation 4-10 shows that $D_{\text{H}^+}^{\text{eff}}$ approaches D_{H^+} as $f_v \rightarrow 1$ (that is, as water content is sufficiently high per Eq. 4.11) (and f_{v0} is much smaller than 1, for example, 0.06 as measured by Morris and Sun [13] and used in their correlation; see discussion below in the next subsection on other models and correlations), indicating that D_{H^+} can be taken as the proton diffusivity at infinite dilution in water, $D_{\text{H}^+}^{0,w}$. Substituting Eqs. 4-9 and 4-10 into Eq. 4.7 and replacing D_{H^+} by $D_{\text{H}^+}^{0,w}$ yields.

$$\kappa = \frac{F^2}{RT} \frac{D_{\text{H}^+}^{0,w}}{\bar{V}_m (1 + \alpha \lambda)} (f_v - f_{v0})^q. \tag{Eq. 4-12}$$

The value of proton diffusivity at infinite dilution in water is well documented in the open literature and text books; for example, Newman and Thomas-Alyea [9] reports a value of $9.312 \times 10^{-5} \text{ cm}^2 \text{ s}^{-1}$ for $D_{\text{H}^+}^{0,w}$ at 25°C.

Equation 4.12 can be derived by an approach similar to that taken by Thampan et al. [10]. In this approach, the molar concentration of protons in Eq. 4.7 can be evaluated using the volume of pore solution, as follows:

$$c_{\text{H}^+} = \frac{1}{\lambda \bar{V}_w}. \tag{Eq. 4-13}$$

Because c_{H^+} is now defined on the basis of per unit volume of pore solution, D_{H^+} can be evaluated accordingly by taking into account the proton interactions with water and membrane when migrating through the pores as follows:

$$\frac{1}{D_{H^+}} = \frac{1}{D_{H^+}^w} + \frac{1}{D_{H^+}^m}, \qquad \text{(Eq. 4-14)}$$

where $D_{H^+}^w$ and $D_{H^+}^m$ are diffusivities for proton interactions, respectively, with water and membrane matrix. Re-writing Eq. 4-14 yields

$$\frac{1}{D_{H^+}} = \frac{1}{D_{H^+,w}}\left(1 + \frac{D_{H^+}^w}{D_{H^+}^m}\right). \qquad \text{(Eq. 4-15)}$$

Now, taking $D_{H^+}^w/D_{H^+}^m$ to be inversely proportional to water content with $\bar{V}_m/\bar{V}_w = \alpha^{-1}$ as the proportionality constant, the limit expressed in Eq. 4-16 is reached:

$$\frac{D_{H^+}^w}{D_{H^+}^m} = \frac{1}{\alpha\lambda}. \qquad \text{(Eq. 4-16)}$$

Substituting Eq. 4-16 into Eq. 4-15 yields an expression for D_{H^+} shown in the following equation:

$$D_{H^+} = \frac{\alpha\lambda}{1+\alpha\lambda}D_{H^+}^w. \qquad \text{(Eq. 4-17)}$$

It is interesting to note from Eq. 4-17 that D_{H^+} approaches $D_{H^+}^w$ as λ becomes sufficiently large (that is, $\lambda \gg 1/\alpha$), which implies that $D_{H^+}^w$ represents the value of proton diffusivity at infinite dilution in water, $D_{H^+}^{o,w}$. Substituting Eq. 4-17 into Eq. 4-10 gives

$$D_{H^+}^{eff} = D_{H^+}^w \frac{\alpha\lambda}{1+\alpha\lambda}(f_v - f_{v0})^q. \qquad \text{(Eq. 4-18)}$$

Substituting Eqs. 4-13 and 4-18 into Eq. 4-7 yields

$$\kappa = \frac{F^2}{RT}\frac{\alpha}{\bar{V}_w}\frac{D_{H^+}^w}{(1+\alpha\lambda)}(f_v - f_{v0})^q. \qquad \text{(Eq. 4-19)}$$

We know the equivalencies shown as follows:

$$\frac{\alpha}{\bar{V}_w} = \frac{\dfrac{\bar{V}_w}{\bar{V}_m}}{\bar{V}_w} = \frac{1}{\bar{V}_m}. \qquad \text{(Eq. 4-20)}$$

Upon substituting Eq. 4-20 into Eq. 4-19 and replacing $D_{H^+}^w$ by $D_{H^+}^{o,w}$, one obtains

$$\kappa = \frac{F^2}{RT} \frac{D_{H^+}^{0,w}}{\bar{V}_m (1+\alpha\lambda)} (f_v - f_{v0})^q, \qquad \text{(Eq. 4-21)}$$

which is the same as Eq. 4.12. In essence, the same constitutive model is derived in two different ways.

It should be pointed out that a key difference between the derivation presented in Eq. 4-12 and the model of Thampan et al. [10] for Eq. 4-21 lies in the treatment of $D_{H^+}^w / D_{H^+}^m$. Thampan et al. considered it to be a constant parameter (though they do expect it to increase as the water content of the membrane decreases). However, in the derivation presented in Eq. 4-12, it is taken to be inversely proportional to water content such that D_{H^+} recovers the upper limiting value of proton diffusion at infinite dilution in water and reasonably represents the lower limiting values at vanishingly small water content.

Setting f_{v0} to 0.06 and q to 1.5 and substituting Eq. 4-11 in either Eq. 4-12 or Eq. 4-21 yields a new constitutive model, given in Eq. 4-22, for proton conductivity in the membrane, which depends on the molar volumes of the dry membrane and water content but otherwise requires no adjustable parameters:

$$\kappa_{30} = \frac{F^2}{RT} \frac{D_{H^+}^{0,w}}{\bar{V}_m (1+\alpha\lambda)} \left(\frac{\alpha\lambda}{1+\alpha\lambda} - 0.06 \right)^{1.5}, \qquad \text{(Eq. 4-22)}$$

where the subscript 30 indicates that conductivity computed from this model are those at 30°C, and $D_{H^+}^{0,w}$ again refers to the value of proton diffusivity at infinite dilution in water. To account for temperature effects, a standard Arrhenius relationship with an activation energy of 2.26 kcal mol^{-1} can be used to arrive at the constitutive model, shown in Eq. 4-23, which relates conductivity to membrane water content and temperature:

$$\kappa = \kappa_{30} \exp\left[1137 \left(\frac{1}{303} - \frac{1}{T} \right) \right]. \qquad \text{(Eq. 4-23)}$$

(3) Other Models or Correlations

Hsu et al. [12] proposed the power-law model of Eq. 4-24, based on the heuristic percolation theory, to describe proton conductivity in Nafion:

$$\kappa = \kappa_0 (f_v - f_{v0})^q \text{ and } f_v = \frac{\lambda \bar{V}_w}{\bar{V}_m + \lambda \bar{V}_w} = \frac{\alpha\lambda}{1+\alpha\lambda}. \qquad \text{(Eq. 4-24)}$$

Using an AC impedance technique with frequencies between 10^0 and 10^7 Hz and a constant measuring current density of 1.25 mA cm^{-2}, Hsu et al. obtained limited conductivity data using a range of equivalent weights (1050, 1100, 1350, and 1500) of Nafion (Na$^+$ form) for small aqueous-phase volume fractions (from about 0.1 to about 0.2) to obtain values for the three constants: $q = 1.5$, $f_{v0} = 0.10$, and $\kappa_0 = 0.16$ S cm^{-1}. Hsu et al.'s power-law conductivity model with these constants significantly under-estimates the conductivity of Nafion membrane as compared with the experimental data of Zawodzinski et al. [4, 7]. For example, at $\lambda = 10$, conductivity estimated by Hsu et al.'s model is 0.0091 S cm^{-1}, whereas Zawodzinski et al. [4, 7] reported a measured value of 0.0484 S cm^{-1}.

Morris and Sun [13] used data they measured for Nafion 117 (H$^+$ form) to obtain the power-law conductivity model as follows:

$$\kappa = 0.125(f_v - 0.06)^{1.95}. \qquad \text{(Eq. 4-25)}$$

Similarly, Morris and Sun's model underestimates the conductivity of Nafion by almost an order of magnitude for $\lambda = 10$, conductivity estimated by Morris and Sun's model is 0.0048 S cm^{-1} as compared to 0.0484 S cm^{-1} reported by Zawodzinski et al. [4, 7].

More recently, Weber and Newman [11] presented the semiempirical conductivity model shown in Eq. 4-26 that is based on heuristic percolation theory and similar to that of Hsu et al. [12] and Morris and Sun [13]:

$$\kappa = 0.5(f_v - 0.06)^{1.5}. \qquad \text{(Eq. 4-26)}$$

Though Weber and Newman claim that their model fits well the majority of the Nafion-conductivity data in the literature, conductivity from their power-law model increases with water content without any upper limit–this is not borne out by the data presented by Gottesfeld and Zawodzinski [15], which shows a peak conductivity at a water content around 22. Springer et al.'s correlation (Eq. 4-2) suffers from the same drawback.

Thampan et al. [10] developed a constitutive conductivity model for Nafion or like membranes based on the dusty-fluid model for transport and the percolation model for structural aspects. Their model also includes the thermodynamics of dissociation of the acid group in the presence of polar solvents such as water. Though the constitutive model of Thampan et al. is relatively comprehensive, it requires four adjustable parameters, which may explain why this model has not received much attention from other researchers as compared with empirical correlation of Springer et al. [6]. For comparison, Thampan et al.'s model is presented as follows:

$$\kappa = \frac{F^2}{RT} \frac{D^e_{12}}{(1+\delta)} \frac{1}{\lambda \bar{V}_w} \bar{\alpha}(f_v - f_{v0})^{1.5}, \qquad \text{(Eq. 4-27)}$$

where D^e_{12} is the effective mutual diffusion coefficient of hydronium ion (H$_3$O$^+$), δ ($\equiv D^e_{12} / D^e_{1m}$) is the ratio of the effective diffusivity of H$_3$O$^+$ interaction with water to that with membrane, which is considered to be an adjustable parameter, and $\bar{\alpha}$ is the degree of acid-group dissociation given by

$$\bar{\alpha} = \frac{(\lambda+1) - \sqrt{(\lambda+1)^2 - 4\lambda(1-1/K_{A,C})}}{2(1-1/K_{A,C})}, \qquad \text{(Eq. 4-28)}$$

with $K_{A,C}$ being the acid dissociation constant, which in turn depends on temperature:

$$K_{A,C} = K_{A,C,298} \exp\left[-\frac{\Delta H_0}{RT}\left(\frac{1}{T} - \frac{1}{298}\right)\right],$$ (Eq. 4-29)

Moreover in Eq. 4-27, the percolation threshold volume fraction f_{v0} is considered to be a fitted parameter, whereas $D^e{}_{12}$ is taken to be the diffusion coefficient of H_3O^+ at infinite dilution in water. In short, the model of Thampan *et al.* [10] requires four adjustable parameters: f_{v0}, $K_{A,C,298}$, ΔH_0, and δ.

4.2.3 IONIC CONDUCTIVITY MEASUREMENT

Experiments can be designed to measure proton conductivity. A method, to be introduced as follows, measures that of Nafion 117 using four-probe electrochemical impedance spectroscopy (EIS) and a Solartron 1260 frequency response analyzer coupled with a Solartron 1287 potentiostat [15]. A schematic diagram of the membrane-conductivity measuring cell is shown in Figure 4-4. The outer electrodes of the testing cell are connected to the working and counter electrodes on the 1287 potentiostat, and the two inner electrodes are connected to the reference electrodes. The method is targeted at the conductivity of Nafion membranes in the in-plane direction. Through-plane measurement is much more difficult and susceptible to interfacial phenomena interfering with the measurement of true membrane properties.

From the amplitude and phase lag of the current response, a complex number, the impedance, can be computed, composed of a real component Z' and an imaginary component Z". An example of raw impedance data plotted on a complex plane is shown in Figure 4-5. To obtain the data as shown in Figure 4-5, the EIS was performed on water-immersed samples by imposing a relatively small (10 mV amplitude) sinusoidal (AC signal) voltage across the membrane sample at frequencies between 100 kHz and 100 Hz (scanning from high to low frequencies), and the resultant current response was measured. An alternating voltage was used in order to

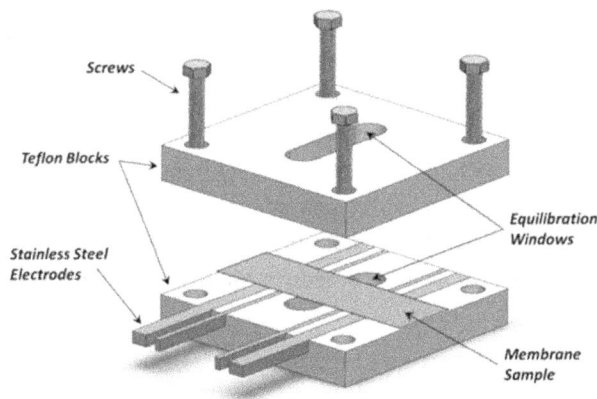

Figure 4-4. Schematic of a four-point membrane-conductivity measuring cell.

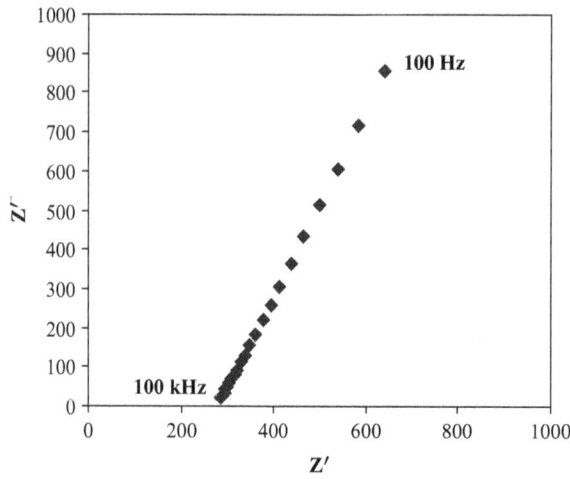

Figure 4-5. Impedance response of a typical proton-conducting membrane between 100 kHz and 100 Hz.

avoid electrolysis or other effects, for example, diffusivity. High frequency can interfere with the ionic conductivity, known as the Debye–Falkenhagen effect. The effect shows an increase in the conductivity when the applied voltage has a sufficiently high frequency [16]. To compute the membrane proton conductivity from the impedance response, the impedance line is extrapolated to the horizontal axis. The extrapolated value of the real impedance where the imaginary response is zero is taken as the membrane resistance. The membrane proton conductivity is then computed from

$$\kappa = \frac{L}{Z'A},$$

(Eq. 4-30)

where L is the length between the reference electrodes, Z' is the real part of the impedance response (extrapolated to $Z'' = 0$), and A is the area available for proton conduction.

For data presented in Figure 4-6, measurement was conducted on Nafion 117 (H$^+$ form) membranes immersed in liquid water for temperatures between 30°C and 80°C. The method of Zawodzinski et al. (see, Zawodzinski et al. [4,7], and Gottesfeld and Zawodzinski [14]) can be used to generate Nafion 117 membranes with high λ values whereby the membranes were swollen in hot glycerol solutions. After swelling in hot glycerol, the membranes are rinsed well and soaked in de-ionized water for at least 48 h to remove all glycerol from the membrane.

The water content or λ values can be determined by gravimetric analysis, in which the membrane samples are weighed when wet at 30°C, dried for at least 72 h at 150°C, and then reweighed. The λ values of each membrane sample can then be calculated as follows:

$$\lambda = \frac{\dfrac{m_w}{M_w}}{m_{m,0} \cdot \text{IEC}},$$

(Eq. 4-31)

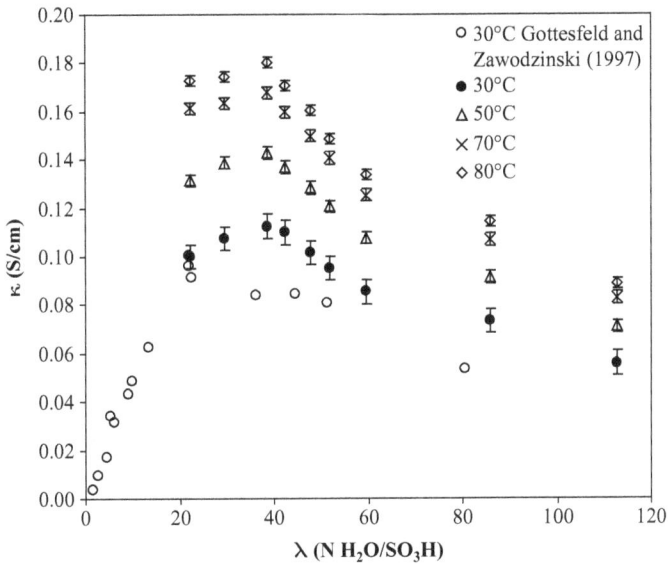

Figure 4-6. Proton conductivity (κ) of Nafion 117 as a function of water content (λ).

where m_w is the mass of water absorbed (mass of the wet membrane minus the mass of the dry membrane), M_w is the molecular weight of water, $m_{m,0}$ is the mass of the dry membrane, and IEC refers to the ion exchange capacity of the membrane (usually expressed in units of milli-equivalents of acid per gram of dry polymer). IEC and equivalent weight (EW) are inversely proportional to each other:

$$IEC = \frac{1000}{EW}, \qquad \text{(Eq. 4-32)}$$

with IEC in milli-equivalents per gram and EW in grams of polymer containing one equivalent of acid group or SO_3H. It should be noted that the λ values of the membrane samples equilibrated in liquid water did not change between 30°C and 80°C.

To study temperature effects, conductivities of the membrane samples can be measured in liquid water at various temperatures, for example, between 30°C and 80°C, as presented in Figure 4-6. For each data point shown in Figure 4-6, the dimensions of the membrane were measured to ensure an accurate determination of the swollen membrane's conductivity. Figure 4-6 shows proton conductivity of Nafion 117 (as measured using a four-point probe) as a function of water content; the error bars in Figure 4-6 indicate uncertainties in the measurements. For comparison, experimental data reported by Zawodzinski *et al.* (1991) and Gottesfeld and Zawodzinski [14] at 30°C are also included in Figure 4-6.

As shown in Figure 4-6, the conductivity of all membrane samples increases with temperature. The average activation energy for proton conduction was found to be around 2 kcal mol^{-1}, with the unswollen Nafion 117 membrane showing an activation energy of 2.3 kcal

mol^{-1} and the highly swollen Nafion 117 membranes (swollen in 200°C glycerol) having an activation energy of 1.9 kcal mol^{-1}. A maximum conductivity at all temperatures was observed for samples with a λ value around 39. These swollen membranes with $\lambda = 38.7$ have conductivities about 7% higher on average than non-swollen Nafion 117 membranes with a λ value around 22. This increase in conductivity for slightly swollen membranes is a result of increased water mobility within the pores, which results in slightly higher proton conductivities for these samples. For λ values greater than 39, the conductivity of the swollen membranes steadily decreases with increasing λ due to increased dilution of the acid groups, which occurs with large membrane swelling.

Figure 4-7 compares the conductivity of Nafion membrane at 30°C as predicted by Eq. 4-23 and other representative correlation and models with experimental data of Gottesfeld and Zawodzinski [14], Zawodzinski *et al.* [4, 7], and Chen *et al.* (denoted as the present work in Figure 4-7) [15]. The parameter values used in computing conductivity from Eq. 4-22 are as follows:

- $D_{H^+}^{o,w} = 10.571 \times 10^{-5}$ cm^2 s^{-1} (diffusivity of proton at infinite dilution in water at 30°C)
- $\bar{V}_m = 550$ cm^3 mol^{-1}
- $\bar{V}_w = 18.0783$ cm^3 mol^{-1}
- $\alpha = 0.03287$

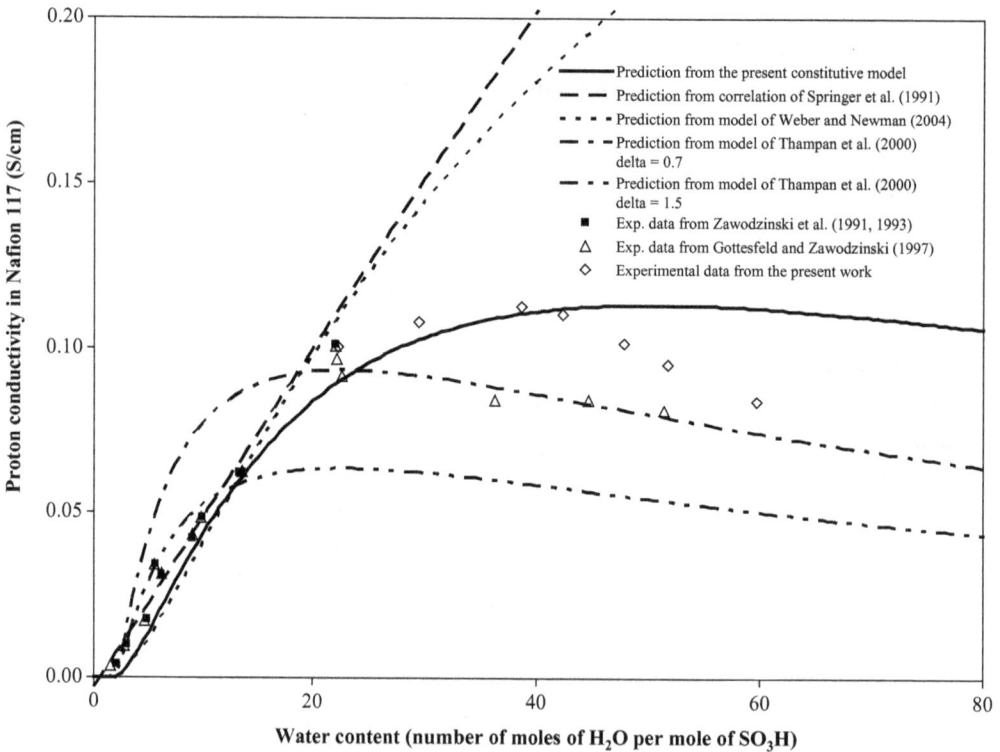

Figure 4-7. Proton conductivity in Nafion membrane at 30°C as a function of water content: comparison between predictions computed by various models and experimental data. The present constitutive model refers to Eq. 4-23, and present work refers to Ref. [15].

Values for the four adjustable parameters used in Thampan *et al.*'s model (Eq. 4-21) are as follows:

- $K_{A,C,298} = 6.2$
- $\Delta H_0 = -52.3$ kJ mol^{-1}
- $f_{v0} = 0.06$
- $\delta = 0.7$ and 1.5, respectively.

As shown in Figure 4-7, conductivity predicted by Eq. 4-23 agrees very well with experimental data for water content from 1.5 to about 45, covering a wide range of water contents. In contrast, conductivities predicted by Springer *et al.*'s [6] and Weber and Newman's model [11] agree with experimental data for water content up to about 22, but then they diverge from each other as water content increases beyond. As for conductivity predicted by Thampan *et al.*'s model [10], with $\delta = 0.7$ good agreement is achieved for water content up to about 10 whereas with $\delta = 1.5$ the model significantly over-predicts conductivity for water contents below about 22; in short, good agreement cannot be achieved using Thampan *et al.*'s model for a wide range of water contents when a single value of δ is used – this was actually recognized by Thampan *et al.* themselves. Although some of the models give good agreement with the experimental data over a limited range of λ values, conditions and membrane water content can vary widely within an operating fuel cell.

4.3 WATER TRANSPORT IN POLYMER ELECTROLYTE MEMBRANES

Water plays an important role in membrane properties, as seen in the previous section. Weber and Newman [17] proposed a physical model as depicted in Figure 4-8, to study the effect of water content on the membrane microstructure. As illustrated in Fig. 4-8, a dry membrane initially absorbs water to solvate its sulfonic acid groups, which builds strong bonds of water-acid groups. The addition of extra water loosens the bonds, leading to the micelles formation in the polymer matrix (Fig. 4-8 b). As more water is absorbed, the clusters become larger, eventually interconnecting with each other. The cluster-channel network forms a transport pathway. The percolation threshold occurs around $\lambda = 2.0$. Figure 4-8c shows a microstructure of the membrane being in contact with saturated water vapor, which forms a complete cluster-channel network. Weber and Newman indicated that in a liquid-water environment, structural reorganization occurs due to that the liquid water repels the fluorocarbon-rich skin of the ionomer and thus allows the liquid infiltration and channel expansion. This will lead to the pore-like microstructure, as shown in Figure 4-8d.

The governing equation describing water transport can be developed based on the conservation principle of species mass as outlined in Chapter 3, by accounting for the storage or accumulation of the bulk region or domain, transfer through the boundary surfaces, and production/consumption, as follows:

$$\varepsilon_m \frac{\partial C_w^m}{\partial t} = \nabla \cdot (D_w^{m,\text{eff}} \nabla C_w^m + \vec{G}_{w,\text{perm}}) - \frac{1}{F} \nabla \cdot \left(n_d \vec{i}_e\right), \qquad \text{(Eq. 4-33)}$$

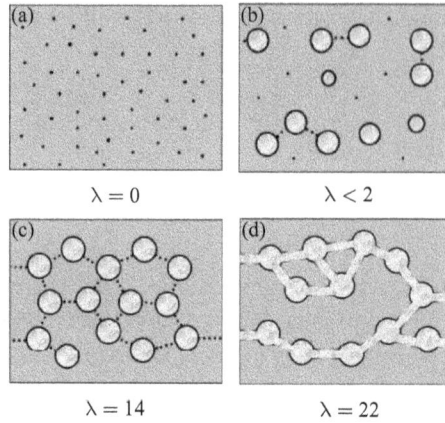

Figure 4-8. Schematic of the Weber and Newman's membrane model: change of the membrane microstructure when altering its water content. The gray area is the fluorocarhon matrix, the black is the polymer side chain, and the dotted line is a collapsed channel. [17] (Courtesy of the Electrochemical Society).

where C_w^m is the equivalent water concentration in the membrane, and related to the water content as follows:

$$C_w^m = \frac{\rho_m \lambda}{\text{EW}}$$ (Eq. 4-34)

where ρ_m and EW are the density and equivalent molecular weight of the membrane, respectively. The conservation equation, Eq. 4-33, from the left to right describes the water behaviors of water holding capacity, diffusion, permeation, and electro-osmotic drag, respectively.

4.3.1 TRANSPORT MECHANISMS

(1) Diffusion

A common mechanism for species transport is diffusion, which is realized through the random motion of molecular particles from regions of high concentrations to those of lower concentrations. The water diffusivity in the membrane depends on water content. The following correlation was proposed by Motupally *et al.* based on experimental data [18]:

$$D_w^m = \begin{cases} 3.1 \times 10^{-3} \lambda \left(e^{0.28\lambda} - 1 \right) \cdot e^{[-2436/T]} & \text{for } 0 < \lambda \leq 3 \\ 4.17 \times 10^{-4} \lambda \left(1 + 161 e^{-\lambda} \right) \cdot e^{[-2436/T]} & \text{otherwise} \end{cases} . \quad \text{(Eq. 4-35)}$$

Another correlation widely used in PEM fuel cell modeling and analysis was developed by Springer *et al.* [6], as follows:

$$D_w^m = \begin{cases} 2.693 \times 10^{-10} \text{ for } \lambda \leq 2 \\ (2.08\lambda - 3.29) \times 10^{-10} e^{(2416/303.15 - 2426/T)} & \text{for } 2 < \lambda \leq 3 \\ (6.84 - 1.3\lambda) \times 10^{-10} e^{(2416/303.15 - 2426/T)} & \text{for } 3 < \lambda \leq 4 \\ (2.563 - 0.33\lambda + 0.0264\lambda^2 - 0.000671\lambda^3) \times 10^{-10} e^{(2416/303.15 - 2426/T)} & 4 > \lambda. \end{cases}$$

(Eq. 4-36)

As for Gore™ membranes that contain Nafion and other reinforcing materials (e.g., to enhance the membrane's mechanical properties), diffusivity correlations can be modified by accounting for the volume fraction of Nafion ionomer ε_m, as follows:

$$D_w^{m,\text{eff}} = \varepsilon_m^{\tau_m} D_w^m.$$

(Eq. 4-37)

As shown in Figure 4-9, strong dependence of diffusivity on membrane water content indicates the nonlinearity of water diffusion in electrolyte membranes.

(2) Water Electro-Osmotic Drag

Another major mechanism for water transport through the membrane is *electro-osmotic flow*, which is the motion induced by an applied potential across a material such as membrane, or other fluid conduits. In polymer electrolyte membranes, protons, driven by the migration force (the electrolyte-phase potential gradient), carry water molecules when moving from the anode to the cathode. The flux of water due to electro-osmotic flow or drag is taken to be proportional to the protonic current density in the membrane as indicated in Eq. 4-33 with the proportionality constant being called the electro-osmotic drag coefficient, which can be experimentally determined. The following correlation developed by Zawodzinski *et al.* is frequently adopted in PEM fuel cell modeling [19]:

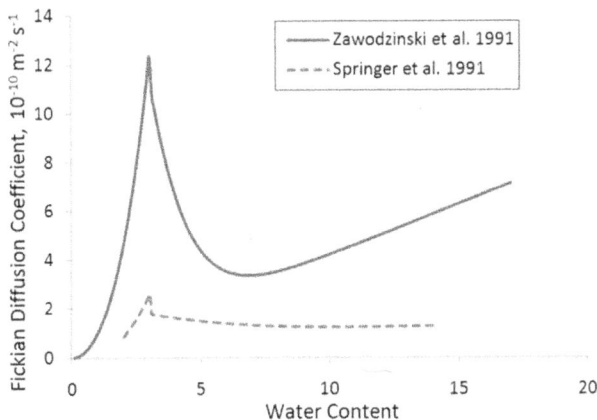

Figure 4-9. Fickian diffusion coefficients of water in Nafion® membranes at 80°C estimated using two correlations [4, 6].

$$n_d = \begin{cases} 1.0 & \text{for} \quad \lambda \le 14 \\ \dfrac{1.5}{8}(\lambda - 14) + 1.0 & \text{otherwise.} \end{cases} \qquad \text{(Eq. 4-38)}$$

Another popular correlation relating the electro-osmotic drag coefficient to water content, shown in Eq. 4-39, was proposed by Springer *et al.* [6]

$$n_d = \frac{2.5\lambda}{22}. \qquad \text{(Eq. 4-39)}$$

In order to compute the water flux due to electro-osmotic drag, the protonic current needs to be known, which can be determined by Ohm's law as follows:

$$\vec{i_e} = -\kappa \nabla \Phi^m, \qquad \text{(Eq. 4-40)}$$

where Φ^m is the electrical potential in the electrolyte membrane and κ is the proton conductivity discussed previously.

(3) Hydraulic Permeation

When liquid water emerges, it can permeate through the electrolyte membrane; this phenomenon is called hydraulic permeation and is driven by the liquid-pressure difference between the two sides of a membrane. The hydraulic-permeation flux is determined by the membrane's intrinsic permeability K_m and liquid pressure gradient, as follows [20]:

$$\vec{G}_{w,\text{perm}} = -\frac{K_m}{M_w \nu^l} \nabla P^l. \qquad \text{(Eq. 4-41)}$$

(4) Net Water Transport Coefficient α

The mechanisms of water transport discussed above determine the water balance between the anode and cathode sides; this balance is important to maintaining anode hydration and shaping the water profile across a polymer membrane which determines overall membrane ionic conductivity. The net water transport coefficient α is defined to characterize the combined effect of these mechanisms: a positive value means net water flow from the anode to cathode. Using a rough definition by considering only diffusion/hydraulic permeation and electro-osmotic drag, the coefficient can be given by

$$\alpha = n_d - \frac{FG}{I}, \qquad \text{(Eq. 4-42)}$$

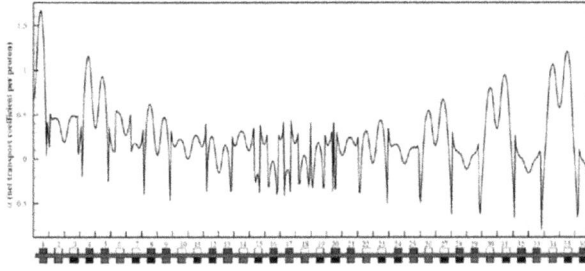

Figure 4-10. The net water transport coefficient per proton at the mid-length cross-section at 0.65 V and 0.91 A cm^{-2}. In the cross-section of flow field at the bottom, the same color represents the same flow path [21].

where G is the magnitude of the water flux, either by diffusion or hydraulic permeation, from the cathode to anode. Figure 4-10 shows a profile of α in a PEM fuel cell with a flow field featured by two parallel channels on each side, indicating highly fluctuating spatial variation with its value mostly in the range from -0.5 to 1.5. Peak values of α appear near the cathode dry inlet channels, Channels 1, 4, 34, and 35, while near the cathode outlet channels, such as Channels 2, 3, and 36, α is less than unity as a result of strong back-diffusion. In addition, under the land, such as the one between Channels 33 and 34, α dramatically changes from a positive to a negative value. This can be explained by the high water concentration under the cathode lands, yielding strong back-diffusion under lands.

Due to the water-content dependence of diffusivity and drag coefficient, water transport in the membrane is highly nonlinear. Büchi and Scherer experimentally measured the water profiles across a membrane, as shown in Fig. 4-11 [22] which indicates nonlinear profiles. Thus, a detailed model with the capability of predicting transport in multiple dimensions is essential to precisely describing the physical process in membranes.

4.3.2   WATER HOLDING CAPACITY

Nafion-based membranes exhibit a large capacity of water uptake. The transport equation of membrane electrolyte can be stated, providing that the velocity is zero in the solid membrane and thus no convective flux is presented, as follows:

$$\varepsilon_m \frac{\partial C_w^m}{\partial t} = \nabla \cdot \left(D_w^{m,\text{eff}} \nabla C_w^m \right) - \frac{1}{F} \nabla \cdot \left(n_d \vec{i}_e \right), \qquad \text{(Eq. 4-43)}$$

where ε_m is the ionomer volume fraction in the electrolyte membrane and the last term on the right accounts for the electro-osmotic drag effect. Here, C_w^m is the equivalent water concentration in the membrane defined as follows:

$$C_w^m = \frac{\rho_m \lambda}{\text{EW}}, \qquad \text{(Eq. 4-44)}$$

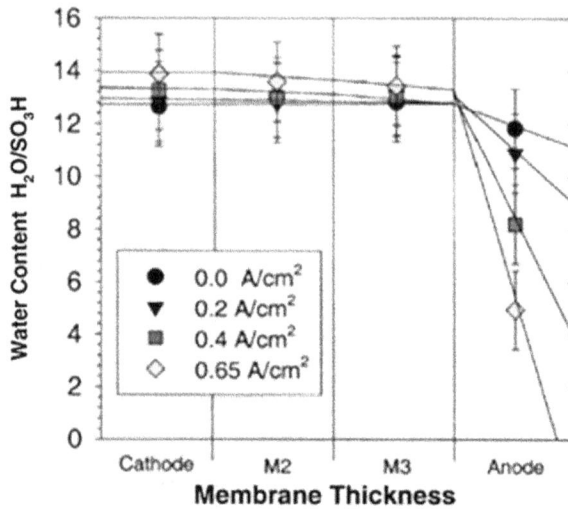

Figure 4-11. Transversal membrane hydration at different current densities at 72°C with both gas flows humidified at 80°C [22] (Courtesy of the Electrochemical Society).

where ρ_m and EW are the density and equivalent molecular weight of the membrane, respectively. EW (equivalent weight) refers to the weight of Nafion per mole of sulfonic acid group. The water content λ is defined as the number of water molecules per sulfonic acid group. It is an important parameter on which the proton conductivity and transport properties of a membrane are based, calculated from an empirical correlation given in the following equation (also see Fig. 4-2):

$$\lambda = \begin{cases} 0.043 + 17.81a - 39.85a^2 + 36.0a^3 & \text{for } 0 < a \leq 1 \\ 14 + 1.4(a - 1) & \text{for } 1 < a \leq 3 \end{cases} , \qquad \text{(Eq. 4-45)}$$

where the water activity a is defined as follows:

$$a = \frac{C_w RT}{p_{w,sat}}. \qquad \text{(Eq. 4-46)}$$

Equations 4-45 and 4-46 present the well-known "Schroeder's paradox," which shows different water contents when the membrane is exposed to saturated vapor ($a = 1$) and liquid water ($a > 1$). This phenomenon was originally reported in 1903 by von Schroeder, who showed a difference in solvent uptake (i.e., swelling) by a solid polymer, for example, gel or membrane, when the polymer was exposed to saturated vapor and pure liquid, respectively [23].

Physically, Eq. 4-43 indicates that the rate of the membrane water storage is equal to the sum of the net flow rates of diffusion and electro-osmotic drag. Using the equilibrium water concentration C_w in the surrounding as the variable, Eq. 4-43 can be rearranged, by incorporating the above three equations, as follows:

$$\varepsilon_m \frac{\partial C_w^m}{\partial t} = \varepsilon^{\text{eff}} \frac{\partial C_w}{\partial t} = \nabla \cdot \left(D_w^{\text{eff}} \nabla C_w \right) - \frac{1}{F} \nabla \cdot \left(n_d \vec{i}_e \right), \qquad \text{(Eq. 4-47)}$$

where the effective factor ε^{eff} is defined as [24]

$$\varepsilon^{\text{eff}} = \varepsilon_m \frac{dC_w^m}{dC_w} = \varepsilon_m \frac{dC_w^m}{d\lambda} \frac{d\lambda}{da} \frac{da}{dC_w} = \varepsilon_m \frac{1}{EW} \frac{RT \rho_m}{p_{w,\text{sat}}} \frac{d\lambda}{da}. \qquad \text{(Eq. 4-48)}$$

For a dry membrane hydrated by product water at a constant current density of I, the time constant τ_m for membrane hydration can be estimated by equating the water storage capability to the ORR's water production, as follows:

$$\tau_m = \frac{\dfrac{\rho_m \delta_m \Delta \lambda}{EW}}{\dfrac{I}{2F}}. \qquad \text{(Eq. 4-49)}$$

For Nafion® 112 at 30°C, $\Delta\lambda = 14$, and $I = 1$ A cm^{-2}, τ_m is about 25 s. This immediately points out the importance of the transient term in Eq. 4-43.

It is of interest to compare the water uptake by a membrane to that by reactant gases within a PEM fuel cell. This ratio can be defined as follows:

$$\frac{\varepsilon^{\text{eff}} V_m}{V_{a,h} + V_{c,ch}}, \qquad \text{(Eq. 4-50)}$$

where V_m, $V_{a,ch}$, and $V_{c,ch}$ represent the volumes of the membrane, anode gas channel, and cathode gas channels, respectively. Figure 4-12 plots the contours of ε^{eff} in the membrane under the steady-state conditions of 0.65 V and RHa/c = 100%/0%, indicating that ε^{eff} ranges from 100 to 1000. Though the volume of the gas-channel space is usually 10–100 times larger than that of the membrane, Figure 4-12 indicates that a membrane holds 10–100 times more water than the vapor holding capacity of the reactant flow channel. Thus, membrane hydration/dehydration play a critical role in transient water management.

4.4 WATER QUANTIFICATION USING NEUTRON RADIOGRAPHY

Because the attenuation of neutron beam is extremely sensitive to the presence of liquid water (which effectively reduces the intensity of the transmitting neutron beam) but not to other fuel cell materials, neutron radiography is a good choice for probing liquid water inside a PEM fuel cell. As shown in Figure 4-13, this is achieved by sending a beam of collimated neutrons through operating fuel cells and measuring the attenuation of the transmitted neutron beam. The spatial and temporal resolutions are determined by the neutron source, detector, and imaging set-up. The NCNR (NIST Center for Neutron Research where NIST refers to National Institute of Standard

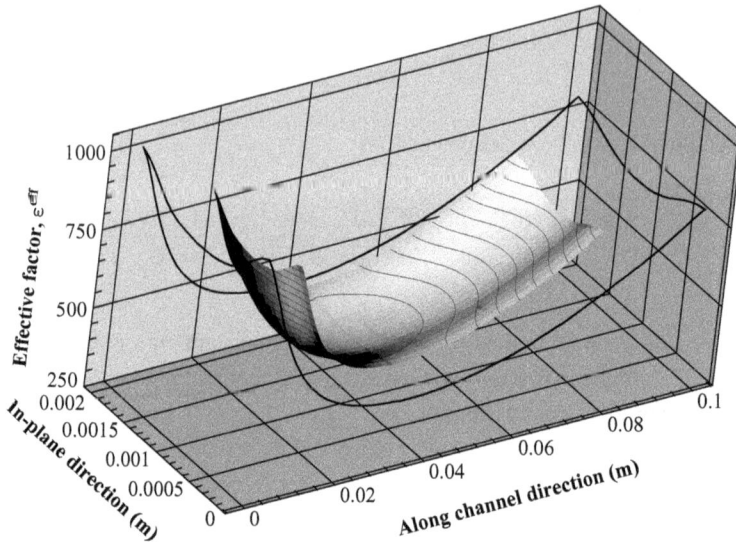

Figure 4-12. The effective factor ε^{eff} in the middle of the membrane under 0.65 V and RHa/c =100%/0% [24].

and Technology) facility uses thermal neutrons which, compared to cold neutrons, require a lower neutron flux to create sufficient contrast to measure water content. Figure 4-13 shows a schematic of neutron radiograph over a PEM fuel cell.

Imaging analysis is usually based on the Lambert–Beer law of attenuation as follows:

$$I = I_0 e^{-N\sigma T}, \qquad \text{(Eq. 4-51)}$$

where I is the intensity of the transmitted neutrons, I_0 is the intensity of the incident neutrons beam, N is the atomic density of the material through which they pass; σ and T_h are, respectively, the cross-section area and the thickness of the neutron beam attenuating medium. For a medium consisting of several materials, the final attenuation is expressed by accounting for the attenuation by all individual materials as shown in the following equation:

$$I = I_o e^{-\sum_i (N\sigma T_h)_i}, \qquad \text{(Eq. 4-52)}$$

where i is the index for each material. When liquid water is present in the fuel cell that neutron beams are transmitted through, the attenuation due to its presence can be isolated from the rest by using the intensity ratio of the radiograph with water (I_{wet}) to that without (I_{dry}):

$$\frac{I_{\text{wet}}}{I_{\text{dry}}} = e^{-(N\sigma T_h)_{\text{H}_2\text{O}}}. \qquad \text{(Eq. 4-53)}$$

Defining $\mu_w = N\sigma T_h$ as the water attenuation coefficient, the above equation is rearranged to

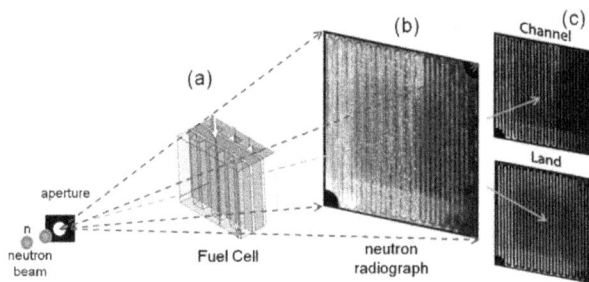

Figure 4-13. Schematic of neutron radiography: neutron beams are directed through an operating fuel cell (a) toward a CCD where a neutron radiograph is recorded (b). Masks can be applied to analyze water contents in the projected areas of channels and lands (c).

$$T_h = \frac{-1}{\mu_w} \ln\left(\frac{I_{\text{wet}}}{I_{\text{dry}}}\right).$$

(Eq. 4-54)

The water attenuation coefficient is determined experimentally by using a specially manufactured coquette of known water thicknesses and measuring the incident neutron intensity. Figure 4-14 displays sample neutron imaging data of water thickness across a PEM fuel cell, along with model prediction.

4.5 ION TRANSPORT IN CATHODE CATALYST LAYERS

Catalyst layers, coated on the electrolyte membrane surface, contain the electrolyte ionomer phase to conduct protons. The ionomer content usually ranges from 0.1 to 0.4, forming a proton conducting network. The ionic resistance of a catalyst layer may not be small, yielding a certain amount of ohmic loss. In catalyst layers, water is transported in both void space and electrolyte ionomer. The layer's ionic conductivity is determined by water content in the ionomer phase.

4.5.1 VARIATION IN WATER CONTENT IN CATALYST LAYERS

(1) Water Vapor Concentration

Without liquid water, diffusion is a major mechanism for water transport in both ionomer and gaseous phases. At phase equilibrium between the ionomer and gas, an effective diffusion coefficient can be defined as follows:

$$D_w^{\text{eff}} = D_w^{g,\text{eff}} + D_w^{m,\text{eff}} = \varepsilon^{\tau_d} D_w^g + \varepsilon_m^{\tau_m} D_w^m \frac{dC_w^m}{dC_w^g},$$

(Eq. 4-55)

where C_w^m is the water molar concentration in the ionomer phase, and $D_w^{m,\text{eff}}$ is the effective water diffusivity in the ionomer phase when expressed in terms of the gradient in the gas-phase

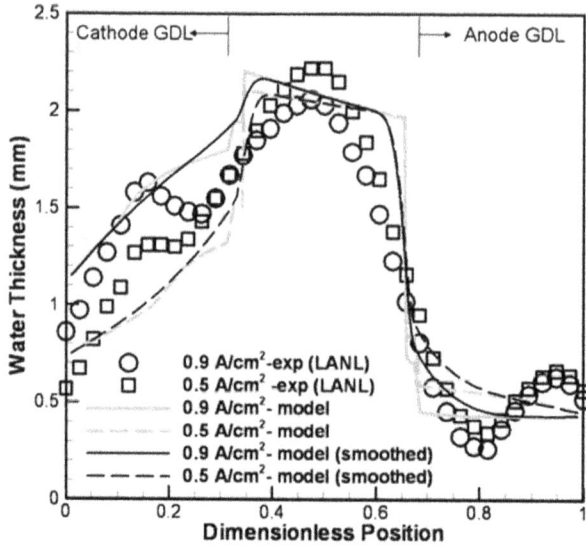

Figure 4-14. Through-plane water profiles in a PEM fuel cell under RHa/c = 100%/50% and 40°C. Experimental data are from LANL (Las Alamos National Laboratory). The membrane resides between the anode and cathode GDLs [25] (Courtesy of the Electrochemical Society).

molar concentration. The Bruggeman correlation is adopted to account for tortuosity. Neglecting convection, the water spatial variation can be evaluated using its diffusion resistance as follows:

$$\Delta C_w = \frac{(1+2n_d)I}{4F}\frac{\delta}{D_w^{\text{eff}}}. \qquad \text{(Eq. 4-56)}$$

Applying the common values of the parameters (e.g., the electro-osmotic drag coefficient $n_d \sim 1.0$) in Eq. 4-56 yield ΔC_w of ~0.08 mol m^{-3} at 1.0 A cm^{-2}, which leads to a fairly small spatial-variation of the ionic conductivity σ_m^{eff}.

(2) Liquid Water Saturation

Liquid water emerges in the situation when the vapor phase is saturated. At equilibrium between gaseous and liquid phases, vapor-phase diffusion becomes zero under the isothermal condition. Liquid flow in void space can be exploited to remove water product, primarily driven by capillary action, which can be treated as a "diffusive" process with a coefficient, as follows:

$$D_c^l = -\frac{k_{rl}}{\nu_l}\sigma\cos\theta_c(K\varepsilon)^{1/2}\frac{dJ(s)}{ds}, \qquad \text{(Eq. 4-57)}$$

where k_{rl} is the relative permeability for liquid and σ is the surface tension. Physically, it describes the extent to which liquid flow is hindered by the presence of gaseous flow. The variation in liquid saturation Δs across the cathode catalyst layer is estimated by

$$\Delta s = \frac{(1+2n_d)I}{4F} \frac{\delta}{D_c^l} M_w.$$

(Eq. 4-58)

In the range of $0.05 < s < 0.8$, using typical values of the parameters (e.g., K of 10^{-14} m^2, θ_c of $120°$, and $k_{rl} = s^3$), we will obtain $\Delta s \sim 1\%$ at 1 A cm^{-2}, which is small and negligible. Thus, the effective exchange current density $ai_{0,T}^{\mathrm{ref},c}$ can be assumed uniform across the cathode electrode if other factors such as Pt loading are uniform. Likewise, the ionic conductivity σ_m^{eff} varies to a small degree across the catalyst layer.

4.5.2 PROTON TRANSPORT IN CATHODE CATALYST LAYERS

Proton behaviors follow the governing equation of ion transport, as detailed in Appendix IV.B. Within catalyst layers, protons are transported via the ionomer network toward the reaction sites. At the macroscopic level, the catalyst layers are treated as ionic conductive media by defining effective ionic conductivity. Assuming uniform oxygen gas content and ionic conductivity, the following one-dimensional (1D) electrolyte-phase potential equation can be set up:

$$\frac{d^2}{dx^2}\Phi^m = \frac{a\, i_{0,T}^{c,\mathrm{ref}}}{\sigma_m^{\mathrm{eff}}} \frac{C_{O_2}}{C_{O_2}^{\mathrm{ref}}} \exp\left(-\frac{\alpha_c F}{R_g T}\cdot(\Phi^s - U_o - \Phi^m)\right).$$

(Eq. 4-59)

Herein, following the original work of Wang and Feng [26] we use σ_m to denote the ionic conductivity. The boundary conditions for the above 1D problem are as follows:

$$\Phi^m = \Phi_\delta^m \text{ and } \frac{d}{dx}\Phi^m = 0 \text{ at } x = \delta.$$

(Eq. 4-60)

The electrolyte phase potential Φ^m can be explicitly solved from Eq. 4-59 and Eq. 4-60, which can be used to obtain local electrochemical reaction rate as follows:

$$j(\bar{x}) = j_\delta\left[\Pi(\Delta U^{j\delta},\bar{x})+1\right] \text{ or } \frac{j(\bar{x})-j_\delta}{j_\delta} = \Pi(\Delta U^{j\delta},\bar{x}).$$

(Eq. 4-61)

Here $\Pi(\Delta U^{j\delta},\bar{x})$ is a function that is defined as follows:

$$\Pi(\Delta U^{j\delta},\bar{x}) = \left[\tan\left(\pm\left[\frac{\alpha_c F}{2R_g T}\Delta U^{j\delta}\right]^{1/2}(1-\bar{x})\right)\right]^2$$

where $\bar{x} = \dfrac{x}{\delta}, I^{j\delta} = -j_\delta \delta, R_\delta = \dfrac{\delta}{\sigma_m^{\text{eff}}}$, and $\Delta U^{j\delta} = R_\delta I^{j\delta}$.   (Eq. 4-62)

Through the explicit solution of Eq. 4-61, a dimensionless parameter $\hbar$ can be defined to quantify the ORR's nonuniformity, as follows [26]:

$$\hbar = \frac{|j_0 - j_\delta|}{\displaystyle\int_0^1 j(\bar{x})d\bar{x}} = \frac{|j_0 - j_\delta|}{\dfrac{I}{\delta}}.$$ (Eq. 4-63)

The smaller the value of $\hbar$, the more uniform the reaction distribution. By substituting Eq. 4-61 into Eq. 4-63, one obtains

$$\hbar = \frac{I\delta}{2\dfrac{\sigma_m^{\text{eff}} R_g T}{\alpha_c F}}.$$ (Eq. 4-64)

It is seen that $\hbar$ is a function of the average current, electrode thickness, ionic conductivity, and temperature. For fixed temperature and thickness, the parameter is only determined by the layer's ionic conductivity and applied average current. A lumped parameter ΔU can be introduced, following Eq. 4-62, as given in the following:

$$\Delta U = \frac{I\delta}{\sigma_m^{\text{eff}}} = IR_\delta.$$ (Eq. 4-65)

Physically, ΔU measures the magnitude of the electrolyte potential drop across the catalyst layer. It lumps the catalyst layer's ionic resistance and average current density. With ΔU defined above, Eq. 4-64 changes to yield

$$\hbar = \frac{\Delta U}{2\dfrac{R_g T}{\alpha_c F}}.$$ (Eq. 4-66)

Figure 4-15 plots the dimensionless transfer current density or local reaction rate profiles and shows that the degree of $j(\bar{x})$ variation increases with $\hbar$. One can directly evaluate the value of $\hbar$ through this figure through the difference of the dimensionless $j(\bar{x})$ s at the two boundaries, that is, $\bar{x} = 0$ and 1.

With Figure 4-15, one can further explore the significance of the lumped parameter $\hbar$ in electrode fundamentals. Physically, the integral of the y-axis variable, that is, $\dfrac{|j(\bar{x})|\delta}{I}$ from 0 to $\bar{x}$ (or the area between the curve and x axis), represents the portion of current contributed by the part of the electrode from 0 to $\bar{x}$. Thus, the reaction current produced by the half of the catalyst layer near the electrolyte membrane reads $\dfrac{\delta}{I}\displaystyle\int_0^{0.5}|j(\bar{x})|d\bar{x}$, whereas the other half is $\dfrac{\delta}{I}\displaystyle\int_{0.5}^1|j(\bar{x})|d\bar{x}$. Through linear approximation, the two integrals can be rearranged, given that $j(\bar{x})$ is negative, as shown in the following equation:

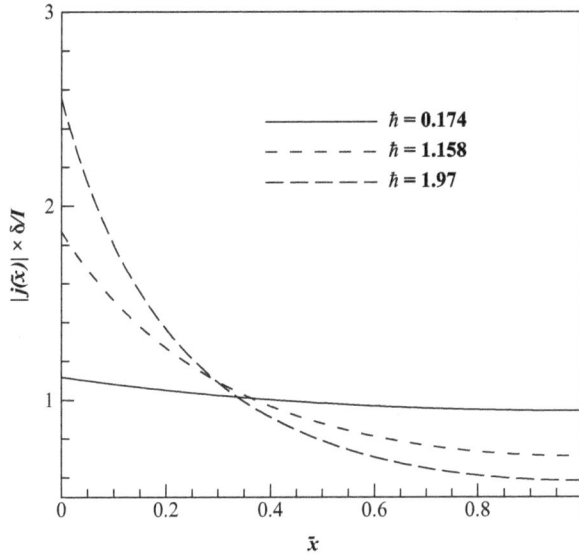

Figure 4-15. Dimensionless reaction current profiles in cathode catalyst layers [26] (Courtesy of the Electrochemical Society).

$$\frac{\delta}{I}\int_0^{0.5}|j(\overline{x})|d\overline{x} \approx \frac{\delta}{4I}|j(0) + j(0.5)| \text{ and } \frac{\delta}{I}\int_{0.5}^1|j(\overline{x})|d\overline{x} \approx \frac{\delta}{4I}|j(0.5) + j(1)|. \text{ (Eq. 4-67)}$$

Note that $j(0) = j(\overline{x} = 0) = j_0$ and $j(1) = j(\overline{x} = 1) = j_\delta$. $\hbar$ can then be approximated by

$$\hbar = \frac{\delta|j_0 - j_\delta|}{I} = \frac{\delta}{I}|j(0) + j(0.5)| - \frac{\delta}{I}|j(0.5) + j(1)| \approx \frac{4\delta}{I}\int_0^{0.5}|j(\overline{x})|d\overline{x} + \frac{4\delta}{I}\int_{0.5}^1|j(\overline{x})|d\overline{x}$$

(Eq. 4-68)

The sum of the two terms in Eq. 4-67 equals to 1, which, along with Eq. 4-68, yields

$$\frac{\delta}{I}\int_0^{0.5}|j(\overline{x})|d\overline{x} \approx 0.5 + \frac{\hbar}{8} \text{ and } \frac{\delta}{I}\int_{0.5}^{1.0}|j(\overline{x})|d\overline{x} \approx 0.5 - \frac{\hbar}{8}. \quad \text{(Eq. 4-69)}$$

Using $\hbar = 1.97$ in Figure 4-15 as an example, the current contribution from the part of $\overline{x} = (0, 0.5)$ is ~75%, whereas the other half is about 25%. In other words, if the electrode is divided into two, the value of $\hbar$ is directly related to the performance of each half of the electrode. Note that the above conclusion is drawn with the linear approximation for integral, which is valid at low $\hbar$. As $\hbar$ approximates 4, Eq. 4-69 provides a rough evaluation.

Furthermore, Figure 4-15 clearly shows the physical meaning of $\hbar$, that is, the degree of spatial variation in the local ORR. In practice, $\hbar$ can be used to guide electrode optimization. For example, for sufficiently small $\hbar$, for example, low current or high humidity, the reaction is almost uniform across the electrode. Thus, a single-layer configuration is sufficient.

For electrodes designed for large $\hbar$ which is encountered under high current or low humidity, multiple-layer configuration can be developed to improve catalyst utilization.

4.5.3 MULTIPLE-LAYERED CATHODE CATALYST LAYERS

Most conclusions from the analysis of single-layered catalyst layers can be extended to multiple-layered catalyst layers. In this configuration, each sublayer is configured separately with loadings of ionomer and catalyst. As an example, Figure 4-16 shows a two sublayer configuration. The governing equations for each sub-layer are written as follows:

$$\text{for Layer I: } \frac{d^2}{dx^2}\Phi_I^m = \frac{\left(a\, i_{0,T}^{c,\text{ref}}\right)_I}{\left(\sigma_m^{\text{eff}}\right)_I}\frac{C_{O_2}}{C_{O_2}^{\text{ref}}}\exp\left(-\frac{\alpha_c F}{R_g T}\cdot(\Phi^s - U_o - \Phi_I^m)\right), \qquad \text{(Eq. 4-70)}$$

$$\text{for Layer II: } \frac{d^2}{dx^2}\Phi_{II}^m = \frac{\left(a\, i_{0,T}^{c,\text{ref}}\right)_{II}}{\left(\sigma_m^{\text{eff}}\right)_{II}}\frac{C_{O_2}}{C_{O_2}^{\text{ref}}}\exp\left(-\frac{\alpha_c F}{R_g T}\cdot(\Phi^s - U_o - \Phi_{II}^m)\right) \qquad \text{(Eq. 4-71)}$$

The boundary and interfacial conditions are as follows:

$$\Phi_I^m = \Phi_\delta^m \text{ and } \frac{d}{dx}\Phi_I^m = 0 \text{ at } x = \delta \qquad \text{(Eq. 4-72)}$$

$$\Phi_I^m = \Phi_{II}^m \text{ and } \left(\sigma_m^{\text{eff}}\right)_I \frac{d}{dx}\Phi_I^m = \left(\sigma_m^{\text{eff}}\right)_{II}\frac{d}{dx}\Phi_{II}^m \text{ at } x = \delta/2 . \qquad \text{(Eq. 4-73)}$$

The analytical solution can be explicitly obtained as

$$\text{in Layer I: } \Phi_I^m - \Phi_\delta^m = \frac{R_g T}{\alpha_c F}\ln\left\{\Pi(\Delta U^{j\delta}, \bar{x}) + 1\right\},$$

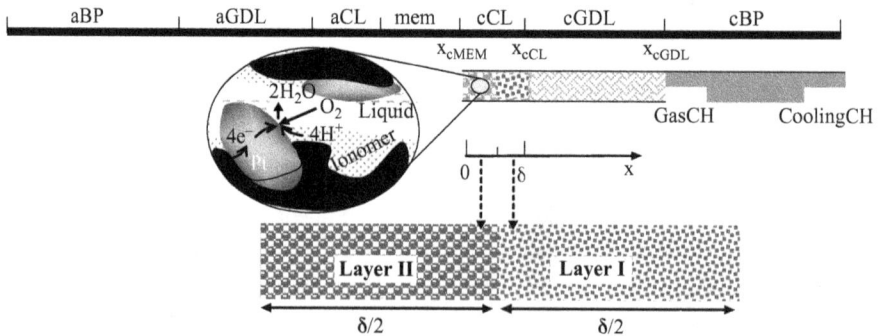

Figure 4-16. Schematic of a dual-layer cathode electrode.

in Layer II: $\Phi_{II}^m - \Phi_\delta^m = \dfrac{R_g T}{\alpha_c F} \ln \left\{ \dfrac{\Psi\left[\Delta U^{j\delta}, \dfrac{\left(\sigma_m^{eff}\right)_I}{\left(\sigma_m^{eff}\right)_{II}}, \dfrac{\left(a\, i_{0,T}^{c,ref}\right)_I}{\left(a\, i_{0,T}^{c,ref}\right)_{II}}\right] \cdot}{\left[II\left(\Delta U^{j\delta}, \dfrac{\left(\sigma_m^{eff}\right)_I}{\left(\sigma_m^{eff}\right)_{II}}, \dfrac{\left(a\, i_{0,T}^{c,ref}\right)_I}{\left(a\, i_{0,T}^{c,ref}\right)_{II}}, \bar{x}\right)+1\right]} \right\}$, (Eq. 4-74)

where the functions are defined as follows:

$$II = \left[\tan\left(\sqrt{\dfrac{\alpha_c F \Delta U^{j\delta}}{2 R_g T} \dfrac{\left(\sigma_m^{eff}\right)_I}{\left(\sigma_m^{eff}\right)_{II}}\dfrac{\left(a\, i_{0,T}^{c,ref}\right)_{II}}{\left(a\, i_{0,T}^{c,ref}\right)_I}} \Psi\left[\Delta U^{j\delta}, \dfrac{\left(\sigma_m^{eff}\right)_I}{\left(\sigma_m^{eff}\right)_{II}}, \dfrac{\left(a\, i_{0,T}^{c,ref}\right)_I}{\left(a\, i_{0,T}^{c,ref}\right)_{II}}\right] \cdot \left(\bar{x}-\dfrac{1}{2}\right)} \right.$$
$$\left. - \tan^{-1}\dfrac{II(\Delta U^{j\delta}, \tfrac{1}{2})}{\sqrt{\Psi\left[\Delta U^{j\delta}, \dfrac{\left(\sigma_m^{eff}\right)_I}{\left(\sigma_m^{eff}\right)_{II}}, \dfrac{\left(a\, i_{0,T}^{c,ref}\right)_I}{\left(a\, i_{0,T}^{c,ref}\right)_{II}}\right]}} \dfrac{\left(\sigma_m^{eff}\right)_I}{\left(\sigma_m^{eff}\right)_{II}}\dfrac{\left(a\, i_{0,T}^{c,ref}\right)_I}{\left(a\, i_{0,T}^{c,ref}\right)_{II}}\right]^2 \quad \text{and}$$

$$\Psi = II(\Delta U^{j\delta}, \tfrac{1}{2})\left[1 - \dfrac{\left(\sigma_m^{eff}\right)_I}{\left(\sigma_m^{eff}\right)_{II}}\dfrac{\left(a i_{0,T}^{c,ref}\right)_I}{\left(a i_{0,T}^{c,ref}\right)_{II}}\right] + 1,$$

$$II = \left[\tan\left(\pm\left[-\dfrac{\alpha_c F}{2 R_g T}\Delta U^{j\delta}\right]^{1/2}\cdot(1-\bar{x})\right)\right]^2. \quad \text{(Eq. 4-75)}$$

Here, $\Delta U^{j\delta}$ is defined the same way as Eq. 4-62 but is based on the properties of Layer I, as shown below:

$$\Delta U^{j\delta} = R_{\delta,I} I^{j\delta} \text{ where } I^{j\delta} = -j_\delta \delta, \ R_{\delta,I} = \dfrac{\delta}{\left(\sigma_m^{eff}\right)_I} \text{ and } j_\delta = j(x=\delta). \text{ (Eq. 4-76)}$$

Figure 4-17 displays the analytical reaction current density at $\Delta U^{j\delta} = 0.1$ V, showing that the reaction rate is continuous at the interface of the two sublayers of difference only in the ionic conductivity. In contrast, remarkable discontinuity occurs upon varying $a i_{0,T}^{c,ref}$. When lowering the ionic conductivity of Layer II, as in the case of $\dfrac{\left(\sigma_m^{eff}\right)_I}{\left(\sigma_m^{eff}\right)_{II}} = 1.25$, the reaction rate in Layer II increases due to the enlarged electrolyte-phase potential variation. When increasing $a i_{0,T}^{c,ref}$ in Layer II, the reaction rate jumps at the interface from Layers I to II.

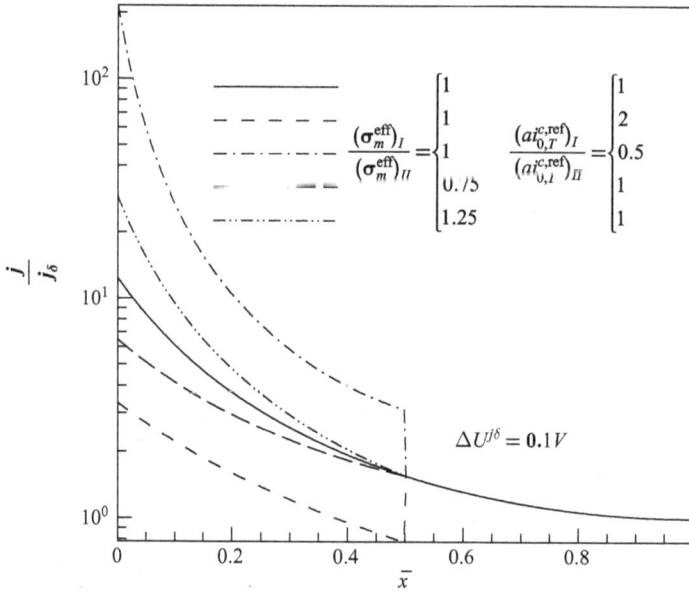

Figure 4-17. The profiles of dimensionless transfer current density under $\Delta U^{j\delta} = 0.1$ V [27] (Courtesy of the Electrochemical Society).

The dual-layer solution can be readily extended to more-layered configurations. Expression of such a solution is tedious, therefore not presented here. Interested readers are referred to Ref. [28] for details. Figure 4-18 displays the sub-layer's performance for three-layered electrodes. The average loadings in Layers 2 and 3 are 0.95 for all the considered cases; that is, the overall loading amount of a material (either the ionomer or Pt) is the same in these two sublayers. However, their performances are different under the higher $\Delta U^{j\delta}$. This result implies that in a dual-layer electrode, one sublayer can be further optimized through multilayered configuration; that is, the more the number of sublayers, the better the optimization can achieve. Detailed discussion on $\Delta U^{j\delta}$ is provided in Ref. [26]. Figure 4-18 also shows that under high $\Delta U^{j\delta}$ both $aj_{0,T}^{c,\mathrm{ref}}$ and σ_m^{eff} in Layer 3 considerably affect the current production of this layer as well as the overall current, whereas the properties of Layer 2 have a relatively small influence.

4.6   CHAPTER SUMMARY

This chapter provides an overview of water and proton transports that occur in the polymer electrolyte membrane and catalyst layers of a PEM fuel cell.

The key points of this chapter are summarized as follows:

- Ionic or protonic conductivity is the most important property of polymer electrolyte membranes and directly determines the ohmic voltage loss. Several protonic conductivity correlations are discussed and compared; a constitutive model of protonic conductivity is derived from ion transport fundamentals; and an experimental method of measuring membrane ionic conductivity is described.
- Membrane water content strongly affects the protonic conductivity and water diffusivity, which are essential to achieving and maintaining high fuel cell performance. Three

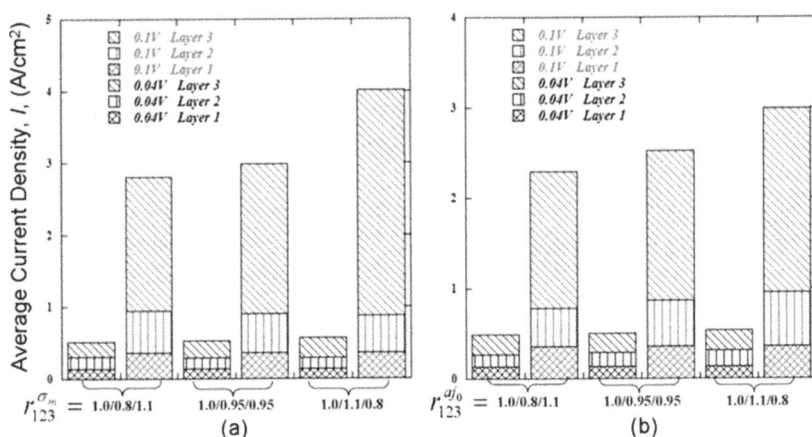

Figure 4-18. Each sub-layer's performance at (a) varying $r_{123}^{\sigma m}$ and (b) varying $r_{123}^{aj_0}$ where r_{123} denotes the ratio of Layers I, II, and III [28].

major mechanisms of water transport in the membrane (diffusion, permeation, and electro-osmotic drag) are identified and discussed; polymer electrolyte membranes possess large water uptake capacity, relative to the gas phase; and membrane water content can be probed experimentally by high-resolution neutron radiography.

• The ionomer electrolyte phase in catalyst layers is a key constituent material for proton transport. The ionic resistance in catalyst layers can be significant and lead to spatial variations in the electrochemical-reaction rates; variations in water vapor concentration and liquid saturation across the catalyst layers are negligibly small; loadings of Nafion and catalyst significantly affect catalyst layer's performance; and a dimensionless group is defined to quantify the non-uniformity of the electrochemical reaction.

REFERENCES

1. Heitner-Wirguin, C. Recent advances in perfluorinated ionomer membranes: structure, properties and applications. *J. Membrane Sci. 120* (1996): 1–33. doi: http://dx.doi.org/10.1016/0376-7388(96)00155-X

2. Mauritz, K.A., and Moore, R.B. State of understanding of Nafion. *Chem. Rev. 104* (2004): 4535–4585. doi: http://dx.doi.org/10.1021/cr0207123

3. Hsu, W.Y., and Gierke, T.D. Ion transport and clustering in Nafion perfluorinated membranes. *J. Membr Sci. 13* (1983): 307. doi: http://dx.doi.org/10.1016/S0376-7388(00)81563-X

4. Zawodzinski, T.A., Neeman, M., Sillerud, L.O., and Gottesfeld, S. Determination of water diffusion coefficients in perfluorosulfonate ionomeric membranes. *J. Phys. Chem. 95* (1991): 6040. doi: http://dx.doi.org/10.1021/j100168a060

5. Hinatsu, J.T., Mizuhata, M., and Takenaka, H. Water uptake of perfluorosulfonic acid membranes from liquid water and water vapor. *J. Electrochem. Soc. 141* (1994): 1493–1498. doi: http://dx.doi.org/10.1149/1.2054951

6. Springer, T.E., Zawodinski, T.A., and Gottesfeld, S. Polymer electrolyte fuel cell model. *J. Electrochem. Soc. 138* (1991): 2334. doi: http://dx.doi.org/10.1149/1.2085971

7. Zawodzinski, T.A. *et al.* A comparative study of water uptake by and transport through iono-meric fuel cell membranes. *J. Electrochem. Soc. 140* (1993): 1981–1985. doi: http://dx.doi.org/10.1149/1.2220749

8. Wang, Y., Mukherjee, P.P., Mishler, J., Mukundan, R., and Borup, R.L. Cold start of polymer electro-lyte fuel cells: three-stage startup characterization. *Electrochimica Acta* **55** (2010): 2636–2644. doi: http://dx.doi.org/10.1016/j.electacta.2009.12.029

9. Newman, J., Thomas-Alyea, K.E. *Electrochemical Systems*, 3rd Ed. Wiley-Interscience, (2004).

10. Thampan, T., Malhotra, S., Tang, H., and Datta, R. Modeling of conductive transport in proton-exchange membranes for fuel cells. *J. Electrochem. Soc. 147* (2000): 3242–3250. doi: http://dx.doi.org/10.1149/1.1393890

11. Weber, A., and Newman, J. Modeling transport in polymer-electrolyte fuel cells. *Chem. Rev. 104* (2004): 4679. doi: http://dx.doi.org/10.1021/cr0207291

12. Hsu, W.Y., Barkley, J.R., and Meakin, P. Ion percolation and insulator-to-conductor transition in Nafion perfluorosulfonic acid membranes. Macromolecules, *13* (1980) 198–200. doi: http://dx.doi.org/10.1021/ma60073a041

13. Morris, D.R., and Sun, X. Water sorption and transport properties of Nafion 117 H. *J. Appl. Polym. Sci. 50* (1993): 1445–1452. doi: http://dx.doi.org/10.1002/app.1993.070500816

14. Gottesfeld, S., and Zawodzinski, T.A. Polymer electrolyte fuel cells. In Alkire, R.C., Gerischer, H., Kolb, D.M., and Tobias, C.W. (Eds.). *Advances in Electrochemical Science and Engineering*, Vol. 5. Wiley-VCH, (1997). doi: http://dx.doi.org/10.1002/9783527616794.ch4

15. Chen, K.S., Hickner, M.A., Siegel, N.P., Noble, D.R., Pasaogullari, U., and Wang, C.Y., Final Report on LDRD Project: Elucidating Performance of Proton-Exchange-Membrane Fuel Cells via Com-putational Modeling with Experimental Discovery and Validation, Sandia Technical Report, Sandia National Laboratories, SAND2006-6964 (2006).

16. Anderson, J.E. The Debye–Falkenhagen effect: experimental fact or fiction? *J. Non-Crystal. Solids* 172–174 (Part 2) (1994): 1190–1194.

17. Weber, A.Z., and Newman, J. Transport in polymer-electrolyte membranes: II. Mathematical model. *J. Electrochem. Soc. 151* (2004): A311. doi: http://dx.doi.org/10.1149/1.1639157

18. Motupally, S., Becker, A.J., and Weidner, J.W. Diffusion of water through Nafion® 115 membranes. *J. Electrochem. Soc. 147* (2000): 3171. doi: http://dx.doi.org/10.1149/1.1393879

19. Zawodzinski, T.A., Davey, J., Valerio, J., and Gottesfeld, S. The water-content dependence of electroos-motic drag in proton-conducting polymer electrolytes. *Electrochim. Acta 40* (1995): 297. doi: http://dx.doi.org/10.1016/0013-4686(94)00277-8

20. Wang, Y. Modeling of two-phase transport in the diffusion media of polymer electrolyte fuel cells. *J. Power Sources 185* (2008): 261–271. doi: http://dx.doi.org/10.1016/j.jpowsour.2008.07.007

21. Wang, Y., and Wang, C.Y.. Simulations of flow and transport phenomena in a polymer electrolyte fuel cell under low-humidity operations. *J. Power Source 147* (2005): 148–161. doi: http://dx.doi.org/10.1016/j.jpowsour.2005.01.047

22. Büchi, F., and Scherer, G.G. Investigation of the transversal water profile in Nafion electrolyte fuel cells. *J. Electrochem. Soc. 148* (2001): A183. doi: http://dx.doi.org/10.1149/1.1345868

23. Vallieres, C. *et al.* On Schroeder's paradox. *J. Membrane Sci. 278* (1–2) (2006): 357–364. doi: http://dx.doi.org/10.1016/j.memsci.2005.11.020

24. Wang, Y., and Wang, C.Y. Transient analysis of polymer electrolyte fuel cells. *Electrochimica Acta 50* (2005): 1307–1315. doi: http://dx.doi.org/10.1016/j.electacta.2004.08.022

25. Wang, Y., and Chen, K.S. Through-plane water distribution in a polymer electrolyte fuel cell: com-parison of numerical prediction with neutron radiography data. *J. Electrochem. Soc. 157* (2010): B1878–B1886. doi: http://dx.doi.org/10.1149/1.3498997

26. Wang, Y., and Feng, X. Analysis of reaction rates in the cathode electrode of polymer electrolyte fuel cells: Part I. Single-layer electrodes. *J. Electrochem. Soc. 155* (2008): B1289–B1295. doi: http://dx.doi.org/10.1149/1.2988763

27. Wang, Y., and Feng, X. Analysis of reaction rates in the cathode electrode of polymer electrolyte fuel cells: Part II. Dual-layer electrodes. *J. Electrochem. Soc. 156* (2009): B403–B409. doi: http://dx.doi.org/10.1149/1.3056057

28. Feng, X., and Wang, Y. Multi-layer configuration for the cathode electrode of polymer electrolyte fuel cell. *Electrochimica Acta 55* (2010): 4579–4586. doi: http://dx.doi.org/10.1016/j.electacta.2010.03.013

CHAPTER 5

VAPOR-PHASE WATER REMOVAL AND MANAGEMENT

Water is produced in the oxygen reduction reaction (ORR) within the cathode catalyst layers (CLs) and removed via the gas diffusion layers (GDLs) and gas flow channels (GFCs). Before the local partial vapor pressure reaches its saturated value, there is no tendency to form liquid water. Thus, water exists in the gas phase as vapor, and as a result the electrolyte membrane is partially hydrated. PEM fuel cells are subjected to a high ohmic resistance under severely dry operating conditions. Product water can be used to hydrate the electrolyte membrane, before being removed through the vapor phase. In the lack of sufficient vapor removal, excess water vapor condenses to liquid water, leading to two-phase flow. Therefore, it is important to understand vapor-phase water removal and management. This chapter covers the fundamentals of convection and diffusion of water vapor and gaseous reactants in the absence of liquid water or ice formation. In addition, Appendix V.A lists transport properties of typical gases at atmospheric pressure that closely relate to PEM fuel cell operation.

5.1 MASS TRANSPORT OVERVIEW

Convection (which refers to transport by bulk motion) and diffusion (as a result of the random walk of the diffusing particles which occurs without requiring bulk motion) are the two common transport mechanisms that underlie numerous transport processes. The general species-mass conservation equations, accounting for convection, diffusion, and sources as well as storage are given by

$$\varepsilon^{\text{eff}} \frac{\partial C_k}{\partial t} + \nabla \cdot (\vec{u} C_k) = \nabla \cdot (D_k^g \nabla C_k^g) + S_k, \qquad \text{(Eq. 5-1)}$$

where ε^{eff} is the effective porosity of a porous medium; C_k and S_k are, respectively, the molar concentration and the rate of generation or consumption of species k; and $\vec{u}$ is the mass-averaged velocity. In a PEM fuel cell, the effective porosity is unity for the gas flow channels and less than unity for porous components such as GDLs and CLs. The first term on the left-hand side in Eq. 5-1 measures the storage or accumulation rate of species k. The second term represents convection, a major transport mechanism in gas flow channels (GFCs). The flow field $\vec{u}$ follows the Navier–Stokes equations for GFCs or Darcy's law for GDLs. The first term on the right-hand side of Eq. 5-1 accounts for Fickian diffusion, which is a major mechanism of reactant/water transport in GDLs and catalyst layers, and water transport in electrolyte membranes.

5.2   DIFFUSION

Diffusion transports matter as a result of the random walks of diffusing particles. It can be explained through the molecular view: the constituent molecules move constantly, usually referred to as "random walks," driven by thermal energy or the particle's kinetic energy. At room temperature, the intermolecular distance of air or hydrogen gas, which is an ideal gas, is around 10 times larger than the molecular diameter; consequently constituent molecules travel randomly from higher to lower concentration regions.

5.2.1   DIFFUSIVITY

Another common process, as a result of random molecular motions, is heat transfer by conduction. There is an obvious analogy between diffusion and conduction, as illustrated by the lattice model presented in Chapter 3. The analogy was recognized by Fick (1855), who developed a quantitative basis for diffusion by adopting the mathematical equation of heat conduction derived by Fourier (1822). Fick's law applies to the binary diffusion with the diffusive flux proportional to the gradient of species concentration:

$$\vec{G}_k = -D_k^g \nabla C_k^g, \tag{Eq. 5-2}$$

where D_k^g is the diffusion coefficient of species k. For isotropic transport media with transport properties in the neighborhood of any point being the same in all directions, D_k^g is a scalar. In fibrous GDLs which are anisotropic media, diffusivity varies with directions; the coefficient is a tensor describing diffusivity in different directions with the diffusive flux written as

$$\vec{G}_k = -\vec{D}_k^g \cdot \nabla C_k^g \text{ where } \vec{D}_k^g = \left\{ D_k^g \right\}_{3x3}. \tag{Eq. 5-3}$$

The reactant gas streams in PEM fuel cells are gaseous mixtures, consisting of several components. Thus, multicomponent diffusion takes place, which can be described by the well-known Maxwell–Stefan flux equations:

$$\nabla x_i = RT \sum_{\substack{j=1 \\ i \neq j}}^{n} \frac{x_i x_j}{D_{ij}} (\vec{v}_j - \vec{v}_i), \tag{Eq. 5-4}$$

where x_i is the gas-phase mole fraction of species i, and $\vec{v}_i$ is the diffusive velocity of species i averaged over a differential volumetric element. Table 5-1 lists the binary diffusion coefficients

Table 5-1. Diffusion coefficients in air at 1 atm [1, 2]

	Binary diffusion coefficient ($m^2/s \times 10^4$)				
$T(K)$	O_2	CO_2	CO	H_2	H_2O (v)
200	0.095	0.074	0.098	0.375	0.1095
300	0.188	0.157	0.202	0.777	0.2538
400	0.325	0.263	0.332	1.25	0.4606
500	0.475	0.385	0.485	1.71	0.6983
600	0.646	0.537	0.659	2.44	0.9403
700	0.838	0.684	0.854	3.17	1.2093
800	1.05	0.857	1.06	3.93	1.5037
900	1.26	1.05	1.28	4.77	1.8225
1000	1.52	1.24	1.54	5.69	2.1644

in the air. Owing to the importance of water vaporair mixtures in PEM fuel cells, the following empirical correlation for water diffusivity in the air $D_{H_2O,air}$ is widely used:

$$D_{w,air} = 1.97 \times 10^{-5} \left(\frac{P_0}{P}\right)\left(\frac{T}{T_0}\right)^{1.685} \quad [m^2/s]; \ 273K < T < 373K, \qquad \text{(Eq. 5-5)}$$

where $P_0 = 1$ atm and $T_0 = 256$ K. The following correlation is also frequently used [2]:

$$D_{w,air} = 1.87 \times 10^{-10} \frac{T^{2.072}}{P} ; \ 280K < T < 450K$$

$$= 2.75 \times 10^{-9} \frac{T^{1.632}}{P} ; \ 450K < T < 1070K, \qquad \text{(Eq. 5-6)}$$

where P is in atmospheres and T is in Kelvins.

For general species, diffusivity and its relevance to molecular movement can be described by using a lattice model, as explained in Chapter 3. The diffusivity in a low-density A-B mixture can be approximated by

$$D_{AB} = \frac{ul}{6} = \frac{1}{3\pi^{3/2}} \frac{(RT)^{3/2}}{N_{AV} d^2 M^{1/2} p}$$

where

$$d = \frac{d_A + d_B}{2} \text{ and } M = 2\left(\frac{1}{M_A} + \frac{1}{M_B}\right)^{-1}, \qquad \text{(Eq. 5-7)}$$

where N_{AV} is the Avogadro's number, d is the average molecular diameter, and M is the average molecular weight.

Equation 5-7 shows the dependency of diffusivity on temperature and pressure. Table 5-2 lists the diffusion coefficients of typical species in PEM fuel cells.

5.2.2 MOLECULAR VERSUS KNUDSEN DIFFUSION

Depending on the scale of interest, two types of diffusion are encountered in PEM fuel cells: one is the molecular (or Fickian) diffusion, which takes place when the mean free length of molecules is relatively large as opposed to the diffusion passage dimension. Under this circumstance, the molecules collide with each other in their passage through the pores; consequently, the molecular interaction is the limiting factor for diffusion. In PEM fuel cell modeling and analysis, the following correlation of diffusion coefficient as a function of temperature and pressure is frequently adopted:

$$D_i^M = D_{0,i}^M \left(\frac{T}{353}\right)^{3/2} \left(\frac{1}{P}\right). \qquad \text{(Eq. 5-8)}$$

The other type of diffusion is called Knudsen diffusion, which dominates when gas molecules collide more frequently with passage walls than with each other. This type of diffusion is important when the mean free path of molecules is on the order of the characteristic length scale of diffusion passage. In catalyst layers, the mean radius of the pores is around 0.1 μm. The mean free path of molecules can be evaluated by

$$\lambda_{\text{molecule}} = \frac{8RT}{\sqrt{2}\pi d^2 N_{AV} P}, \qquad \text{(Eq. 5-9)}$$

which is ca 0.1 μm under 80°C and 2 atm; therefore, the Knudsen diffusion is important for gas species transport in the catalyst layer. Its coefficient in a long straight pore can be derived from the kinetic theory of gases:

$$D_i^K = \frac{2r_{\text{pore}}}{3} \sqrt{\frac{8RT}{\pi M_i}}. \qquad \text{(Eq. 5-10)}$$

The value of D_i^K is around $2.64 \times 10^{-5} \text{m}^2/\text{s}$ at $r_{\text{pore}} = 0.1$ μm under typical PEM fuel cell operating conditions, which is comparable to that of the molecular diffusion coefficient. An

Table 5-2. Diffusion coefficients of hydrogen, oxygen and water vapor in PEM fuel cells under 80°C and 1 atm [3]

H_2 diffusivity in anode gas, D_{o,H_2}	$1.1 \times 10{-}4 \text{ m}^2/\text{s}$
H_2O diffusivity in anode gas, $D_{o,w}^a$	$1.1 \times 10{-}4 \text{ m}^2/\text{s}$
O_2 diffusivity in cathode gas, D_{o,O_2}	$3.24 \times 10{-}5 \text{ m}^2/\text{s}$
H_2O diffusivity in cathode gas, $D_{o,w}^c$	$3.89 \times 10{-}5 \text{ m}^2/\text{s}$

important parameter is the Knudsen number (Kn), defined as $\dfrac{\lambda_{molecule}}{L}$, which distinguishes the molecular and Knudsen diffusion regimes. In the catalyst layer, the harmonic mean can be taken to calculate the average diffusivity D^{eff}:

$$\frac{1}{D^{eff}} = \frac{1}{D^M} + \frac{1}{D^K}$$

(Eq. 5-11)

Figure 5-1 displays the diffusion coefficients in a multilength scale porous medium with the two distinct diffusion regimes, namely the Knudsen and molecular regimes. In between, both diffusions are important and thus need to be taken into account. In addition to the two diffusion regimes discussed previously, when the average pore size is on the order of the molecular diameter, the diffusivity should be determined from a molecular configurational model [5], in which the crystalline structure of the medium, the molecular dimension of diffusing species, and the pore geometry and microstructure all become important.

5.2.3 DIFFUSION IN GDLS

GDLs bridge the diffusive passages between the catalyst layer and gas flow channel. In this component, diffusion is important to species transport in both the through-plane and in-plane

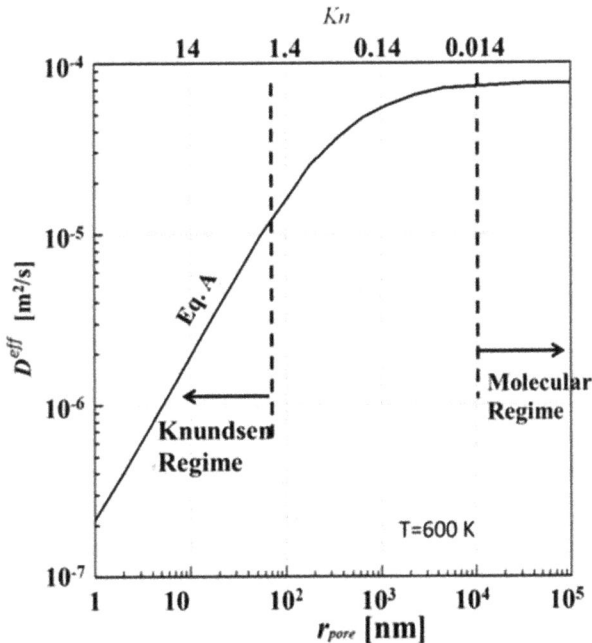

Figure 5-1. The mass diffusivity with respect to pore size or Knudsen number (Kn). Equation A refers to Eq. 5-11 (redrawn from [4] with modifications).

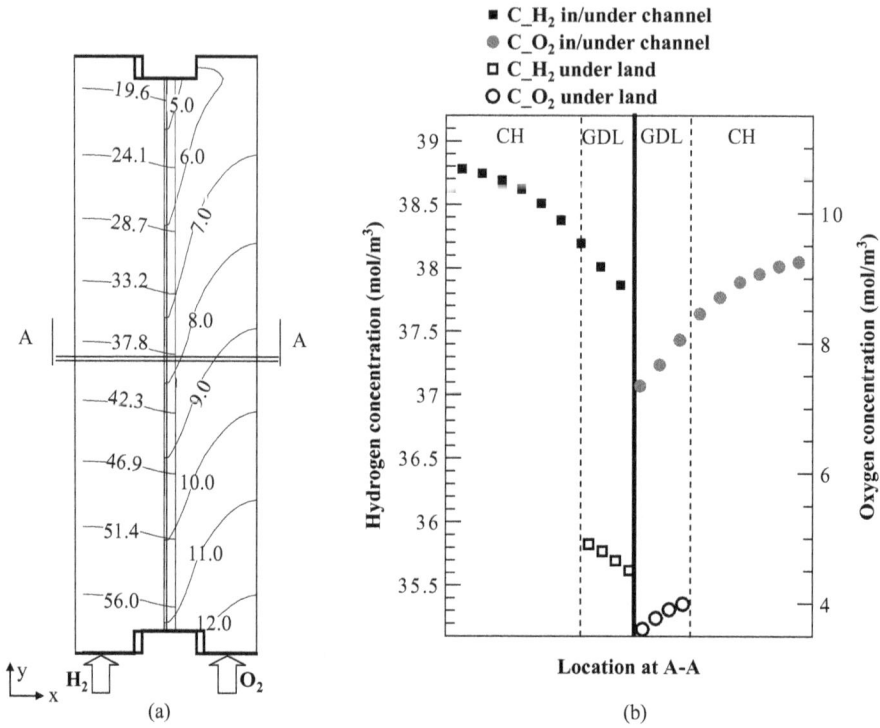

Figure 5-2. (a) Hydrogen and oxygen concentration contours in a PEM fuel cell and (b) at the A–A location (operating conditions: anode/cathode pressure, Pa/c = 2 atm, cell temperature, T_{cell} = 80°C, anode/cathode RH, RHa/c = 66%/66%, and anode/cathode stoich, stoicha/c = 1.5/2).

directions. Figure 5-2 displays the oxygen and hydrogen concentration contours on the cathode and anode sides, respectively. Both H_2 and O_2 concentrations decrease toward the catalyst layer and down the stream, as a result of the HOR and ORR electrochemical reactions, which consume H_2 and O_2, respectively.

Lateral diffusion is critical to oxygen supply and water removal under lands. Figure 5-3 presents the oxygen and water distributions in the GDL under a land, which clearly show lateral gradients of the H_2O and O_2 concentrations between the two adjacent channels. A gradient in the water concentration toward the dry inlet channel (Channel 1) is evident. Figure 5-2(b) shows that the lowest O_2 concentration appears under the land due to the local ORR.

Figure 5-4 presents the water diffusive flux across the two surfaces of a cathode GDL. The areas under the curves, that is, the integrals of the curves, represent the amount of water across each face of the GDL per unit length. Because there is no water production/consumption in the GDL under the selected conditions, the area between the two curves across the two surfaces of the GDL represents the net water flux along the lateral direction. It is seen that the amount of water into the gas channel is almost twice of that across the interface between the catalyst layer and GDL under Channel 1, whereas the water fluxes across the two GDL surfaces are almost equal under Channel 2. The difference in the amount of product water captured by the inlet Channel 1 and outlet Channel 2 indicates that internal humidification exists, imposing a

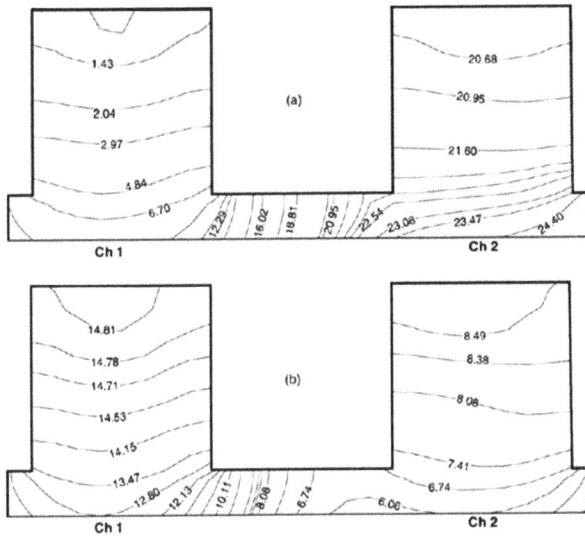

Figure 5-3. (a) H_2O and (b) O_2 concentration [mol/m³] distributions, respectively, in the cathode GDL at the mid-length cross-section between Channels 1 and 2 with no convection and $V_{cell} = 0.65$ V and $I = 0.91$ A/cm². Channel 1 is the inlet of a flow path; Channel 2 is the outlet of the other flow path. The entire flow-field configuration is shown in Figure 5-4.

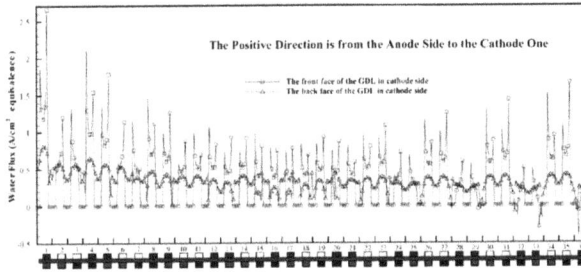

Figure 5-4. Water flux across the cathode GDL surfaces at the mid-length cross-section with no convection in the GDL and under 0.65 V and 0.91 A/cm². The cathode flow field is a dual serpentine channel configuration with the two paths in parallel and counter flow pattern [6].

significant effect on humidifying the dry inlet flow. The same trend is also seen in the channels close to the air inlet such as Channels 4, 5, 34, and 35, and the channels close to the exhaust outlet such as Channels 3, 6, 33, and 36. In Channel 36, there exists a negative flux across the front face of the GDL, which means that water diffuses from Channel 36 into the GDL and is transported laterally through the GDL to Channel 35. Similar phenomena also appear in the outlet.

5.3 SPECIES CONVECTION

5.3.1 FLOW MODELING WITH CONSTANT-FLOW ASSUMPTION

In order to account for convection, flow fields must be known first. Flows encountered in PEM fuel cells are usually in the laminar regime due to the relatively low magnitude of velocity and small physical dimension. In general, Newtonian fluid flows in the laminar regime are well described by the Navier–Stokes equations. In porous components such as GDLs and catalyst layers, Darcy's law is frequently adopted to relate momentum flux to pressure gradient. In PEM fuel cell studies reported in the open literature so far, flow fields were poorly described in many cases and the coupling with electrochemical reactions were neglected to some degree; for example, the mass exchanges between the anode and cathode sides were not taken into account. One popular modeling approach is the constant flow assumption, which neglects the mass exchange and assumes constant density of the gas mixture, yielding the following standard flow equations:

continuity: $\nabla \cdot \vec{u} = 0$,

momentum: $\dfrac{1}{\varepsilon^2} \nabla \cdot (\rho \vec{u} \vec{u}) = -\nabla p + \nabla \cdot \vec{\tau} + S_u$. (Eq. 5-12)

Appendix V.A provides transport properties of typical gases in PEM fuel cells, such as density and viscosity. The last term in the momentum equation accounts for Darcy's force in porous media, and vanishes in gas flow channels (or GFCs). Because Eq. 5-12 results in significant model simplification and fast convergence in numerical simulation, and more importantly acceptable approximation in terms of fuel cell performance prediction, at present this constant gas-density approach is widely adopted in fuel cell modeling and simulation.

5.3.2 FLOW FORMULATION WITHOUT THE CONSTANT-FLOW ASSUMPTION

In PEM fuel cells, mass is exchanged between the anode and cathode as a result of the electrochemical reactions and transport through the electrolyte membranes. The mass exchange can considerably alter gas composition and fluid flow. Herein, we term the formulation without the constant-flow assumption the variable-flow model. This variable-flow formulation describes a more realistic flow field. The general governing equations of steady-state fluid flows, which account for mass addition or consumption, are expressed by

continuity: $\nabla \cdot (\rho \vec{u}) = S_m$

momentum: $\dfrac{1}{\varepsilon^2} \nabla \cdot (\rho \vec{u} \vec{u}) = -\nabla p + \nabla \cdot \tau + S_u$, (Eq. 5-13)

where $\vec{u}$ is the superficial velocity in porous media and the variable gas density is a function of molar concentrations of the constituent species as described below:

$$\rho = \sum_k C_k M_k. \tag{Eq. 5-14}$$

Another way to derive the continuity equation for the gas mixture in Eq. 5-13 is by summing up all the species equations: that is, performing the operation of $\sum_k (M_k \times \text{Eq. 5-14})$ yields

$$\nabla \cdot \left(\vec{u} \sum_k C_k M_k \right) = \nabla \cdot \left(\sum_k D_k^{\text{eff}} \nabla M_k C_k \right) + \sum_k S_k M_k \tag{Eq. 5-15}$$

or

$$\nabla \cdot (\rho \vec{u}) = \nabla \cdot \left(\sum_k D_k^{\text{eff}} \nabla M_k C_k \right) + \sum_k S_k M_k. \tag{Eq. 5-16}$$

Comparing Eq. 5-16 with the continuity equation in Eq. 5-13 yields

$$S_m = \nabla \cdot \left(\sum_k D_k^{\text{eff}} \nabla M_k C_k \right) + \sum_k S_k M_k. \tag{Eq. 5-17}$$

In the above, the diffusion terms contain contributions from both gas and membrane phases. Assuming the summation of multicomponent gas diffusion terms is equal to zero, and hence separating out the term on water diffusion through the membrane phase, one arrives at [7]

$$S_m = M_w \nabla \cdot (D_w^m \nabla C_w^m) + \sum_k S_k M_k = M_w \nabla \cdot (D_{w,m} \nabla C_w) + \sum_k S_k M_k, \tag{Eq. 5-18}$$

where C_w^m is the molar concentration of water in the membrane phase, and $D_{w,m}$ is the modified diffusion coefficient for water in the ionomer if expressed in terms of the gradient in the gas phase molar concentration. Details on water transport in membranes are given in Chapter 4.

Figure 5-5 displays the mass source S_m in the anode catalyst layer along the channel, scaled by the anode inlet flow rate. The total mass source consists of the contributions from hydrogen consumption, water electro-osmotic drag, and water back diffusion, as seen in Eq. 5-18. The integral of each curve represents the ratio of the total mass source/sink in the catalyst layer to the anode inlet flow rate. For an anode stoichiometric flow ratio of 2 (or stoich$_a$ = 2.0), the anode loses a large amount of mass in the first quarter of the channel due to water electro-osmotic drag, and it gradually receives mass through the back-diffusion downstream. Compared with the mass source related to water, hydrogen contribution is negligibly small. In addition, the water electro-osmotic drag and diffusion almost cancel each other in terms of mass contribution, starting from the dimensionless distance of ~25% from the inlet; in the remaining three-quarters of the channel, a small total mass source is generated. In the case of stoich$_a$ = 1.2, the anode loses mass to the cathode in the entire region and the magnitude of the total mass source is significant not only in the first quarter but also in the last quarter of the channel.

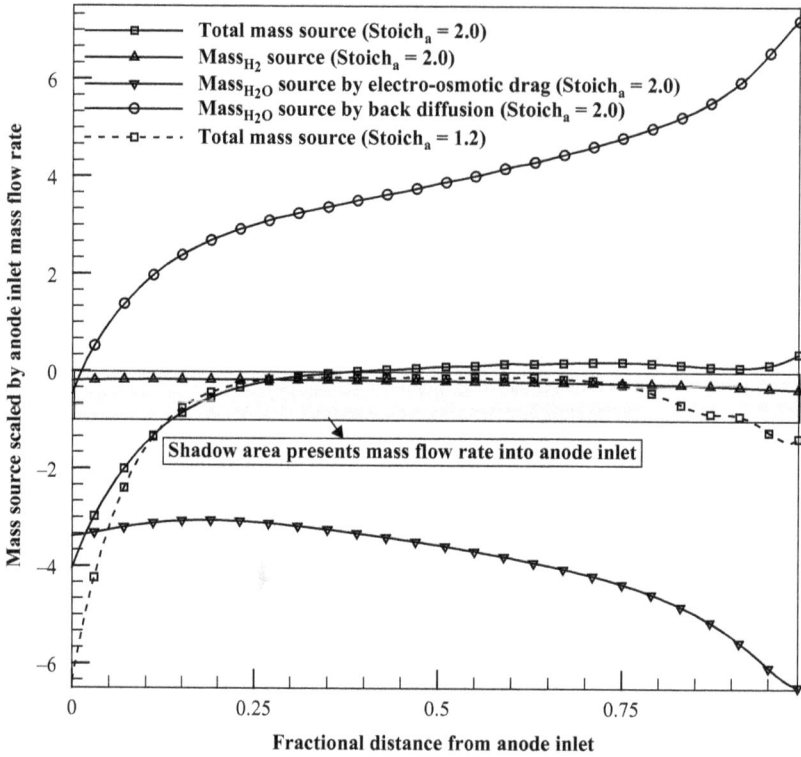

Figure 5-5. Mass sources in the anode catalyst layer (scaled by the mass injection rate in anode) predicted by the variable-flow model for stoich$_a$ = 2.0 (at 0.625 V and 0.45 A/cm^2) and stoich$_a$ = 1.2 (at 0.61 V and 0.5 A/cm^2), respectively [7] (Courtesy of the Electrochemical Society).

The partial differential equation of Eq. 5-1 can be solved for water and oxygen molar concentrations with the flow field computed by Eq. 5-13, whereas the hydrogen and nitrogen concentrations can be obtained using the ideal gas law:

$$C_{H_2/N_2} = \frac{p}{RT} - \sum_{k=\text{others}} C_k. \tag{Eq. 5-19}$$

This approach of indirectly solving for the hydrogen concentration is valid for a binary gas (H$_2$/H$_2$O) anode. In this situation, the H$_2$O species equation along with the ideal gas law (i.e., Eq. 5-19) provides a sufficient number of equations for two unknowns: H$_2$ and H$_2$O molar concentrations.

It is instructive to estimate the transverse gas velocity induced by the mass source/sink and the ensuing convection effect on species transport. In the dimension across a GDL or a catalyst layer, integration of the continuity equation in the through-plane direction across the anode results in

$$\rho u = M_{H_2} \frac{I}{2F} + M_w \alpha \frac{I}{F} = \left(\frac{M_{H_2}}{2} + M_w \alpha \right) \frac{I}{F}. \tag{Eq. 5-20}$$

Here, u is the velocity component in the through-plane direction and α the net water flux per proton through the membrane. If we define the Peclet number (Pe) as a parameter to measure the relative importance of this transverse convection to molecular diffusion, it follows:

$$Pe = \frac{u\delta_{GDL}}{D} = \frac{\left(\dfrac{M_{H_2}}{2} + M_w\alpha\right)\dfrac{I}{F}\delta_{GDL}}{\rho D} = 0.03(1+18\alpha) \approx 0.1 \qquad \text{(Eq. 5-21)}$$

for $I = 1.0$ A/cm^2, $\delta_{GDL} = 0.3$ mm, and $\alpha = 0.1$. This indicates that the convective effect due to the transverse flow is small, relative to the diffusive transport. Figure 5-6 presents the cross-sectional velocity vectors and water concentration contours at the mid-length of the cell. At this location, strong back-diffusion of water occurs, as evident from the higher water concentration

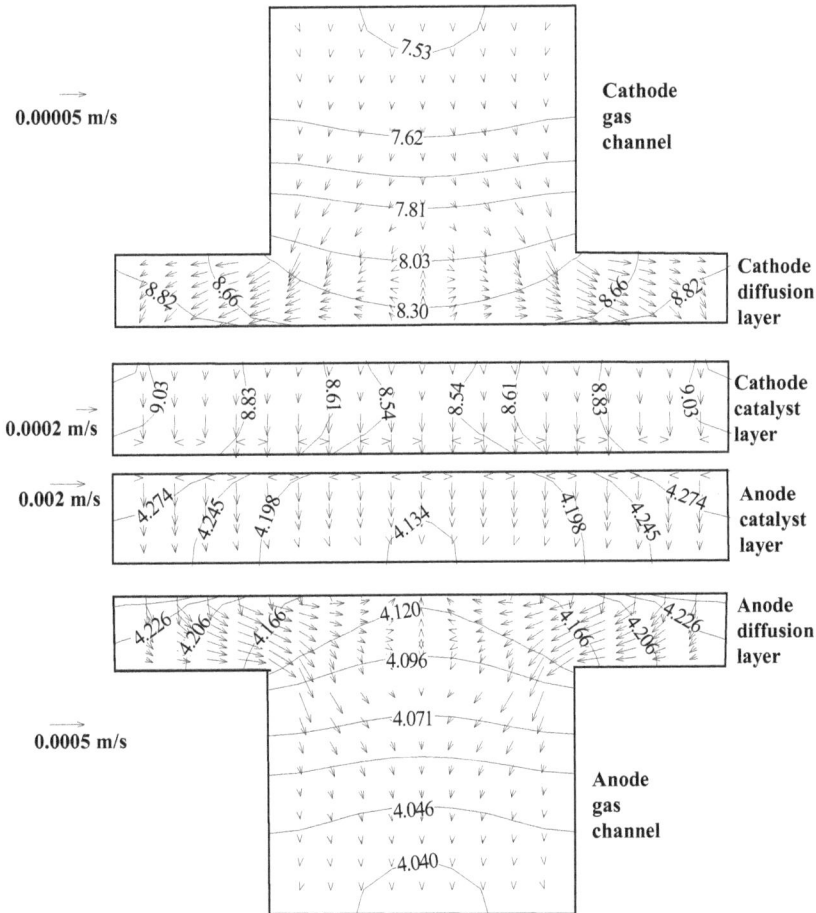

Figure 5-6. Flow field and water concentration (mol/m^3) at the mid-length cross-section of a PEM fuel cell (i.e., $y = Ly/2$), predicted by the variable-flow model with stoich$_a$ = 2.0 and 0.625 V.
Note: the catalyst layers are expanded in the scale for clarity [7] (Courtesy of the Electrochemical Society).

on the cathode side. The back diffusion offsets the electro-osmotic drag, resulting in a small transverse flow.

Gas reactant flows in channels are strongly affected by the mass exchange and gas density variation. Figure 5-7 shows streamwise variations of the average axial velocity and flow density in gas flow channels predicted by the variable-flow model for stoich$_a$ = 2.0. The predicted density in the cathode decreases only by 4% along the channel, whereas that in the anode deceases by 12% over the first quarter of the length, followed by an increase of ~30% from the lowest mixture gas density. The negligible variation in the cathode gas density is expected because nitrogen is a dominant species in the gas mixture. The large density variation in the anode stems from the large molecular-weight contrast of hydrogen gas to water vapor. In addition to the gas density, the velocity variation along the fuel cell length is more dramatic in the anode. Figure 5-7 shows that the anode average velocity decreases by nearly 50%, whereas the cathode velocity changes by mere 9%. Again, the relatively small variation observed in the cathode velocity can be easily explained by the much larger density of the cathode gas stream.

Species convective transport is altered appreciably as a result of the varying flow velocity. Figure 5-8 compares the average water/hydrogen molar concentrations along the gas channel for stoich$_a$ = 2.0, predicted by the variable- and constant-flow models. The remarkable difference appears in the hydrogen profile. The variable-flow model predicts an almost flat profile; that is, the H$_2$ molar concentration remains nearly constant. In contrast, the constant-flow model predicts a rapidly declining H$_2$ profile. This is because in the constant-flow model, the flow rate

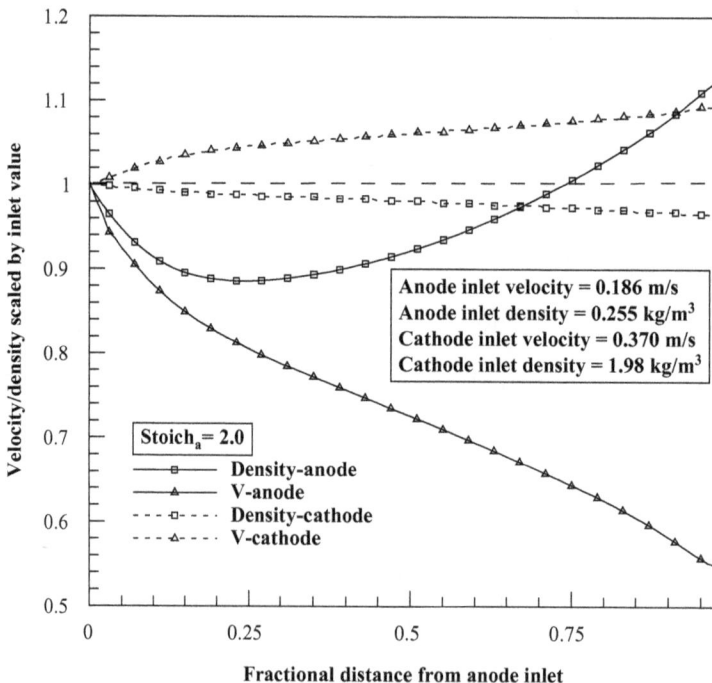

Figure 5-7. Average axial velocity and gas density in gas channels predicted by the variable-flow model for stoich$_a$ = 2 and 0.625 V [7] (Courtesy of the Electrochemical Society).

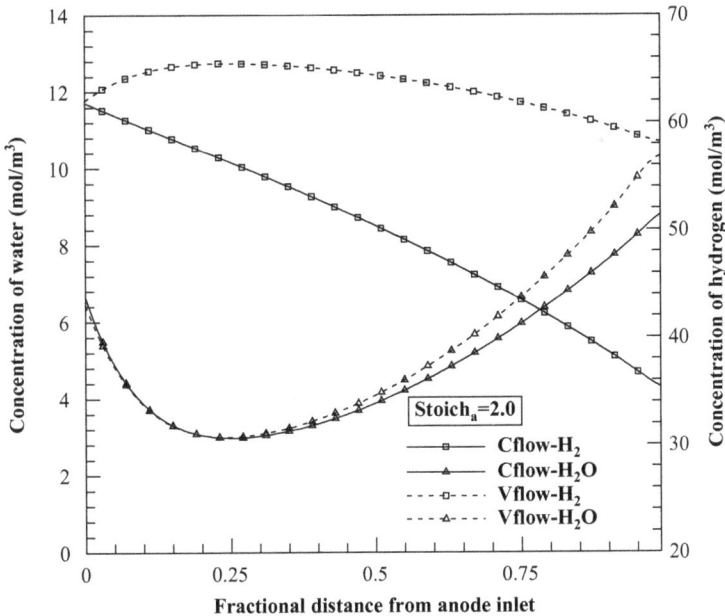

Figure 5-8. The average water/hydrogen molar concentrations in the anode gas channel predicted by the constant-flow and variable-flow models for stoich$_a$ = 2.0 and 0.625 V [7] (Courtesy of the Electrochemical Society).

does not decrease as in the variable-flow model, and consequently, the hydrogen concentration must be lowered in order to satisfy its consumption by the reaction. This dramatic difference in hydrogen concentration, however, does not affect the overpotential for HOR as the reaction is sufficiently facile under both concentrations. As compared with hydrogen, the water profiles predicted by both models are much closer; that is, water concentration is less likely affected by the flow treatment. They look similar also in the sense that the water concentration decreases near the inlet due to the dominant electro-osmotic drag and then increases as back diffusion becomes appreciable.

Figure 5-9 shows the hydrogen through-plane profiles for stoich$_a$ = 2. At the mid-length of the fuel cell, both curves under the land and channel exhibit a decline from the GDL toward the catalyst layer, indicative of a typical diffusion-reaction process. In addition, lower hydrogen concentration appears under the land than under the channel due to the HOR's consumption under the land. However, near the inlet region, both curves under the land and channel reveal a totally opposite trend: hydrogen concentration increases from the channel, to the GDL, and then to the catalyst layer, whereas the concentration under the land is higher than that under the channel. The reverse profile results clearly from the strong transverse flow existing in the region. Nonetheless, the weak effect of convection only slightly raises the H$_2$ concentration.

Figure 5-10 attempts to establish the validity range of the constant-flow model. First, comparison is made in Figure 5-10 between the two models for stoich$_a$ = 2 and a current density of 0.5 A/cm^2 with full humidification in both anode and cathode. No appreciable difference is observed, demonstrating that the constant-flow assumption is physically sound and computationally advantageous for modeling PEM fuel cells that operate under high to full humidification. For low-humidity

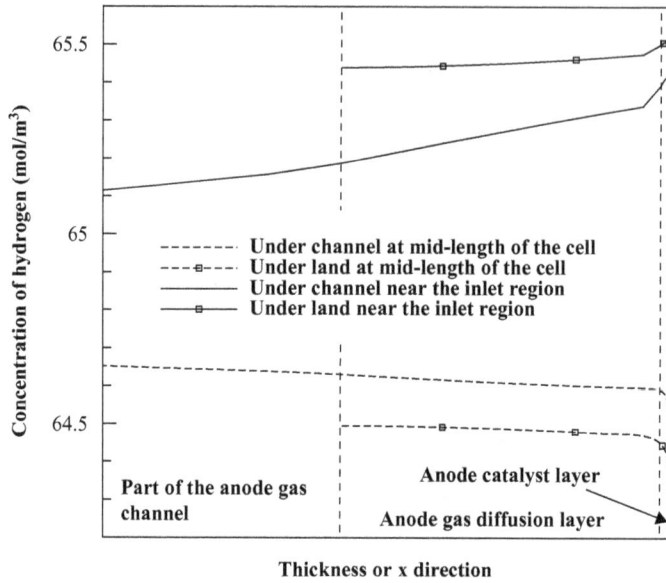

Figure 5-9. Through-plane profiles of hydrogen concentrations in the anode for stoich$_a$ = 2 and 0.625 V, predicted by the variable-flow model [7] (Courtesy of the Electrochemical Society).

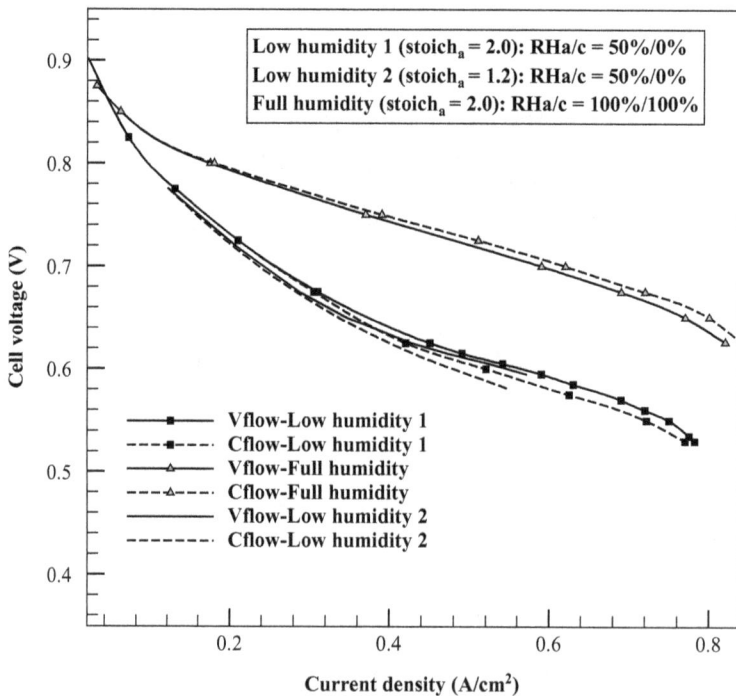

Figure 5-10. Comparison of polarization curves computed by constant- and variable-flow models [7] (Courtesy of the Electrochemical Society).

operation (e.g., $RH_{a/c} = 50\%/0\%$), the differences in the average current are found to be less than 10% and 14%, respectively, for the anode stoichiometric flow ratios of 1.2 and 2.0. The two polarization curves at $stoich_a = 1.2$ are terminated before 0.6 A/cm^2, which is the maximum current density possible for the H_2 stoichiometric flow ratio of 1.2 and current density of 0.5 A/cm^2.

5.3.3 CONVECTION IN GDLs

Diffusion is generally regarded as a major mechanism for gas transport in conventional flow fields, in particular, in GDLs and CLs. In interdigitated flow fields, strong flow is promoted in GDLs thus convection dominates species transport. Strong flow may be promoted in common serpentine configurations, for example, in the presence of appreciable pressure difference between adjacent channels, by-pass flow in the GDL develops from the higher pressure channel toward the lower one. The by-pass flow can be evaluated using Darcy's law, and it alters the lateral convection: promoting species convection from higher-pressure channel to low-pressure one.

Figure 5-11 shows the gas pressure profile in the cathode GDL at the mid-length cross-section in a PEM fuel cell with its flow field shown in the bottom of the figure. A large pressure drop is seen between the adjacent channels at the two ends of the fuel cell. Between Channels 1 and 2, c.a. 1.5 kPa pressure difference is present. According to Darcy's law, a by-pass flow or a lateral velocity of about 0.15 m/s will be generated in the GDL with a permeability of 10^{-12} m^2 from the dry Channel 1 to the wet Channel 2. As a result, there is a certain amount of inlet fresh air bypassing through the GDL and directly flowing into the channel in the other flow path toward the exit. This gas bypass, or "short circuit", greatly reduces reactant utilization of the fuel cell. To quantify the bypass, the average gas velocity in the cathode gas channels is displayed in Figure 5-11 as well, showing that the bypass causes severe leakage of injected flow. For Path 1 (shown in dark in the upper channels), from the inlet Channel 1 to the middle of the fuel cell, for example, Channel 17, the mass of flow in the channel decreases by about 80%. From the middle to the outlet Channel 36, the flow acquires four times mass from the other flow path.

The effect of convection on species transport is indicated in Figure 5-12, which superposes H_2O/O_2 concentration contours with the velocity vector plot in the cathode side between

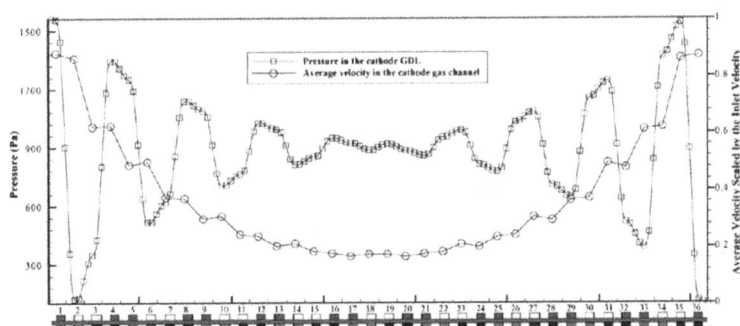

Figure 5-11. Pressure distribution in the cathode GDL and average velocity in the channel (scaled by cathode inlet velocity) at the mid-length cross-section with flow present in the GDL.

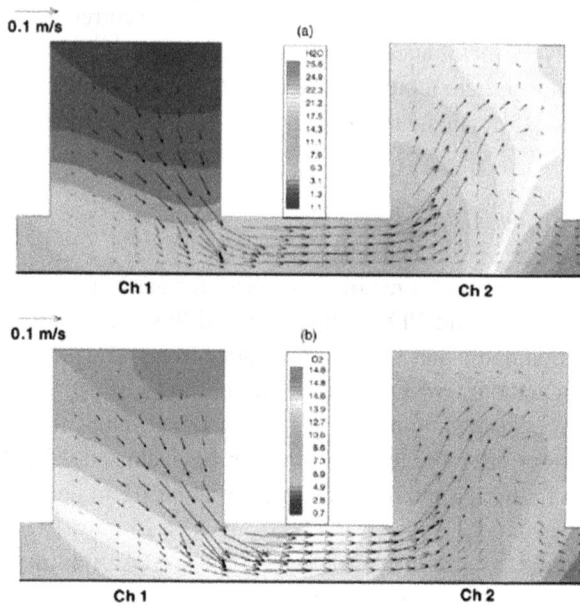

Figure 5-12. (a) H_2O and (b) O_2 distributions and velocity on the cathode side at a cross-section between Channels 1 and 2 under 0.65 V and 0.88 A/cm^2.

Channels 1 and 2. It is seen that about 0.1 m/s velocity in the porous GDL is induced by the pressure difference between the two channels, consistent with the estimate made earlier. The Peclet number (Pe) can be used to measure the relative strength of convection to molecular diffusion:

$$Pe = \frac{u\delta_{land}}{D_{GDL}^{eff}}.$$

(Eq. 5-22)

For $D_{GDL}^{eff} \approx 10^{-5}$ m^2/s and $\delta_{land} = 0.001$ m, Pe is about 10, demonstrating that convection dominates the lateral species transport in the GDL.

5.4 PORE-SCALE TRANSPORT

Continuum macroscopic approach is a preferred method to study transport phenomena in porous media. Fully accounting for the wide-spanning length scale and real pore structures is in general computationally costly. Macroscopic methods use several effective parameters to account for the information of structure feature; thus, these parameters must be known or can be predetermined. Oftentimes, these parameters are determined only experimentally, especially in situations where the pore structure is complex. Another important issue in continuum modeling is the proper choice of scales. Adopted averaged parameters contain implicitly specified length scales; thus, the model must operate over similar scales. Parameters, averaged over too small a length scale, are not meaningful.

Pore-scale modeling provides a link between the medium's microscopic properties and the transport behavior at a larger scale. Pore-scale studies are an important complement to the macroscopic approach because it can be used to predict macroscopic parameters and provide a means for studying scaling issues. A number of issues must be addressed, however, before the pore-scale modeling approach becomes practical and useful. One issue is the ability to predict consistent transport parameters using basic morphological information. This capability has been demonstrated for many different types of specific media such as sphere packs and fibrous media. Another issue is upscaling from the pore to the macroscopic level. Both the mathematical and numerical problems can be quite challenging because of computational burdens, morphological description, numerical convergence, multiscale bridging, and so on. Readers who are interested in detailed discussion on the pore-level studies are referred to consult with reference books [8, 9].

In a PEM fuel cell, both GDLs and catalyst layers are porous materials. Gas flow channels can use porous materials to improve cell performance by taking advantage of porous media's advanced properties. The properties, therefore, are critical to the functions of these components. In the following subsection, we will use carbon papers as an example to show a way of property evaluation using a pore-scale model.

5.4.1 STOCHASTIC MATERIAL RECONSTRUCTION

Carbon papers, a popular GDL substrate, are nonwoven fibrous porous media, where carbon fibers are randomly placed and bonded by binders. Figure 5-13 shows a microscopic image of carbon papers. In operation, the transport of reactants (e.g., hydrogen and oxygen) and product (e.g., water) occurs through this medium due to the electrochemical reactions, such as HOR and ORR. Thus, its transport properties, determined by GDL structure, are critical to fuel cell performance. Digital reconstruction of material structure can be performed through two major methods: one is to create three-dimensional (3D) images using volume imaging techniques, the other to develop stochastic models. The former requires the use of nonintrusive imaging techniques, such as X-ray tomograph or magnetic resonance imaging. The material is repeatedly sectioned and imaged automatically, and the images are used in software constructions of the material's 3D structure. X-ray Tomograph is a very popular imaging method widely used for medical and industrial imaging, to be discussed in Chapter 6. Magnetic resonance imaging (MRI) is a popular imaging method used in medical radiology to visualize a human body's internal structures. It uses the property of nuclear magnetic resonance (NMR) to image nuclei of atoms inside a body. An MRI scanner is a tube where the patient lies inside. The magnetic field is used to align the magnetization of some atomic nuclei in the body with radio frequency fields. The nuclei produce a rotating magnetic field detectable by the scanner and recorded to construct an image of the scanned area. The imaging method can be extremely expensive and time-consuming, and its resolution is limited by the source wavelength and detectors. The latter method is typically based on the material's structural knowledge and stochastic theory. It is not as expensive as the former method and is also faster. Its resolution is generally limited by the reconstruction grids applied to the model, and reconstructed materials can be modified by adjusting model parameters. A stochastic model is presented below for reconstructing the microstructure of a carbon paper GDL [10].

(a)

(b)

Figure 5-13. Microscope images of the carbon paper GDL in PEM fuel cells.

A look at Figure 5-13 leads to an impression that the constituent carbon fibers can be approximated by horizontally orientated straight lines and assuming that their curvatures are negligible, which are dilated with respect to 3D. This can be verified through the production process of carbon papers, in which fibers are disposed "almost" horizontally. In addition, the fiber system can be treated as a stack of several thin sections consisting of such laterally orientated fibers which are allowed to interpenetrate. The fibers within a given thin section are seen as mutually penetrating cylinders or filaments. The fiber system is therefore modeled as a stack of independent and identically distributed layers, where adjacent layers are in touch with each other. In real carbon papers, the fibers in each thin section are crossing each other to a small extent. It has been demonstrated that in spite of these simplifications, the predicted structure and the structure of physically measured 3D data of such a GDL are very similar.

The model for the fibers in each section is based on the Poisson line tessellations (PLT) in a plane. A planar line tessellation is a set of cells $\xi_1, \xi_2, \ldots$ generated by intersecting lines $l_1, l_2, \ldots$ scattered in the plane. A line l_n in the plane can be represented by its normal form, that is, its orthogonal signed distance to the origin x_n and its direction m_n measured as the angle to the x-axis; see Figure 5-14(a). Thus, a line tessellation can be seen as an independently marked point process $(X_1, M_1), (X_2, M_2), \ldots$ where in the case of a PLT the random distances $X_1, X_2, \ldots$ form a homogeneous Poisson point process on the real line with intensity $\lambda > 0$. The random variables $M_1, M_2, \ldots$, describing the directions, are uniformly distributed in the interval $[0, \pi)$ and mutually independent as well as independent from the Poisson point process. The PLT is then the set of random polygons $\Xi_1, \Xi_2, \ldots$ created by the random lines $(X_1, M_1), (X_2, M_2), \ldots$; see Figure 5-14(b). The PLT is completely described by one parameter, the intensity of the Poisson point process $\lambda > 0$, which can be interpreted as the mean total edge length per unit area of the PLT.

The complete multilayer model consists of a stack of such dilated PLTs (see Figure 5-14), where all layers are assumed to be independent and identically distributed. Adjacent layers are just touching each other. The fibers of the carbon paper are adhered by means of a binder, which is introduced in the manufacturing process and can be treated as thin films bounded by fibers. One binder is modeled by an independent filling of cells, that is, a cell of the underlying PLT is chosen to contain binder with probability $p \in [0,1]$. A realization of the complete model is shown in Figure 5-15.

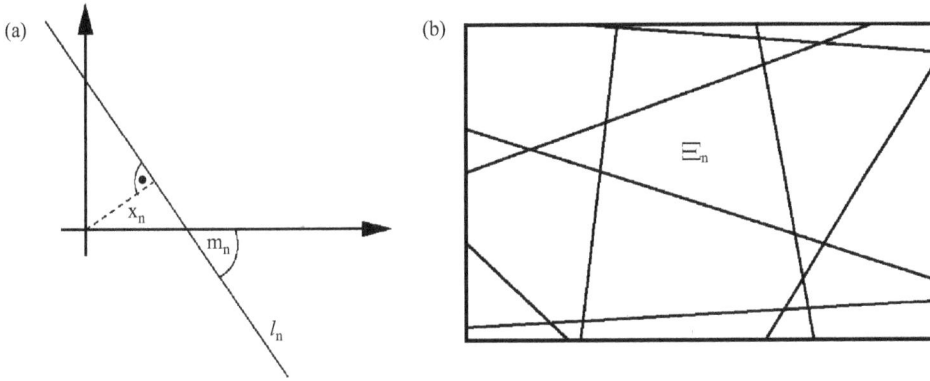

Figure 5-14. (a) Construction of a line; (b) a cell Ξ_n of a PLT (Poisson line tessellation) [10].

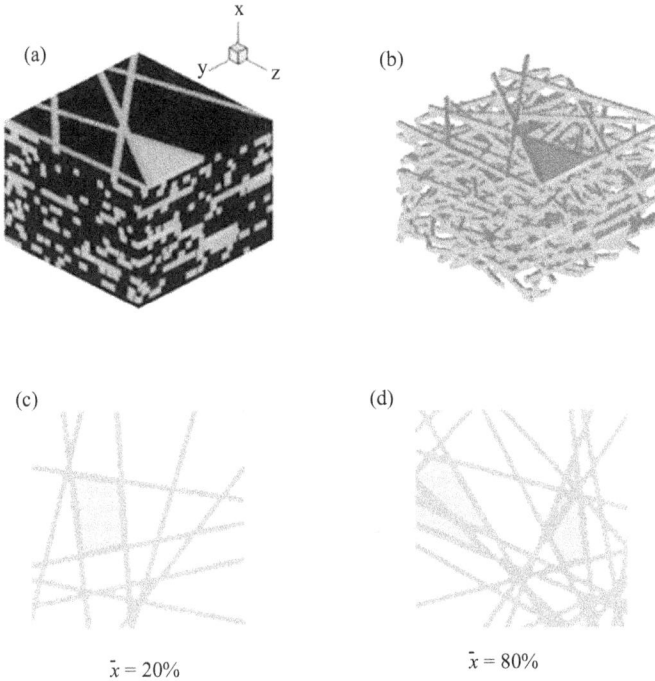

Figure 5-15. (a) A realization of the stochastic model for carbon paper reconstruction; (b) solid matrix; (c) solid matrix cross-section at $\bar{x} = 20\%$; (d) solid matrix cross-section at $\bar{x} = 80\%$.

5.4.2 PORE-SCALE TRANSPORT MODELING

Direct simulation is a powerful tool in fundamental studies of pore-level phenomena. Some studies also term it "direct numerical simulation". The modeling process is based on first principles or conservation laws, for example, the Navier–Stokes equations and energy conservation equations. One issue is that direct simulation usually involves a large number of computational grid-points in order to describe the full morphology of a material's microstructure. Presented in the following is an example of direct simulation on a carbon paper GDL [10].

(1) Void Space

The void space of a carbon paper GDL is mostly interconnected, allowing mass transport. In direct simulation, a sufficiently fine mesh that is capable of distinguishing the void from solid is employed. In void space, there may exist fluid motion, which is described by the Navier–Stokes equations:

$$\frac{\partial \rho_g}{\partial t} + \nabla \cdot \left(\rho_g \vec{u} \right) = 0 \qquad \text{(Eq. 5-23)}$$

$$\left[\frac{\partial \rho_g \vec{u}}{\partial t} + \nabla \cdot \left(\rho_g \vec{u} \vec{u} \right) \right] = -\nabla P_g + \nabla \cdot \vec{\tau}, \qquad \text{(Eq. 5-24)}$$

where the quantities are defined at the pore level; consequently, the effective properties such as porosity and permeability no longer appear in the governing equations. The void network provides species transport pathways. In addition to the convection arising from fluid flow, molecular diffusion is a major mechanism for water and oxygen transport, and the general conservation equation of species reads

$$\frac{\partial C_k}{\partial t} + \nabla \cdot (\vec{u} C_k) = \nabla \cdot (D_k \nabla C_k), \qquad \text{(Eq. 5-25)}$$

where the superscript k denotes species, and the coefficient D_k is a function of temperature and pressure as follows:

$$D_k = D_{0,k} \left(\frac{T}{353} \right)^{3/2} \left(\frac{1}{P} \right). \qquad \text{(Eq. 5-26)}$$

In contrast to macroscopic approaches, the intrinsic diffusivity, instead of an effective property, is used in the direct simulation approach.

(2) Solid Matrix

The solid matrix of a GDL consists of carbon fibers and binder, which enable electron transport. The governing equation of electron transport is derived following Ohm's law:

$$0 = \nabla \cdot (\sigma_s \nabla \Phi_s). \qquad \text{(Eq. 5-27)}$$

(3) Boundary Conditions

As an example, we focus on the phenomena in the through-plane direction (x-axis), specifically from catalyst layers to gas flow channels. In reality, 3D transport occurs as a result of the GDL's random structure.

GDL–Catalyst Layer Interface

The cathode catalyst layer produces water and waste heat through the ORR, with oxygen and electrons consumed. Thus, at the interface of the GDL and catalyst layer, inward flows of oxygen and electrons, and outward flows of water and heat are promoted for the catalyst layer.

Water is also added to the cathode through the electro-osmatic drag. Part of water returns to the anode by way of back-diffusion. Given that the anode conditions are unavailable, the net water transfer coefficient a is adopted to quantify the net water exchange between the anode and cathode sides. Its value is usually positive, indicating water loss from the anode to the cathode. With this parameter, the water flux at this boundary is written as

$$\left(-D_w \frac{\partial C_w}{\partial x} - \vec{u} \cdot \vec{n} C_w\right) = (1 + 2\alpha) \frac{I}{2F}. \qquad \text{(Eq. 5-28)}$$

Oxygen is consumed by the ORR, with its flux expressed by Faraday's law:

$$\left(-D_{O_2} \frac{\partial C_{O_2}}{\partial x} - \vec{u} \cdot \vec{n} C_{O_2}\right) = -\frac{I}{4F}. \qquad \text{(Eq. 5-29)}$$

The electrolyte membranes are good insulators to prevent electrical current shorting. The electronic current flux is then equal to the ORR's consumption rate:

$$-\sigma_s \frac{\partial \Phi_s}{\partial x} = I. \qquad \text{(Eq. 5-30)}$$

Mass flows are promoted in GDLs by the reaction activities, as indicated by the variable-flow approach: three major mechanisms contribute to the mass exchange between the anode and cathodes. This mass source/sink is nonzero in the catalyst layer, expressed in terms of rate per volume:

$$S_m = M_w \nabla \cdot (D_w^{m,\text{eff}} \nabla C_w) + M_{O_2} \frac{j}{4F} - M_w \frac{j}{2F} - M_w \nabla \cdot \left(\frac{n_d}{F} \vec{i}_e\right), \qquad \text{(Eq. 5-31)}$$

where j is the transfer current density (which varies across the catalyst layer) and $\vec{i}_e$ is the protonic current flux. The above equation is integrated across the catalyst layer to obtain the surface mass flux. Again, with the net water transfer coefficient, a, the mass flux is written as follows:

$$-\rho\vec{u}\cdot\vec{n} = \left(\frac{I}{2F}M_{\mathrm{H}_2} + \alpha\frac{I}{F}M_w\right). \qquad \text{(Eq. 5-32)}$$

GDL–Gas Flow Channel Interface

One GDL surface faces the gas flow channel. In the flow channel, the stoichiometric flow ratio is around 1.5–2, leading to a velocity of ranging from ~1 m/s to 10 m/s or higher and low Reynolds laminar flows. The gradients of oxygen concentration and water content along the flow channel are small, as opposed to that in the other two dimensions. In the absence of channel flow conditions, constant values are assumed at the outer surface because of the small dimension of the simulated GDL, relative to the channel length. The boundary conditions are written as

$$\frac{\partial}{\partial n}\begin{pmatrix}\vec{u}\\P\end{pmatrix}=0 \ \text{ and } \ \begin{pmatrix}C_{\mathrm{O}_2}\\C_w\\\Phi_s\end{pmatrix}=\begin{pmatrix}C_{\mathrm{O}_2}^{\mathrm{ch}}\\C_w^{\mathrm{ch}}\\V_{\mathrm{cell}}\end{pmatrix}. \qquad \text{(Eq. 5-33)}$$

Fluid–Solid Interfaces

In the computational domain, the governing equations are applied in each grid according to the phase it represents. In void space fluid flow and species transport equations apply. At the surface of the solid matrix, no-slip and no-flux boundary conditions are imposed. For the electron transport equation in the solid matrix, no-flux condition is applied at the fluid–solid interfaces. As for the energy equation, the temperature and heat flux are set to be continuous across the fluid–solid interface.

5.4.3 PORE-LEVEL PHENOMENA

Figure 5-16 shows the pore-level fluid flows at three locations with their velocity ranging from 0.0001 to 0.001 m/s. Diameters of the carbon fibers (the gray area) are around 10 μm; therefore, the pore-level flow can easily go around the fibers. The binder, however, can be flat in shape bounded by fibers, and exhibit a much larger blocking area than a fiber, forcing the flow to go laterally. Pore-level flows are nonuniform as a result of the stochastic nature of carbon-paper microstructure. Local pore-level flow velocity can reach as high as 0.003 m/s, which is still relatively small, and the resultant species convection is insignificant, as indicated by the local Peclet number:

$$Pe = \frac{u\delta_{\mathrm{GDL}}}{D_g} < 0.1. \qquad \text{(Eq. 5-34)}$$

The pressure data can be used to evaluate the permeability K based on Darcy's law as follows:

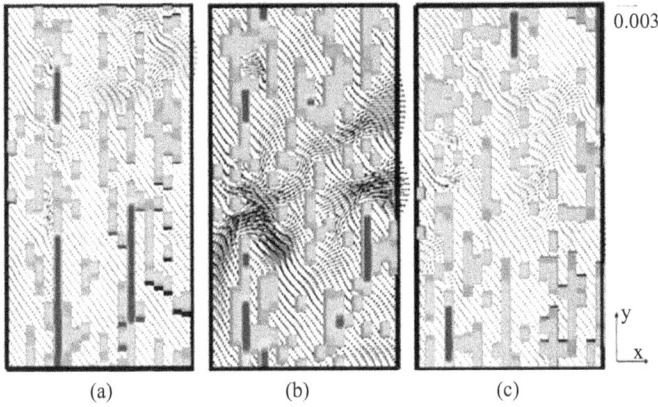

Figure 5-16. Flow fields at three GDL cross-sections at $\bar{z} =$: (a) 0.25, (b) 0.5, (c) 0.75, where $\bar{z}$ is the dimensionless distance in the z direction. The gray region denotes the solid phase with the light gray being fibers and the dark gray being binders [10].

$$\vec{u} = -\frac{K}{\mu}\nabla P, \qquad \text{(Eq. 5-35)}$$

which yields $K\sim 3.1 \times 10^{-12}$ m²; this value is only slightly lower than that given by the Carman–Kozeny model:

$$K = \frac{\varepsilon^3}{180(1-\varepsilon)^2}d^2, \qquad \text{(Eq. 5-36)}$$

where ε is the porosity and d is the mean pore diameter. One major reason for the difference in predicted K arises from the fibrous nature of carbon papers, whereas the Carman–Kozeny model was originally developed for packed beds. The binder's morphological feature, which is flat thus imposes a strong blockage over fluid flow, contributes to the smaller value of permeability by the direct simulation than the Carman–Kozeny model.

Figure 5-17 displays the oxygen contours in void space, indicating a small variation in the oxygen concentration, ~ 1 mol/m³. A low oxygen concentration appears in the lower-left corner of Fig. 5-17 (a), arising from the presence of binders as shown in Fig. 5-16 (a). In the literature, tortuosity is usually described as a property of the porous media [11], which lowers the effective mass transport in the medium:

$$D_k^{g,\text{eff}} = \frac{\varepsilon}{\tau}D_k^g, \qquad \text{(Eq. 5-37)}$$

In terms of the length of transport passage, tortuosity τ is defined as

$$\tau = \frac{L_{\text{actual}}}{\delta_{\text{GDL}}}, \qquad \text{(Eq. 5-38)}$$

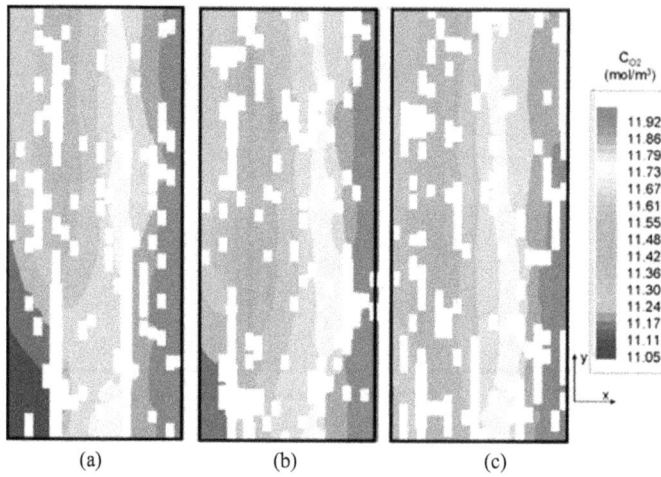

Figure 5-17. Oxygen distributions at three cross-sections at $\bar{z} =$: (a) 0.25, (b) 0.5, (c) 0.75, where $\bar{z}$ is the dimensionless distance in the z direction.

where L_{actual} is the actual length of the diffusion passage and δ_{GDL} is the GDL thickness. Figure 5-18 shows the concept of tortuosity using three examples.

Several empirical and analytical relations between the porosity and tortuosity exist because in contrast to the porosity, the tortuosity cannot be measured directly. In addition to Eq. 5-37, $D_k^{g,\text{eff}} = \dfrac{\varepsilon}{\tau^2} D_k^g$ has been proposed for inclined capillary tubes. Assuming that porosity and tortuosity are the only two factors that determine the effective diffusivity, the following general expression can be set up:

$$D_k^{g,\text{eff}} = \frac{D_k^g}{f(\varepsilon,\tau)}. \qquad \text{(Eq. 5-39)}$$

Using the direct simulation results (to obtain $D_k^{g,\text{eff}}$) and Eq. 5-37, τ is evaluated to be 1.2. Using the well-known Bruggeman equation, τ is determined to be 1.13. Figure 5-19 plots $D_k^{g,\text{eff}}/D_k^g$ as a function of ε for the two cases. In addition, the MacMullin number (N_M), defined as the ratio of resistance of porous media saturated with an electrolyte to the bulk resistance of the same electrolyte, also measures the effectiveness of species transport:

$$N_M = \frac{D_k^g}{D_k^{g,\text{eff}}} = \frac{1}{f(\varepsilon,\tau)}. \qquad \text{(Eq. 5-40)}$$

In Equation 5-37, the MacMullin number is implicitly defined as τ/ε, and its value calculated from the direct simulation is ~1.6 [10]. Table 5-3 lists the expressions of N_M for various porous media as a function of ε. In general porous media, $f(\varepsilon,\tau)$ is determined by the pore structure such as the shape and arrangement. Figure 5-19 displays the variations in the effective

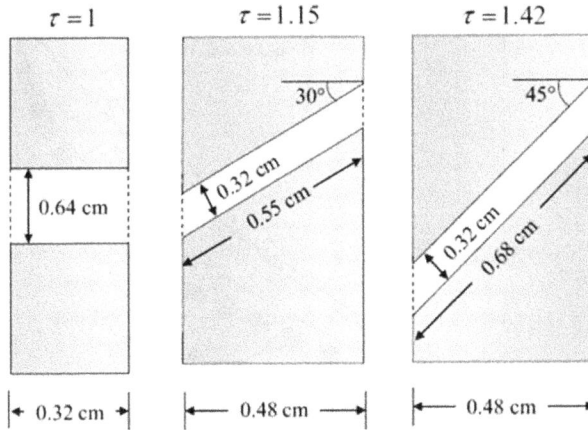

Figure 5-18. Three examples of pore structures at different tortuosities [12] (Courtesy of the Electrochemical Society).

Table 5-3. The MacMullin number (N_M) of a system consisting of a dispersed non-conducting phase in a conductive medium [12] (Courtesy of the Electrochemical Society)

Label	Geometry	Arrangement	Size	Expression
I	Spheres	Random	Uniform	$N_M = \dfrac{(5-\varepsilon)(3+\varepsilon)}{8(1+\varepsilon)\varepsilon}$
II	Spheres	Cubic lattice	Uniform	$N_M = \dfrac{(3-\varepsilon)\left[\dfrac{4}{3}+0.409(1-\varepsilon)^{7/3}\right]-1.315(1-\varepsilon)^{10/3}}{2\varepsilon\left[\dfrac{4}{3}+0.409(1-\varepsilon)^{7/3}\right]-1.315(1-\varepsilon)^{10/3}}$
III	Spheres	Random and ordered	Range	$N_M = \varepsilon^{-1.5}$
IV	Cylinders	Parallel (square array)	Uniform	$N_M = \dfrac{2-\varepsilon-0.3058(1-\varepsilon)^4-1.334(1-\varepsilon)^8}{\varepsilon-0.3058(1-\varepsilon)^4-1.334(1-\varepsilon)^8}$
V	Fibrous material (Cylinders)	Random	–	$N_M = \dfrac{0.9126}{\varepsilon(\varepsilon-0.11)^{0.785}}$

mass diffusivity with respect to the porosity for the tetragonal and hexagonal network models. The result of Neale and Nader [13] are also plotted for comparison. It is seen that the random network exhibits a lower effective mass diffusivity for a given porosity, that is, randomness results in a larger resistance or more tortuous mass transport paths.

Figure 5-20 shows the possible path of the species transport under different orientations in the geometries. Parallel fibers, as in carbon cloth, and random arrangement of spheres with some separation show that species transport can follow a similar curved zig-zag or wave path. This is also true for ordered arrangement of spheres, where the zig-zag path is more symmetric. Consequently, carbon cloth tends to follow the Bruggeman expression. However, fibers with

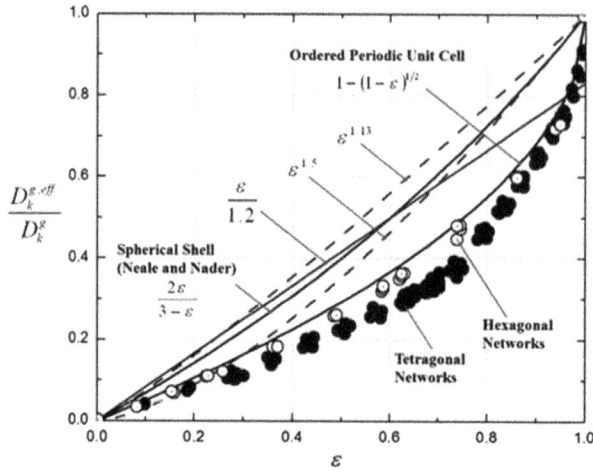

Figure 5-19. The effective diffusivity as a function of porosity for the tetragonal and hexagonal network models (some data are obtained from [4]).

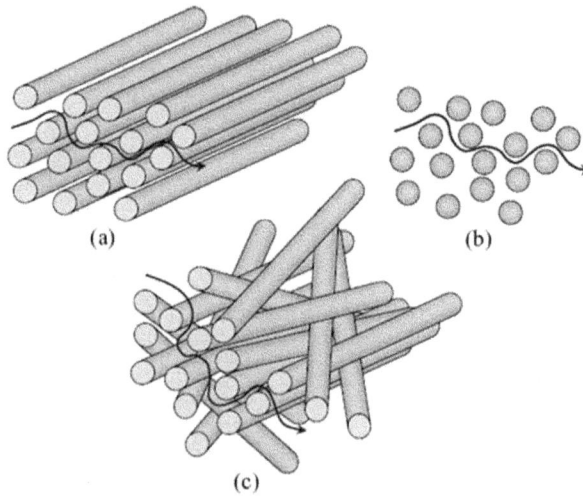

Figure 5-20. Path of the species diffusion for (a) woven carbon-cloth GDL having fibers mostly in order, (b) packed or suspended spheres, and (c) carbon-paper GDL having randomly oriented fibers [12] (Courtesy of the Electrochemical Society).

random orientation, as in carbon-paper GDL, can block immediate paths and create more erratic patterns, thus extending the path length for species transport in void space. Consequently, the tortuosity becomes larger for the carbon-paper GDL. Table 5-4 compares the physical properties between carbon paper and cloth used in PEM fuel cells.

Following the tortuosity defined in the pore network, a similar parameter can be introduced to the solid matrix. Figure 5-21 displays the electric potential distributions in the solid matrix and shows a small potential drop. Though carbon fibers are a good conductor in general, the highly tortuous nature of the solid matrix yields an enlarged apparent ohmic resistance. The

Table 5-4. Physical properties of carbon paper and cloth [12] (Courtesy of the Electrochemical Society)

Label	Description	Thickness (cm)	Samples stacked	ε	Measured N_M	$\tau = \varepsilon N_M$	$\tau = \sqrt{\varepsilon N_M}$
Carbon-paper GDL							
A	Mitsubishi Rayon Pyrofil MFG-070	1.78×10^{-2}	7	0.66	4.69 ± 0.24	3.10	1.76
B	Ballard AvCarb P50	1.78×10^{-2}	4	0.70	3.49 ± 0.15	2.44	1.56
C	Toray TGP-H-120	3.05×10^{-2}	4	0.78	3.65 ± 0.19	2.85	1.69
D	SGL Carbon Sigracet 10BB w/ MPL	4.06×10^{-2}	3	0.84	2.08 ± 0.25	1.75	1.32
Carbon-cloth GDL							
E	Showa Denko K. K. SCT-NF2-1	2.54×10^{-2}	5	0.50	2.76 ± 0.06	1.38	1.17
F	W. L. Gore CARBEL CL w/MPL	4.06×10^{-2}	3	0.74	1.64 ± 0.14	1.21	1.10
G	E-TEK ELAT LT-1400 W w/MPL	4.06×10^{-2}	3	0.74	1.64 ± 0.23	1.21	1.10

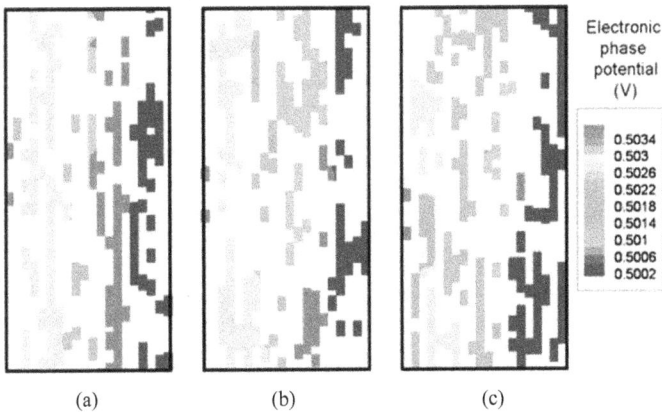

Figure 5-21. Electronic potential distributions at $\bar{z} =$: (a) 0.25, (b) 0.5, (c) 0.75, where $\bar{z}$ is the dimensionless distance in the z direction.

high tortuosity arises from the fact that fibers are arranged laterally, and therefore the through-plane conduction occurs at the contacting points. Following that in the pore network, one can define $\sigma_s^{\text{eff}} = \dfrac{(1-\varepsilon)}{\tau_s}\sigma_s$. The solid matrix's tortuosity τ_s is evaluated to be 13.8 from the simulation result. Readers who are interested in detailed discussion on carbon paper fabrication are referred to consult with Ref. [14].

5.5 TRANSIENT PHENOMENA

5.5.1 TRANSIENT TERMS AND TIME CONSTANTS

In transient operation, species or energy can be stored in or released from bulk materials, which is accounted for by the transient term in the conservation equation. Polymer electrolyte membrane (e.g., Nafion) exhibits a large water holding capacity. Catalyst layers contain several phases including ionomer and gas. Thus, the transient term of the water transport equation is derived to account for their holding capacity of water [15]:

$$\varepsilon_g \frac{\partial C_w}{\partial t} + \varepsilon_m \frac{\partial C_w^m}{\partial t} = \nabla \cdot \left(D_w^{g\,\text{eff}} \nabla C_w \right) + \nabla \cdot \left(D_w^{m\,\text{eff}} \nabla C_w^m \right) - \frac{1}{F} \nabla \cdot \left(n_d \vec{i}_e \right). \quad \text{(Eq. 5-41)}$$

The two terms on the left-hand side represent the rates of water storage in the gas and membrane phases, respectively. Assuming that the waters in the gas and membrane phases are in equilibrium, the above equation is rearranged as

$$\varepsilon^{\text{eff}} \frac{\partial C_w}{\partial t} = \nabla \cdot \left(D_w^{\,\text{eff}} \nabla C_w \right) - \frac{1}{F} \nabla \cdot \left(n_d \vec{i}_e \right), \quad \text{(Eq. 5-42)}$$

where the effective factor ε^{eff} is defined as

$$\varepsilon^{\text{eff}} = \varepsilon_g + \varepsilon_m \frac{dC_w^m}{dC_w} = \varepsilon_g + \varepsilon_m \frac{1}{EW} \frac{RT}{P_{w,\text{sat}}} (\rho + \lambda \frac{d\rho}{d\lambda}) \frac{d\lambda}{da}. \quad \text{(Eq. 5-43)}$$

For Nafion® 11-series membranes having EW (equivalent weight) of 1.1 kg/mol, and membrane density ρ of 1980 kg/m^3 at 80°C, the effective factor ε^{eff} is on the order of 100–1,000 for the water activity less than one or a < 1.

Another transient phenomenon is charging or discharging of the electrochemical double-layers. The double layer, which is formed within a thin layer (with a thickness on the order of nanometer) adjacent to the reaction interface, acts as a capacitor during transient. More specifically, the double layer refers to two parallel layers of charges surrounding the object. It is a structure that develops on the surface of a solid electrode, gas bubble, or even liquid droplet, when being placed in an electrolyte. Because of its importance in interfacial processes, several models were proposed to understand the double-layer dynamics; see Figure 5-22. A detailed discussion on the double-layer model is presented in Appendix IV.B. Similar to porous battery electrodes, the double layer in the catalyst layer can be regarded as being in parallel to a charge transfer reaction resistor. The importance of the double layer in transient can be evaluated by its time constant:

$$\tau_{\text{dl}} = \delta_{\text{CL}}^2 aC \left(\frac{1}{\kappa} + \frac{1}{\sigma} \right). \quad \text{(Eq. 5-44)}$$

In common operation,
    capacity C is around 20 μF/cm^2,
    specific area a is about 10^3/cm,
    catalyst layer thickness δ_{CL} is 10^{-3} cm,

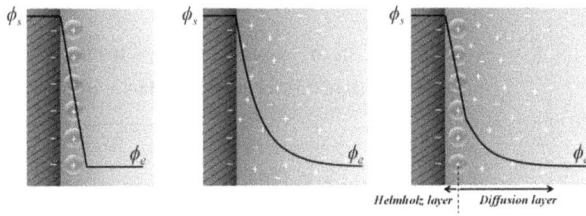

Figure 5-22. Models of the double layer: Helmholtz model (left); Gouy–Chapman model (middle); Stern's model that combines Helmholtz and Gouy–Chapman models (right).

κ is around 0.1 S/cm,

and σ is about 50 S/cm,

which lead to a time constant about 0.2 μs, sufficiently short to be safely ignored for automotive PEM fuel cells.

In comparison, the time constant for species transport (e.g., diffusion) through a GDL can be estimated by

$$\tau_k = \frac{\delta_{GDL}^{\,2}}{D_k^{g,\text{eff}}}. \tag{Eq. 5-45}$$

For PEM fuel cells, D_g^{eff} is around $10^{-5}\,\text{m}^2/\text{s}$ and δ_{GDL} of 0.3 mm, yielding τ_k on the order of 0.01 s.

5.5.2 *TRANSIENT UNDERGOING CONSTANT VOLTAGE OR STEP CHANGE IN VOLTAGE*

Transient operation under constant or varying voltages is frequently encountered in PEM fuel cell testing and diagnostics. Figure 5-23 shows the dynamic response of current density to step change in cathode inlet humidification under given voltages. The step change was set at $t = 0$, starting from a steady state at RHa/c =100%/0% (that is, a fully humidified anode and a completely dry cathode). It is seen that it takes approximately 20 s for the fuel cell to reach another steady state, in accordance with the time constant of membrane water uptake (see Chapter 4). The transition duration at higher voltage is slightly longer because lower current in this condition results in less water production, thus taking longer to hydrate the membrane. Despite that humidified air stream contains less oxygen, the current density continues to increase till another steady state, demonstrating that the operation falls in the regime dominated by the ohmic resistance.

The water profiles shown in Figure 5-24 are of interest to note. At $t = 0$, the gas in the cathode channel is relatively dry and the water generated in the cathode catalyst layer is transported to the gas flow channel; thus, a water concentration gradient develops, directing toward the flow channel. Once the inlet instantly switches to the fully humidified operation, the high humidification front propagates into the middle of the gas channel, reversing the gradient.

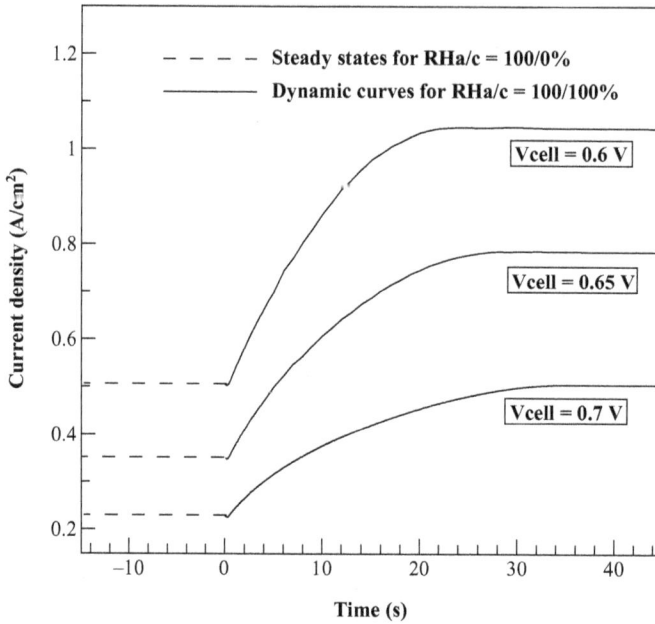

Figure 5-23. Dynamic responses of current density to the step change of the cathode inlet humidification from RH = 0–100%.

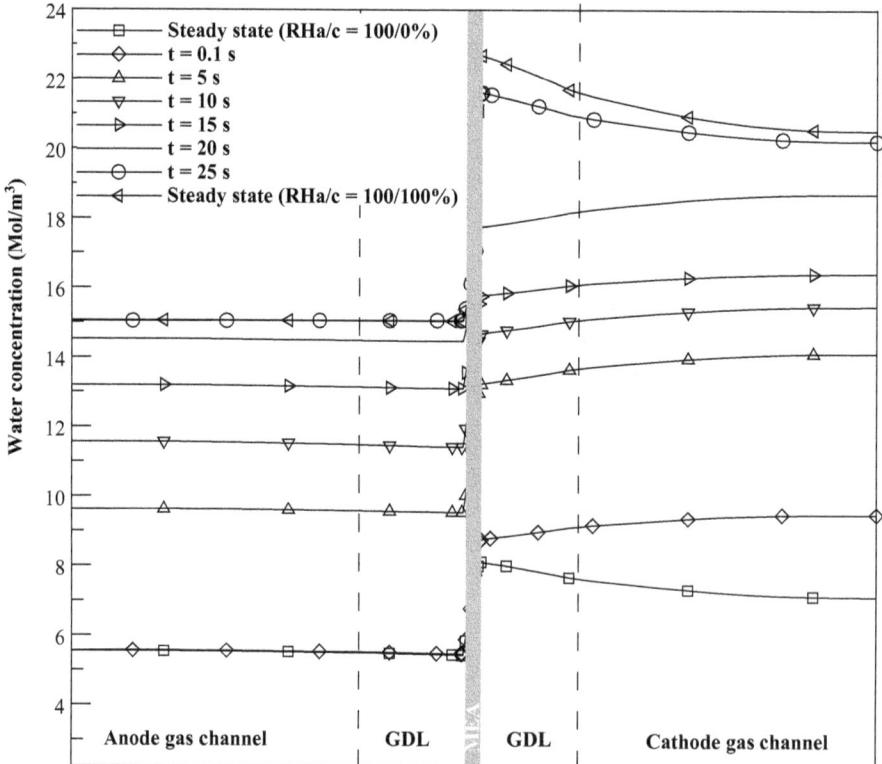

Figure 5-24. Water profiles after the step change of the cathode inlet humidification from RH = 0–100% at 0.65 V.

This means that the membrane takes up water not only from the ORR production but also from the humidified gas stream in the channel. The consequent increase in the membrane water content then gives rise to a higher current density under the constant voltage and hence more water production. The membrane hydration increases with time and eventually reaches another steady state, recovering the trend of the water concentration gradient in the GDL, that is, the initial shape. Figure 5-24 shows that the water concentration profile on the anode side is altered by the change in the cathode humidity condition. This is illustrative of the important role of water back diffusion through the membrane, determined by the cathode and anode water conditions.

The increase of the anode water concentration with time can be readily explained by Figure 5-25, which shows the evolution of water profiles in the membrane. The water content reaches steady state first upstream. After the membrane in the inlet area reaches full humidification within about 10 s, it takes additional 15 s for the downstream membrane to be fully hydrated.

Figure 5-26 displays the water and oxygen concentration profiles, showing that within the initial 5 s the water concentration is substantially reduced on the cathode side, while there is little change on the anode. The cathode gas, however, remains nearly fully humidified (i.e., $C_{sat}(80°C) \approx 15.9$ mol/m^3) during this period. The oxygen concentration increases significantly in the first 0.1 s due to the injection of dry, undiluted gas into the cathode.

Figure 5-27 presents the dynamic response of average current density to a step change in cell voltage and compares a fully humidified case (i.e., RHa/c = 100%/100%) with a dry cathode case (i.e., RHa/c=100%/0%). Times for the onset of step changes are chosen arbitrarily so as to set apart the two curves. The most part of undershoot occurs in a fraction of a second, consistent with the time constant of species diffusion in GDLs. In contrast, under the dry condition the current density

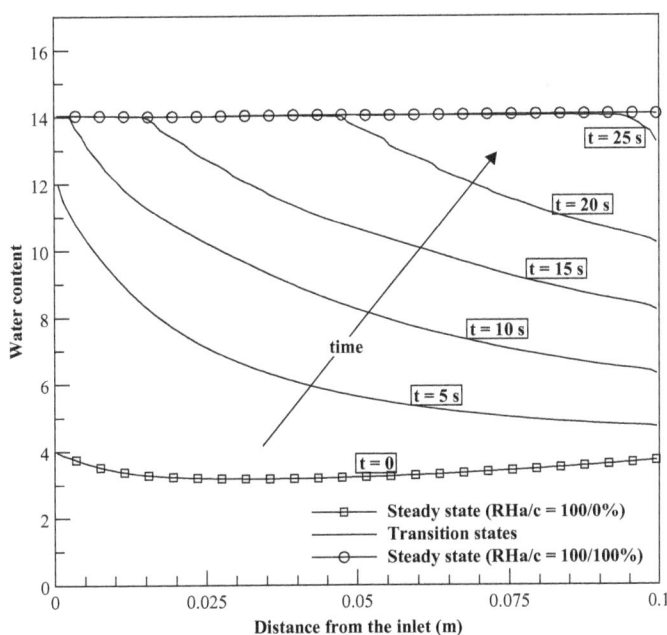

Figure 5-25. Membrane water profiles, when the cathode inlet humidification changes from RH = 0–100% under 0.65 V.

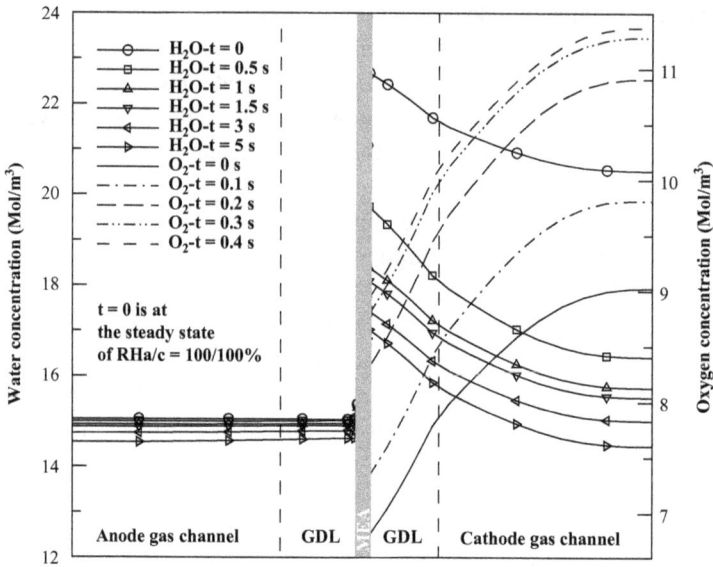

Figure 5-26. Water and oxygen concentration profiles, when the cathode inlet humidification changes from RH = 100–0% under 0.65 V.

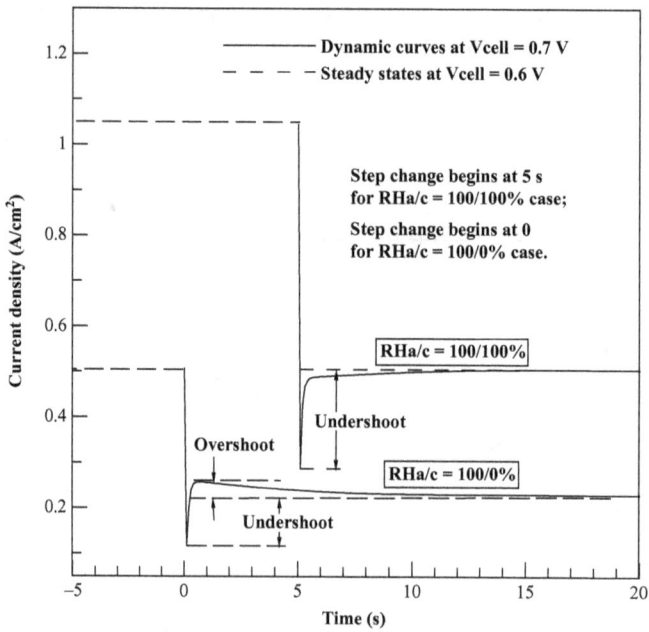

Figure 5-27. Dynamic response of average current density to the step change in voltage from 0.6 to 0.7 V [15].

undergoes undershoot followed by overshoot. The overshoot can be explained by the fact that the response of the membrane hydration lags that of oxygen transport. Also, the transient takes about 10 s, indicative of the dominance of membrane uptake in fuel cell dynamics.

5.5.3 TRANSIENT UNDERGOING CONSTANT CURRENT OR STEP CHANGE IN CURRENT

Control of current density is frequently encountered in testing and diagnostics. It is practically important because load adjustment is generally realized through tuning current density.

Figure 5-28 shows the dynamic responses of cell voltage to various step changes in current density. The voltage responses are nearly instantaneous for the fully humidified case, whereas it takes seconds for the low-humidity fuel cell to reach another steady state. The dynamic behaviors of low-humidity cases exhibit voltage undershoot. The degree of undershoot increases with the magnitude of the imposed current change. When the current changes from 0.1 to 1.0 A/cm^2, the cell voltage drops to almost zero before recovering. Upon a larger increase in current loading (i.e., >1.0 A/cm^2), the fuel cell will likely reverse its voltage during transient.

To further explore the cause for the observed voltage undershoot, Figure 5-29 presents the water profiles in the MEA during a transient. Dashed lines denote the initial water content profile immediately before the step change, and the solid line at $t = 6$ s marks the

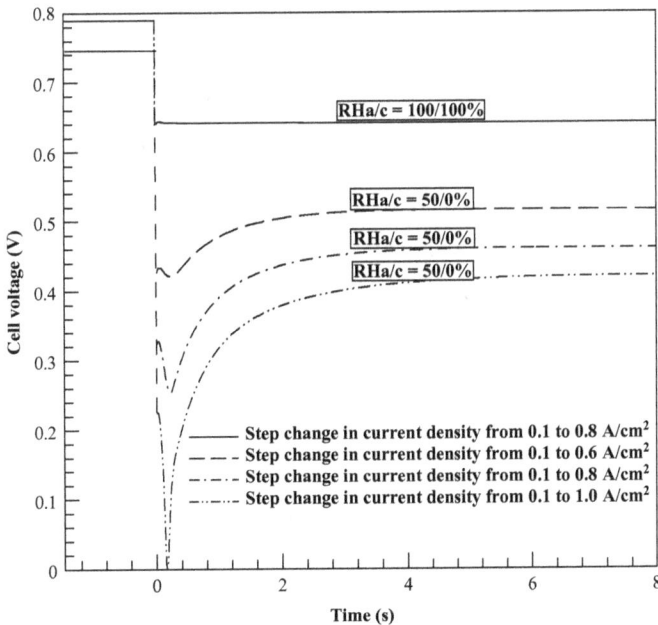

Figure 5-28. Dynamic responses of cell voltage to various step changes in current density for a PEM fuel cell using Gore® 18 membrane [16].

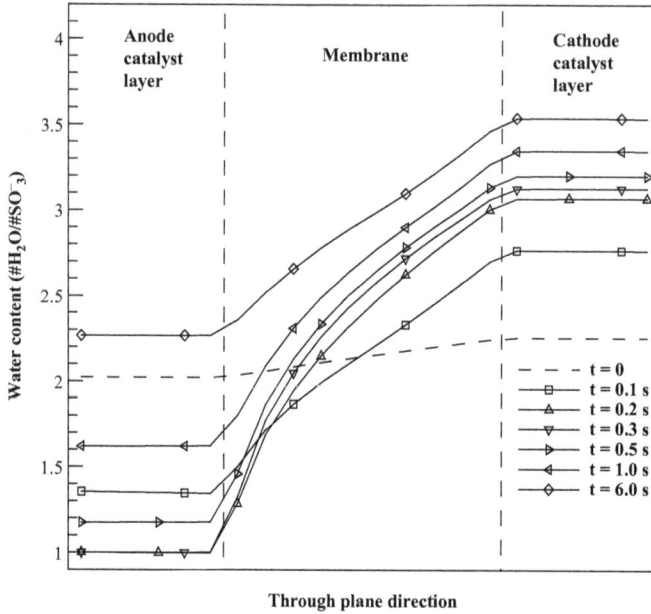

Figure 5-29. Dynamic profiles of the water content in the MEA after a step change in current density from 0.1–0.8 A/cm^2 under dry operation (RHa/c=50%/0%) for a PEM fuel cell using a Gore® 18 membrane.

final profile of the new steady state. Right after the step change, the predicted water content decreases in the anode, whereas it increases in the cathode. This can be explained by the water electro-osmotic drag, which increases almost proportionally to the current density. The anode hydration level reaches its lowest value around 0.3 s after the step change, roughly the same time as the cell voltage reaching the minimum as shown in Figure 5-28. Over the period of that time instant, the anode begins to acquire water through back-diffusion as the water concentration difference between the anode and cathode is further enlarged. It takes about 6 s for the water profile to reach the steady state, which is of the

same order as the time constant of membrane hydration, that is, $\tau_{m,H} = \dfrac{\dfrac{\rho_m \delta_m \Delta \lambda}{EW}}{\dfrac{I}{2F}} \sim 7$ s. In addition,

the time constant of water diffusion across the membrane, that is, $\tau_{m,D} = \dfrac{\delta_m^{\,2}}{D_w^{m,\text{eff}}} \sim 0.2$ s at $\lambda = 3$,

coincides with the time when the cell voltage reaches the minimum, indicating that back-diffusion begins to re-hydrate the anode.

Figure 5-30 displays variations of the water molar concentration profile in the anode channel. Once again, the dashed line represents the initial state, and the solid line at $t = 6$ s denotes the new steady state. It is interesting to note that water concentration in the anode channel is sensitive to the step change in current density. Furthermore, there is severe dryout of the anode downstream during transient. This phenomenon can be explained by the fact that most of the water carried

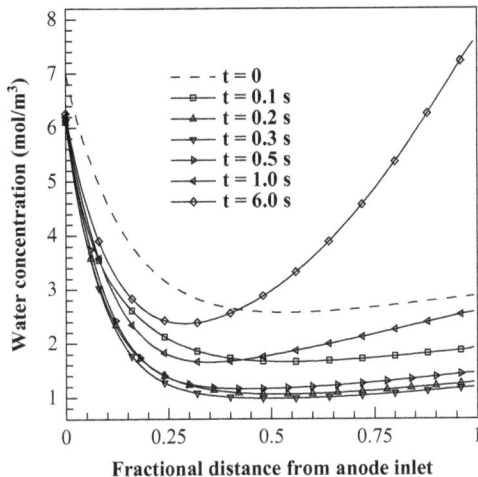

Figure 5-30. Water concentration profiles in the anode gas channel after the step change from 0.1 to 0.8 A/cm^2 under dry operation (RHa/c = 50%/0%).

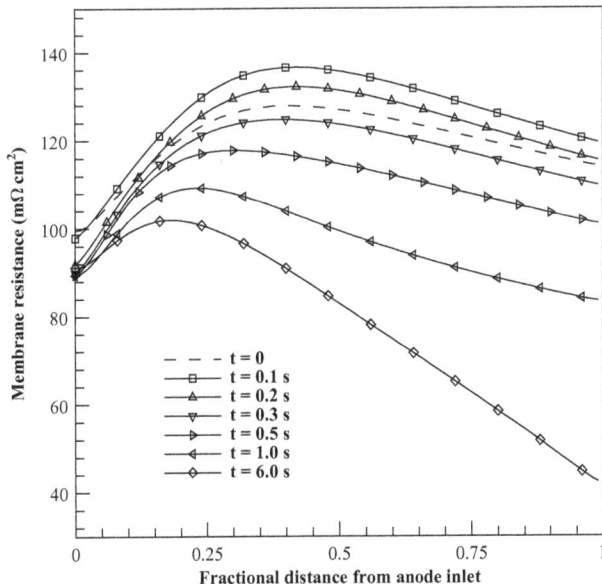

Figure 5-31. Membrane resistance profiles after the step change from 0.1 to 0.8 A/cm^2 under dry operation (RHa/c = 50%/0%).

in partially humidified anode inlet gas is transferred to the cathode upstream due to the fast gas transport across the anode GDL, which occurs in a time constant around 0.01–0.1 s, leaving the downstream region temporarily dry. Once water back-diffusion takes effect, the anode water concentration downstream recovers.

As a result of water redistribution during the transient, the membrane resistance experiences overshoot, as shown in Figure 5-31. As indicated by the dashed line, the initial membrane resistance increases first along the flow, due to the anode dehydration caused by the electro-osmotic drag, and then decreases due to water production. The same spatial trend remains at the final steady state of a higher current density, except that the anode dehydration region is shortened due to the higher water production rate (see the solid line that denotes the membrane resistance at $t = 6$ s). In the transition, the membrane resistance initially increases overall before drop-down, indicating the significant role played by the anode-side membrane dehydration. A considerable decrease in the membrane resistance occurs downstream as a result of higher water production at 0.8 A/cm^2, in accordance with what is shown in Figure 5-30.

5.6   WATER MANAGEMENT BETWEEN A PEM FUEL CELL AND FUEL PROCESSOR

Water product by a PEM fuel cell can be utilized in its fuel process when using hydrocarbon-based fuels. This section is based on Ref. [17] and provides a brief introduction to water management between PEM fuel cells and their fuel processor. Figure 5-32 shows a simplified schematic of a generic fuel cell-fuel processor system. Streams flowing into the system boundary include a generic carbonaceous fuel $(C_nH_mO_p)$ feed to the fuel processor (FP), and air feeds to the (FP) and the fuel cell cathode.

Through the fuel reforming reaction, the carbonaceous fuel, air, and water are converted into a mixture of hydrogen, carbon dioxide, and nitrogen. A generic reaction can be stated as follows:

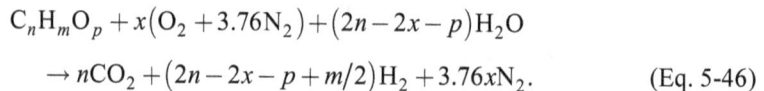

$$C_nH_mO_p + x(O_2 + 3.76N_2) + (2n - 2x - p)H_2O$$
$$\rightarrow nCO_2 + (2n - 2x - p + m/2)H_2 + 3.76xN_2. \qquad \text{(Eq. 5-46)}$$

In typical steam-reforming fuel processors, the reformate and combustion streams would be kept separate. The combustion products (generating the heat for the steam reforming reaction) would then be fed either into the burner or into the radiator/condenser, which can be

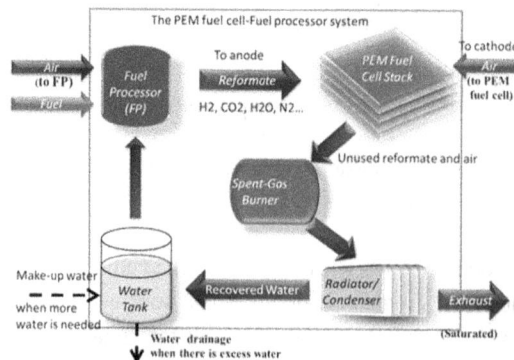

Figure 5-32. Schematic of a fuel cell-fuel processor system with the various streams entering and leaving the system (redraw from [17] with modification).

located after the fuel cell stack to utilize the extra hydrogen gas and oxygen. The heat generated at the burner is reused to preheat the incoming feeds in the fuel processor.

If the total amount of water recovered at the condenser and at other locations is greater than that needed for fuel processing and humidification, the system is a net water producer, and a water tank is added to drain the excess water; see the dotted stream leaving the system at the lower left of Figure 5-32. Alternatively, the excess water can be discharged in the form of vapor by increasing the temperature of the exhaust gas out of the condenser. Conversely, if the amount of water recovered is less than that used in the system, the system is a net water consumer, and make-up water must be added for sustainable operation. In some applications, the option of providing make-up water, however, is undesirable or unacceptable.

5.6.1   WATER BALANCE MODEL

The fuel processor converts a carbonaceous fuel, $C_n H_m O_p$, to hydrogen and carbon dioxide by an overall reforming process represented by Eq. 5-46. The PEM fuel cell operates with no prior humidification of the air fed to the cathode. Thus, the overall system takes in fuel, air, and possibly water, producing electric power and an exhaust gas containing carbon dioxide, water vapor, and nitrogen.

The amount of water needed to convert the input fuel to hydrogen is determined by the fuel composition (i.e., n and p) and the oxygen-to-fuel molar ratio x ($x = O_2: C_n C_H O_p$); see Eq. 5-46. Thus, $2n-2x-p$ is the minimum water-to-fuel molar ratio ($H_2O: C_n C_H O_p$) for the fuel conversion. In terms of the steam-to-carbon ratio Ψ, the minimum ratio Ψ_{min} is expressed by $(2n-2x-p)/n$. In practice, excess of water is usually used (i.e., $\Psi > \Psi_{min}$) to increase the fuel utilization. For the base case analysis discussed below, Ψ of 1.5 was used. For methane fuel, this corresponds to 44% more water than the minimum required. In the case of excess of water, that is, $\Psi > (2n - 2x - p)/n$, the overall fuel processing reaction becomes

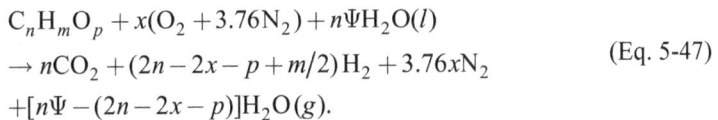

$$C_n H_m O_p + x(O_2 + 3.76 N_2) + n\Psi H_2 O(l)$$
$$\rightarrow nCO_2 + (2n - 2x - p + m/2) H_2 + 3.76x N_2 \qquad \text{(Eq. 5-47)}$$
$$+ [n\Psi - (2n - 2x - p)] H_2 O(g).$$

The oxygen-to-fuel ratio x is calculated such that the reaction is thermo-neutral, that is, the heat of reaction ($\Delta H_{r,298}$) is zero. Therefore, without any heat loss, there is no change in temperature between the incoming reactants and the outgoing products of the fuel processor. The reformate gas from the processor enters the anode channel of the fuel cell stack, and air is fed to the cathode side. The hydrogen and oxygen react electrochemically to produce electricity, heat, and water. Extra hydrogen is burned in the spent-gas burner. By cooling, some water vapor in the burner product gas can be recovered and recycled back to the water tank.

Table 5-5 shows the parameters and a summary of results for a base case using methane. The left side of the table lists the input parameters and assumptions, while the right side presents the results. Additional computational details and results are given in Table 5-6. The methane feed rate is 1 gmol/min. The system is assumed to operate at a pressure of 1 atm absolute and a steam-to-carbon molar ratio of 1.5. For Ψ of 1.5, thermo-neutrality is achieved at the oxygen-to-fuel ratio x of 0.478, corresponding to an air feed rate of 2.27 gmol/min. The product gas from the fuel processor, see Eq. 5-47, contains 3.05 gmol/min

Table 5-5. Water balance in a fuel cell-fuel processor system operating on methane [17] (Courtesy of Elsevier)

Fuel: methane (CH_4); basis: 1 gmol/min of fuel			
Input/assumptions		Calculated parameters	
Molecular weight	16	O_2/fuel molar ratio into FP, x	0.478
$\Delta H_{f,298}$ (kcal/gmol)	−17.9	Air feed into FP (gmaol/min)	2.270
$\Delta H_{c,298}$ (kcal/gmol)	192	O_{air}/C ratio	0.956
System pressure (atm)	1.0	Water feed into FP (gmol/min)	1.500
H_2O/C ratio into FP, Ψ	1.5	Idealized FP products	
Heat of reaction, $\Delta H_{r,298}$ (kcal/gmol)	0	H_2 (gmol/min)	3.05
Fuel (H_2) utilization (%)	80	H_2 conc. In reformate (%-dry)	48.4
O_2 utilization (%)	40	LHV of H_2 (kW(t))	12.3
Ambient temperature, T_{air}	35°C, 95°F	Fuel processor efficiency (%)	91.8
Condenser approach temperature, $T_{approach}$	11°C, 52°F	H_2O in reformate (gmol/min)	0.46
		H_2O conc. In reformate (%-wet)	7.2
		Air stoichiometry in fuel cell	2.50
		Air into cathode (gmol/min)	14.5
		Burner product (gmol/min)	19.3
		H_2O in burner product (gmol/min)	3.5
		H_2O conc. In burner product (%-wet)	18.2
		Exhaust gas temperature, $T_{exhaust} = T_{air} + T_{approach}$	46°C, 115°F
		Saturated moisture content in exhaust gas (%-wet)	10.0
		Recoverable water (gmol/min)	1.75
		Net water produced (gmol/min)	+0.25
		Net water produced (ml/min)	+4.6

of hydrogen (48.4% on a dry basis), with a lower heating value (LHV) of 12.3 kW. Thus, the fuel processor efficiency, defined as the LHV of hydrogen produced relative to that of the fuel, is 91.8%. The reformate gas contains 0.46 gmol/min of water vapor or 7.2% of the reformate stream. For the PEM fuel cell operation at 60°C at which the water vapor pressure is 0.193 atm, no water is condensed out of the reformate when cooled to 60°C.

Assuming fuel and oxidant utilizations of 80% and 40%, respectively, the rate of air fed to the fuel cell cathode is 14.5 gmol/min. After the fuel cell and the spent-gas burner, all of the hydrogen, 3.05 gmol/min, is converted to water. Thus, the total water content in the burner product gas is 3.5 gmol/min, which corresponds to a water concentration of 18.2% in the burner product.

Under an ambient temperature of 35°C (95°F) and a condenser at 11°C (52°F), the exhaust gas can leave the system at 46°C (115°F). The saturation water vapor concentration at this temperature

Table 5-6. Computational details and results for methane [17] (Courtesy of Elsevier)

$C_nH_mO_p$	n	M	p	
	1	4	0	Methane
	Mol. wt.	$-\Delta hf$	Δhc	ΔHc
	(g/mol): 16	(cal/gmol): 17,889	(cal/gmol): 191,758	(Btu/lb): 21,573
Fuel properties				
Fuel feed	1 gmol			
Oxygen	0.478 gmol			
Nitrogen	1.797 gmol			
Required S/C	1.045			
Operating S/C	1.500			
Excess water (%)	43.6			
Water feed	1.500 gmol			
Ideal products				
H_2	3.045 gmol		48.35%-wet	12.3 kW(t)
CO_2	1.000 gmol		15.88%-wet	
N_2	1.797 gmol		28.54%-wet	
H_2O (g)	0.455 gmol		7.23%-wet	
Total	6.297 gmol		100%-wet	
Heat of reaction	0.0 cal		−ve: exo; +ve: endo	
Reactor T estimate for Cp calculations	500°C			
Efficiency (%)	91.8			
Anode gas temperature	60°C		139.4°F	
Saturated vapor pressure	2.833 psia		0.193 atm	19.27%-wet
Fuel utilization stack	80%			
H_2 reacting in stack	2.44 gmol			
Air stoich in stack	2.5			
Oxygen into cathode	3.045 gmol			
Nitrogen into cathode	11.448 gmol			

(Continued)

Table 5-6. Computational details and results for methane [17] (Courtesy of Elsevier) (*Continued*)

$C_nH_mO_p$	n	M	p	
Combined product after burner				
CO_2	1.000 gmol		5.19%-wet	
N_2	13.245 gmol		68.74%-wet	
O_2	1.522 gmol		7.90%-wet	
H_2O	3.500 gmol		18.17%-wet	
Total	19.267 gmol		100.00%-wet	
System pressure	1.0 atm			
Ambient temperature	35.3°C		95°F	
Radiator/condenser approach T	11.1°C		20°F	
Exhaust gas temperature	46.4°C		115°F	
Saturated pressure	1.470 psia		0.100 atm	10.00%-wet
Water recovered at radiator	1.748 gmol			
Radiator exhaust gas				
CO_2	1.000 gmol		5.71%-wet	
N_2	13.245 gmol		75.60%-wet	
O_2	1.522 gmol		8.69%-wet	
H_2O	1.752 gmol		10.00%-wet	
Total	17.519 gmol		100.00%-wet	
Excess water in FP =	0.25 gmol		4.5 ml/(gmol of fuel);	
water recovered – water used			23.4 ml/(Mcal)	

and 1 atm is 10%. Therefore, cooling the burner exhaust to 46°C (115°F) would condense out 1.75 gmol/min of water. Comparing with the amount of water fed to the fuel processor, it is seen that the system recovers 0.25 gmol/min of water more than it consumes. Thus, this system operates as a net water producer.

5.6.2 EFFECT OF THE STEAM-TO-CARBON RATIO

Hydrocarbons that are heavier than methane have a tendency to form coke during processing, which is a highly undesirable byproduct because it can foul and shut down the reactors. To suppress coke formation, relatively high steam-to-carbon ratios are usually used. Table 5-7 shows the results of varying the steam-to-carbon ratio, while other assumptions and input parameters are the same as in Table 5-5.

Table 5-7 shows the results of changing Ψ by 20% from the base case value of 1.5 (i.e., to 1.2 and to 1.8, respectively); the last column shows a much higher Ψ of 3.0, as is done in some fuel processors. To maintain thermo-neutrality, the oxygen-to-fuel ratio, x, needs to be adjusted, as shown in Table 5-7. The amount of oxygen fed to the fuel processor increases in order to generate heat to vaporize the increasing amount of water in the feed. This successively increasing x yields lower hydrogen concentration and higher water content in the reformate, as shown in Table 5-7. The lower hydrogen yields reduced heating values of the reformate gas and fuel processor efficiencies.

Figure 5-33 shows the net water produced and the total system efficiency as functions of percent change in the steam-to-carbon ratio. The efficiency changes only 9% upon a 100%

Table 5-7. Varying the steam-to-carbon ratio Ψ [17] (Courtesy of Elsevier)

Basis: 1 gmol/min of methane		Base case		
H_2O/C ratio into FP, Ψ	1.2 (−20%)	1.5	1.8 (+20%)	3.0 (+100%)
Water feed into FP (gmol/min)	1.2	1.5	1.8	3.0
O_2/fuel molar ratio into FP, x	0.450	0.478	0.523	0.614
Air feed into FP (gmol/min)	2.14	2.27	2.40	2.92
Idealized FP products				
H_2 (gmol/min)	3.10	3.05	2.99	2.77
H_2 conc. In reformate (%-dry)	52.6	48.4	44.6	33.4
LHV of H_2 (kW(t))	12.5	12.3	12.1	11.2
Fuel processor efficiency (%)	93.4	91.8	90.1	83.5
H_2O in reformate (gmol/min)	0.10	0.46	0.81	2.23
H_2O conc. In reformate (%-wet)	1.71	7.2	12.1	26.8
Air into cathode (gmol/min)	14.8	14.5	14.2	13.2
H_2O in burner product (gmol/min)	3.2	3.5	3.8	5.0
H_2O conc. In burner product (%-wet)	16.8	18.2	19.6	24.9
Exhaust gas temperature, $T_{exhaust} = T_{air} + T_{approach}$	46°C, 115°F	46°C, 115°F	46°C, 115°F	46°C, 115°F
Saturated moisture content in exhaust gas (%-wet)	10.0	10.0	10.0	10.0
Recoverable water (gmol/min)	1.43	1.75	2.06	3.32
Net water produced (gmol/min)	+0.23	+0.25	+0.26	+0.32
Net water produced (ml/min)	+4.2	+4.6	+4.8	+5.8

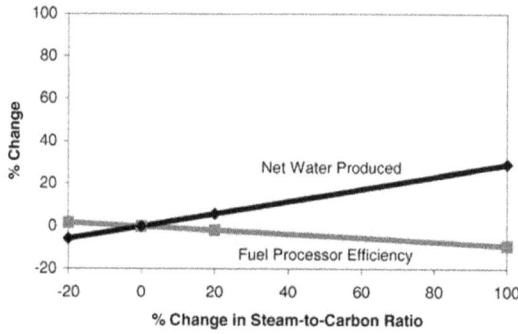

Figure 5-33. Effects of the steam-to-carbon ratio on the net water production and fuel processor efficiency [17] (Courtesy of Elsevier).

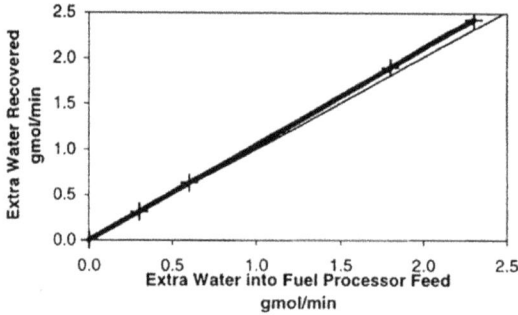

Figure 5-34. Water recovery as a function of the extra water added at the fuel processor. "Extra" refers to the additional water with respect to the base case [17] (Courtesy of Elsevier).

change in Ψ. The responses are in the opposite directions, indicating that net water production is inversely correlated with efficiency. Under the constant exhaust gas temperature at 46°C, the water vapor concentration remains unchanged at 10%. Thus, increasing Ψ improves the amount of water that can be recovered at the condenser. Figure 5-34 shows that the recovered water is greater than the increase of water fed to the fuel processor, and the difference (i.e., surplus) increases with Ψ. The difference between the extra water recovered and the diagonal line can be traced to additional water that would be produced from the extra fuel consumed to compensate for efficiency losses.

5.7 CHAPTER SUMMARY

This chapter discusses the transport phenomena in PEM fuel cell in the absence of liquid water or ice, more specifically, the transport of the gaseous reactants and water vapor. The key points in this chapter are summarized as follows:

- Diffusion and convection are the two major mechanisms for water vapor and gaseous reactant species transport in PEM fuel cells. Convection dominates in gas flow channels (GDCs), whereas diffusion usually dominates in gas diffusion layers (GDLs) and catalyst layers (CLs); both molecular and Knudsen diffusion are important because of the small pore dimension.

- Two approaches, the constant-flow and variable-flow formulations, are presented and discussed for flow modeling. The constant-flow formulation (which takes the gas-mixture density to be constant and neglects the mass exchange between the anode and cathode) leads to simplified governing equations/numerical treatment and thus reduced computational cost; the variable-flow formulation (which describes the real flow conditions without simplifications) better captures the flowfield and yields more accurate predictions of the species concentrations but with much higher computational cost, and for fuel cell performance prediction, the two approaches yield similar results under common operating conditions.

- A pore-level study, along with material stochastic modeling, is presented to shed light on the fluid flow and transport phenomena and evaluate macroscopic material properties. Pore-level simulation results show that the microflow within the pore network of a GDL is random in nature due to the complex structure of the solid matrix and confirm that convection is weak as compared with diffusion; macroscopic properties such as permeability and tortuosity are evaluated from the pore-level study, and various correlations of the effective diffusivity are also discussed.

- Transient analysis of several important processes in PEM fuel cell is performed. The time constants for membrane hydration, oxygen transport through GDL, and electrochemical double-layer charging/discharging are, respectively, around 10–100 s, 0.01–0.1 s, and less than 1 ms; in the transient of step change in the cell voltage under dry operation, current density experiences a undershoot, followed by an overshoot, and in the transient of step change in the current density under dry operation, cell voltage may undergo a deep undershoot, bringing the voltage down to 0 V.

- Water balance between a fuel processor and PEM fuel cell is analyzed. Product water by PEM fuel cells can be recovered for the fuel processing of carbonaceous fuels, whereas the reformate gas from a fuel processor can be fed into the PEM fuel cell for the electrochemical energy conversion.

REFERENCES

1. Mills, A.F. Mass transfer. In Kreith, et al. *Heat and Mass Transfer*. CRC Press, (1999).

2. Marrero, T.R., and Mason, E.A. Gaseous diffusion coefficients. *J. Phys. Chem. Ref. Data. 1* (1992): 3–118. doi: http://dx.doi.org/10.1063/1.3253094

3. Wang, Y., Chen, K.S., Mishler, J., Cho, S.C., and Adroher, X.C. A review of polymer electrolyte membrane fuel cells: technology, applications, and needs on fundamental research. *Appl Energy, 88* (2011) 981-1007. doi: http://dx.doi.org/10.1016/j.apenergy.2010.09.030

4. Mezedur, M.M., Kaviany, M. and Moore, W., Effect of pore structure, randomness and size on effective mass diffusivity, *AIChE J., 48* (2002) 1. doi: http://dx.doi.org/10.1002/aic.690480104

5. Weisz, P.B., Zeolites–New horizons in catalysis, *Chemtec*, 8 (1973) 498.

6. Wang, Y. and Wang, C.Y., Simulations of flow and transport phenomena in a polymer electrolyte fuel cell under low-humidity operations, *J. Power Source*, 147 (2005) 148–161. doi: http://dx.doi.org/10.1016/j.jpowsour.2005.01.047

7. Wang, Y. and Wang, C.Y., Modeling polymer electrolyte fuel cells with large density and velocity changes, J. *Electrochem. Soc.*, *152* (2005) A445–A453. doi: http://dx.doi.org/10.1149/1.1851059

8. Kavianv, M., *Principles of Heat Transfer in Porous Media*, Springer-Veriag (1991).

9. Prat, M., Recent advances in pore-scale models for drying of porous media, *Chemical Engineering Journal*, *86* (2002) 153–164. doi: http://dx.doi.org/10.1016/S1385-8947(01)00283-2

10. Wang, Y., Cho, S.C., Thiedmann, R., Schmidt, V., Lehnert, W. and Feng, X., Stochastic modeling and direct simulation of the diffusion media for polymer electrolyte Fuel Cells, *Int. J. Heat and Mass Transfer*, *53* (2010) 1128–1138. doi: http://dx.doi.org/10.1016/j.ijheatmasstransfer.2009.10.044

11. Shen, L. and Chen, Z., Critical review of the impact of tortuosity on diffusion, *Chem. Eng. Sci.*, *62* (2007) 3748–3755. doi: http://dx.doi.org/10.1016/j.ces.2007.03.041

12. Martínez, M.J., Shimpalee, S. and Van Zee, J.W., Measurement of MacMullin numbers for PEMFC gas-diffusion media, J. *Electrochem. Soc.*, *156* (1) (2009) B80–B85. doi: http://dx.doi .org/10.1149/1.3005564

13. Neale, G.H. and Nader, W.K., Prediction of transport processes within porous media: Diffusive flow Processes within homogeneous swarms of spherical particles, *AIChE J.*, *19* (1973) 112. doi: http://dx.doi .org/10.1002/aic.690190116

14. Mathias, M., Roth, J., Fleming, J. and Lehnert, W., Diffusion media materials and characterization, in Vielstich, W., Gasteiger, H., Lamm, A. (Eds.), *Handbook of Fuel Cells: Fundamentals, Technology and Applications*, vol .3, John Wiley & Sons Ltd, (2003).

15. Wang, Y. and Wang, C.Y., Transient analysis of polymer electrolyte fuel cells, *Electrochimica Acta*, *50* (2005) 1307–1315. doi: http://dx.doi.org/10.1016/j.electacta.2004.08.022

16. Wang, Y. and Wang, C.Y., Dynamics of polymer electrolyte fuel cells undergoing load changes, *electrochimica Acta*, *51* (2006) 3924–3933. doi: http://dx.doi.org/10.1016/j.electacta.2005.11.005

17. Ahmed, S., Kopasz, J., Kumar, R. and Krumpelt, M., Water balance in a polymer electrolyte fuel cell system, *J. Power Sources*, *112* (2002) 519–530. doi: http://dx.doi.org/10.1016/S0378 -7753(02)00452-4

CHAPTER 6

LIQUID WATER DYNAMICS AND REMOVAL

In high humidification operation, liquid water will emerge in the catalyst layers (CLs), gas diffusion layers (GDLs), and gas flow channels (GFCs) of PEM fuel cells. Liquid water, though hydrating the membrane thus improving its ionic conductivity, hinders reactant transport and electrochemical reaction activity, and thus plays a profound role in PEM fuel cell operation. In this chapter, we examine the two-phase flow in these three components of PEM fuel cell and at the GDL–GFC interface.

In addition, two appendices are provided for this chapter: Appendix VI.A summarizes the general governing equations of multiphase flow and heat transfer in porous media; and Appendix VI.B briefly introduces the multiphase mixture model in porous media that is widely applied in the two-phase model of PEM fuel cells.

6.1   MULTIPHASE FLOW OVERVIEW

Multiphase flow refers to fluid flows consisting of two or more phases (solid, liquid, and gas). Many natural phenomena occur with multiphase transport such as clouds, aerosols, waves, and bubble flows. Multiphase flow plays a critical role in various fields of engineering, such as chemical reactors (e.g., continuously stirred tank reactors or CSTRs), heat exchangers, refrigerators, and air conditioners. Classical fluid mechanics focus on single-phase flows. Multiple fluids arise when more than one phases exist in the system. The phases interact with each other at their interfaces, which dramatically increases the complexity of the flow phenomena.

The multiphase phenomena frequently observed in PEM fuel cells is the air–liquid water two-phase flow. Product water accumulates inside a PEM fuel cell upon the lack of efficient capability of water vapor removal. When the partial vapor pressure reaches its saturated value, product water condenses, resulting in two-phase flow. Two-phase flow is one of the most complex and vital phenomena in PEM fuel cells. Liquid water may block the pore paths of mass transport in GDLs and CLs, thereby reducing fuel cell performance and durability. Figure 6-1 shows liquid water formation and breakthrough in a CL, microporous layer (MPL), and GDL. Since liquid water formation depends on the vapor saturation pressure, which is a strong function of temperature, heat transfer and its coupling with water condensation and/or evaporation are critical to the study of two-phase transport and the ensuing cathode flooding in PEM fuel cells, to be discussed in Chapter 9.

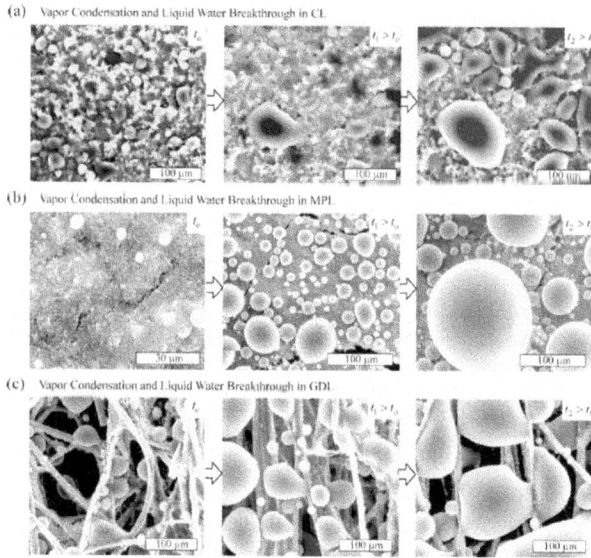

Figure 6-1. Liquid water formation and breakthrough in a CL, microporous layer (MPL), and GDL [1] (Courtesy of Elsevier).

6.1.1 MODELING MULTIPHASE FLOWS

The standard flow equations, that is, the Navier–Strokes equations, which describe the behavior of individual phase flow, are a beginning point for modeling multiphase flows. In this section, we briefly describe several basics in the context of general multiphase flow modeling. Readers who are interested in more detailed discussions of two-phase flow are referred to consult with the reference texts [2–5]. In addition, Appendix VI.A summarizes the general governing equations of multiphase flow and heat transfer in porous media.

A multiphase flow system consists of several components or phases, the mass conservation for each phase N can be expressed as [3]

$$\frac{\partial \rho_k \alpha_k}{\partial t} + \nabla \cdot (\rho_k \vec{j}_k) = I_k, \qquad \text{(Eq. 6-1)}$$

where ρ *is* the fluid density, α is the volume fraction, $\vec{j}_k (= \alpha_k \vec{u}_k)$ is the volumetric flux of a component k, and I_k is the rate of mass added to the phase k from other phases $(\sum_k I_k = 0)$. For regular interfaces such as in stratified flows, a model can be formulated by the above equations for each component, in conjunction with boundary conditions at their interfaces.

In many occasions, however, the interfaces are irregular and constantly change with time. Averaging is an effective method widely applied to simplify the multiphase flow model. An example is dispersed flows: one can define an infinitesimal control volume in which no significant flow-property change is observed but sufficient numbers of the individual phase elements such as bubbles or droplets are contained. Through volume averaging, continuum formulation can be set up. In the mixture approach, a detailed interfacial interaction is approximated by simple or empirical correlations. One way to eliminate the interfacial interaction is through summing the mass conservation equations over all the phases:

$$\frac{\partial}{\partial t}\left(\sum_k \rho_k \alpha_k\right) + \nabla \cdot \left(\sum_k \rho_k \vec{j}_k\right) = 0. \qquad \text{(Eq. 6-2)}$$

Because the interfacial terms between phases are canceled (i.e., $\sum_k I_k = 0$), there is no need to develop extra models for the interfacial mass exchange. The momentum conservation can be developed by accounting for the momentum storage rate, its flux at the boundaries, and forces over the control volume. To simplify the problem, the boundary of the control volume is assumed to be continuous within one phase. Then, the net force on the component N is expressed as

$$\vec{F}_k' = \frac{\partial}{\partial t}(\rho_k \alpha_k \vec{u}_k) + \nabla \cdot (\rho_k \alpha_k \vec{u}_k \vec{u}_k). \qquad \text{(Eq. 6-3)}$$

Separating the pressure and viscous stress along with the gravitational force, the momentum equation becomes

$$\frac{\partial}{\partial t}(\rho_k \alpha_k \vec{u}_k) + \nabla \cdot (\rho_k \alpha_k \vec{u}_k \vec{u}_k) = \alpha_k \rho_k \vec{g} + \vec{F}_k - \delta_k \left(\nabla p - \nabla \cdot \vec{\sigma}_C^d\right), \qquad \text{(Eq. 6-4)}$$

where p and $\vec{\sigma}_C^d$ are pressure and the deviatoric stress, respectively, and $\delta_k = 0$ for the disperse phase (D) and $\delta_k = 1$ for the continuous phase (C). To cancel the forces among phases (i.e., $\sum_k \vec{F}_k = 0$), summation can be taken over all the phases, yielding

$$\frac{\partial}{\partial t}\left(\sum_k \rho_k \alpha_k \vec{u}_k\right) + \nabla \cdot \left(\sum_k \rho_k \alpha_k \vec{u}_k \vec{u}_k\right) = \rho \vec{g} - \nabla p + \nabla \cdot \vec{\sigma}_C^d. \qquad \text{(Eq. 6-5)}$$

6.2 MULTIPHASE FLOW IN GDLS/CLS

Both GDLs and CLs are porous media consisting of the solid matrix and pore network. When liquid water emerges in the pore space, it hinders gaseous reactant supply to the catalyst sites. Under severe conditions, electrode "flooding" (which refers to the pore space being filled up with liquid water) occurs, causing material degradation and efficiency reduction. To facilitate water drainage, the porous materials are usually rendered hydrophobic by adding nonwetting chemicals such as polytetrafluoroethylene (PTFE, a.k.a. DuPont's Teflon™). Figure 6-2 shows the two common GDL materials: carbon paper and carbon cloth, which exhibit different structures of the solid matrix.

Figure 6-3 schematically shows a possible morphology of liquid water in GDLs, which follows the branching-type geometry. The topology consists of large main streams and small branches with them connected to randomly distributed condensation sites. Water vapor condenses at the

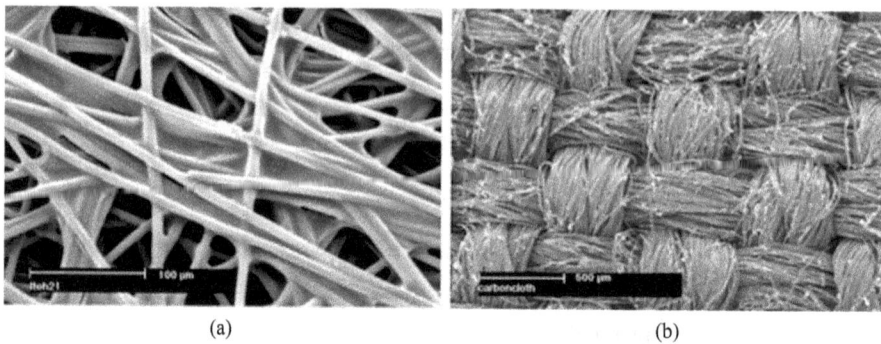

Figure 6-2. Microscopic images of (a) carbon paper and (b) carbon cloth, respectively.

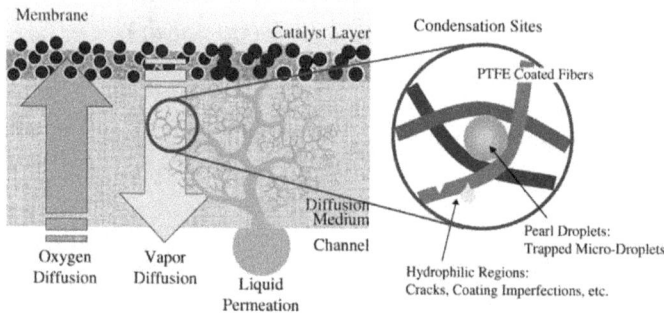

Figure 6-3. A liquid water transport model: branching micro- to macro-transport [5] (Courtesy of Elsevier).

microdroplet surface (rather than uniformly distributed) and intermittently agglomerates to form larger droplets. The major streams extend from the catalyst layer to the gas flow channel and transport liquid water by capillary action. While large streams act as backbones for liquid transport, small streams collect water condensate from microdroplets. As a result, the distribution of liquid water is determined by both condensation (microdroplets) and major streams of liquid-phase capillary flow [5].

Figure 6-4 illustrates the flow patterns in porous media as a function of the capillary number and ratio of viscosities of the invading and receding fluids. Liquid water flow in PEM fuel cells usually falls in the capillary fingering regime in which the displacement is sufficiently slow that the viscous forces are negligibly small, relative to the capillary forces.

6.2.1 EXPERIMENTAL VISUALIZATION

Direct optical visualization of liquid water within a GDL or catalyst layer is difficult: optical access can reach at most a depth of a couple of pores underneath the surface, for example, for GDLs, ~20 μm depth. Confocal microscope was employed in several studies to visualize liquid movement, and it is, however, difficult to apply it for the quantification of liquid content.

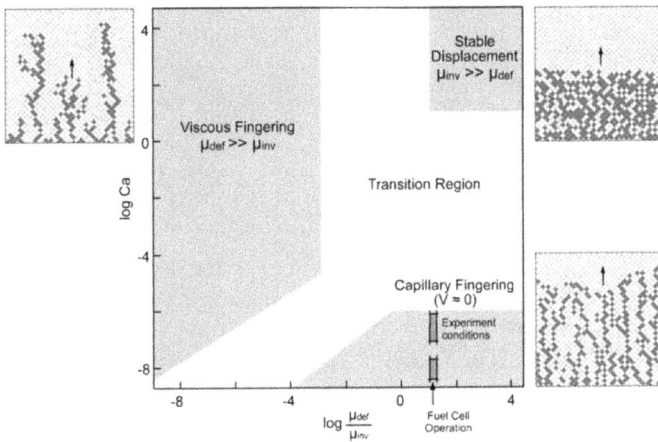

Figure 6-4. Flow map for fluid transport through porous media, along with a representative schematic of the flow patterns in three flow regimes [6–8] (Courtesy of Elsevier).

Two advanced imaging methods, namely (a) X-ray imaging and (b) neutron radiography, are often applied to probe liquid-water contents in GDLs. Other methods such as MRI and supersonic methods were proposed as well, but have received little attention to date.

6.2.1.1 X-ray Imaging

X-ray tomography, which is usually called X-ray computed tomography (CT), is widely used in medical imaging. Through the X-ray attenuation by a sample (or object), a 2D picture of the object at a specified angle can be obtained. By viewing the object at different angles, a 3D image of the internal material structure can be built. Liquid-water content is then obtained by comparing the X-ray absorptions in the absence and presence of liquid phase. Using Beer–Lambert's law which is similar to neutron radiography, liquid content can be obtained. Sinha *et al.* [9] presented pioneering work on applying X-ray imaging to PEM fuel cells. Figure 6-5 shows the liquid water location and profiles in a GDL probed by X-ray imaging [10].

6.2.1.2 Neutron Radiography

Neutron radiography is a powerful tool for *in situ* visualization of liquid water inside PEM fuel cells. This is achieved by sending beams of collimated thermal neutrons through a working fuel cell and measuring the attenuation of the transmitted neutron beams. The spatial and temporal resolutions are determined by the neutron source and the imaging setup. The attenuated beams are captured by a detector. Using Beer–Lambert's law, the liquid-water content can be separated out, as detailed in Chapter 4. Neutron radiography has been employed to quantify liquid-water content in PEM fuel cells in different dimensions [11–13]. A main challenge was to obtain the through-plane water profiles, which was not possible until recently when the high resolution neutron radiography became available. Figure 6-6 shows the water contours, obtained by

(a)

(b)

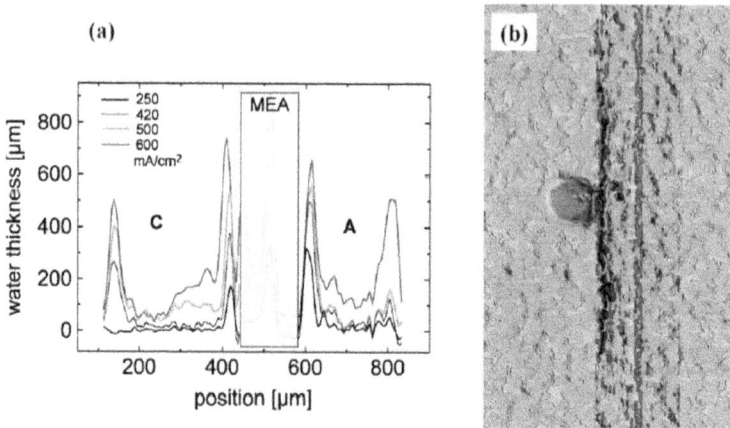

Figure 6-5. (a) Liquid water profile in GDLs. Depending on operating conditions, one or two diffusion barriers formed by liquid water can be detected. (b) Eruptive water transport from the GDL to the gas channel [10] (Courtesy of American Institute of Physics).

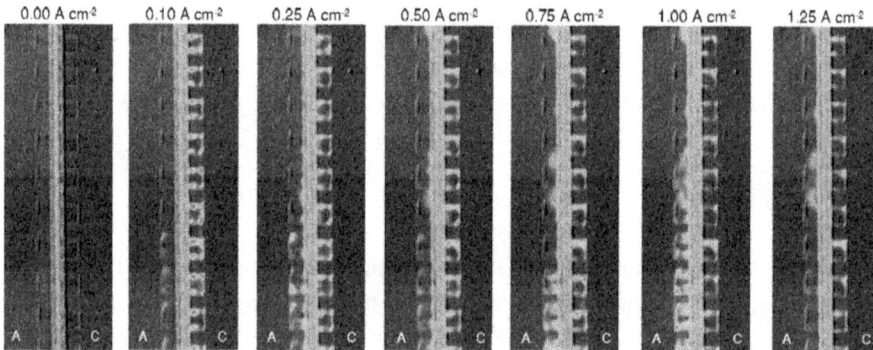

Figure 6-6. High-resolution false color radiographs for a PEM fuel cell at 60°C with 100% RH inlet gas feeds. Red and green represent higher water content; blue and black represent low water content [12] (Courtesy of the Electrochemical Society).

probing with neutron imaging with a pixel size of ~15 μm. Figure 6-7 shows the through-plane water profiles from neutron imaging data.

6.2.2 MULTIPHASE MIXTURE (M²) FORMULATION

A number of macroscopic two-phase flow models have been proposed for PEM fuel cells, particularly for GDLs and CLs. The multiphase mixture (M^2) formulation, originally developed by Wang and Cheng [15], is widely adopted in modeling PEM fuel cells. The formulation defines mixture properties and develops the governing equations for mixture variables, such as the mixture density, velocity, and species concentration. Appendix VI.B introduces the multiphase mixture model in general porous media. A detailed M^2 description of two-phase flows in PEM fuel cells will be presented in the following section.

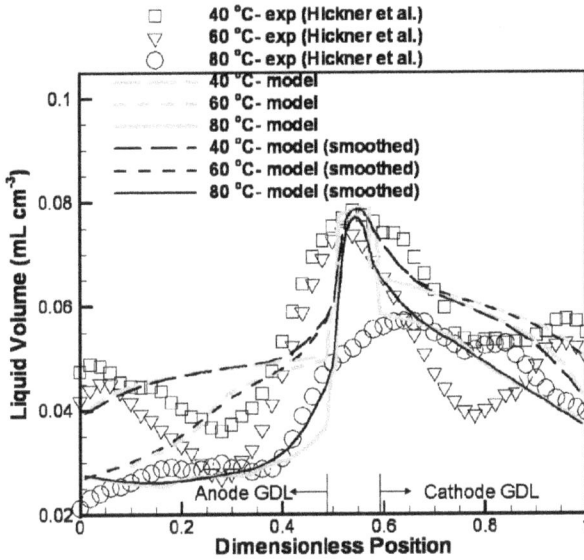

Figure 6-7. Water profiles in the MEA and GDLs from model prediction and experimental data under 0.75 A/cm² and RHa/ c=100%/100% [14] (Courtesy of the Electrochemical Society).

6.2.2.1 Flow Equations

In porous media, the two phases are assumed to be well mixed; thus at each local site mixture variables can be defined based on the liquid and gas fractions, for example, the mixture velocity and density. Assuming that Darcy's law is applicable to individual liquid and gas phases, respectively, the mass conservation and momentum equations are written as

$$\nabla \cdot (\rho \, \vec{u}) = 0, \qquad \text{(Eq. 6-6)}$$

$$\rho \vec{u} = -\frac{K}{\nu} \nabla P, \qquad \text{(Eq. 6-7)}$$

where the two-phase mixture density and velocity are defined as

$$\rho = s\rho_l + (1-s)\rho_g \ \text{ and } \ \rho \vec{u} = \vec{u}_l \rho_l + \vec{u}_g \rho_g. \qquad \text{(Eq. 6-8)}$$

The liquid saturation s, defined as the liquid volume fraction in pore space, is related to the mixture water concentration C_w:

$$s = \begin{cases} 0 & C_w \leq C_{w,\text{sat}} \\[2mm] \dfrac{C_w - C_{w,\text{sat}}}{\rho_l/M_w - C_{w,\text{sat}}} & C_w > C_{w,\text{sat}}\,, \end{cases} \qquad \text{(Eq. 6-9)}$$

where $\rho C_w = C_w^l \rho_l s + C_w^g \rho_g (1-s)$.

The mixture kinematic viscosity is defined by

$$\nu = \left(\frac{k_{rl}}{\nu_l} + \frac{k_{rg}}{\nu_g} \right)^{-1} . \qquad \text{(Eq. 6-10)}$$

The interaction between the flows of two phases is accounted for by the relative permeabilities k_{rl} and k_{rg}, which are dimensionless measure of the effective permeability of that phase. The relative permeability is the ratio of the effective permeability of that phase to the absolute intrinsic permeability of a medium. Physically, these parameters describe the extent to which one fluid is hindered by others in pore spaces and hence can be formulated as a function of the volume fraction of each phase. One formula for the relative permeability frequently used in PEM fuel cell modeling is as follows:

$$k_{rl} = s^3 \text{ and } k_{rg} = (1-s)^3. \qquad \text{(Eq. 6-11)}$$

Figure 6-8 shows the relative permeability of air ($k_{r,a}$) plotted against saturation for the through- and in-plane directions, respectively, for the two GDL materials. In Figure 6-8 (a), the experimental data reported by Koido *et al.* [16] are also plotted. The calculated values of $k_{r,a}$ in the through-plane are found to be unrealistically low. Data in the in-plane direction are more consistent and reliable. For carbon paper, $k_{r,a}$ is found to follow $(1-s^2)^4$, whereas for carbon cloth, the dependency of $k_{r,a}$ on s is more along the line of $(1-s)^3$.

6.2.2.2 Species Transport

Hydrogen gas and oxygen and water vapor follow similar mechanisms of transport, as discussed in the preceding chapter. In liquid water, the transport of reactant hydrogen and oxygen is extremely slow as opposed to that in the gaseous phase, and thus can be neglected. The flowing liquid is essentially pure water, and thus a large amount of water can be stored and transported in the liquid phase. Liquid flow in GDLs and CLs is mainly due to capillary action (i.e., driven by capillary pressure gradients). The general conservation equation of species is given by

$$\varepsilon^{\text{eff}} \frac{\partial C_k}{\partial t} + \nabla \cdot (\gamma_k \vec{u} C_k) = \nabla \cdot (D_k^{g,\text{eff}} \nabla C_k^g) - \nabla \cdot \left[\left(\frac{m f_k^l}{M_k} - \frac{C_k^g}{\rho_g} \right) \vec{j}_l \right] + S_k. \qquad \text{(Eq. 6-12)}$$

Because the presence of liquid narrows down gas transport passages, the diffusivity of gaseous species is modified, following the Bruggeman correlation, as follows, to account for this effect:

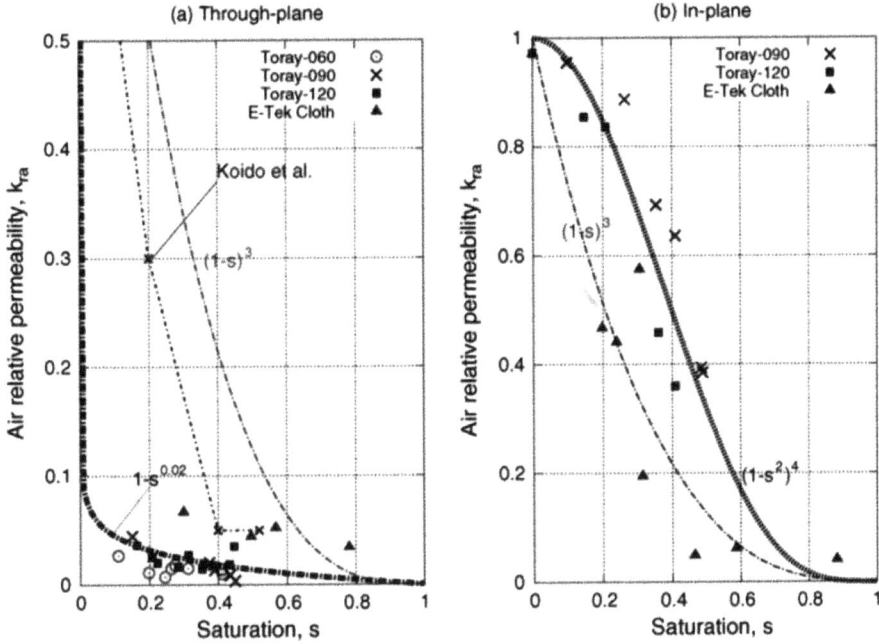

Figure 6-8. Air relative permeability k_{ra} in through- and in-plane directions for carbon paper and carbon cloth GDLs [6]: (a) through plane, (b) in-plane (Courtesy of Elsevier).

$$D_g^{\text{eff}} = [\varepsilon(1-s)]^{\tau d} D_g. \qquad \text{(Eq. 6-13)}$$

In Eq. 6-12, the convection correction factor γ is a function of the liquid saturation, s:

for water: $\gamma_w = \dfrac{\rho}{C_w}(\dfrac{\lambda_l}{M_w} + \dfrac{\lambda_g}{\rho_g} C_{\text{sat}}),$

for oxygen and hydrogen gas: $\gamma_k = \dfrac{\rho \lambda_g}{\rho_g (1-s)}.$ $\qquad \text{(Eq. 6-14)}$

where $\lambda_{l/g}$ is the relative mobilities of individual phases, defined as

$$\lambda_l = \dfrac{k_{rl}}{\nu_l} \nu \text{ and } \lambda_g = \dfrac{k_{rg}}{\nu_g} \nu. \qquad \text{(Eq. 6-15)}$$

Thus,

$$\lambda_l + \lambda_g = 1. \qquad \text{(Eq. 6-16)}$$

The capillary flux $\vec{j_l}$ in the water equation is determined by the capillary pressure gradient and gravity:

$$\vec{j}_l = \frac{\lambda_l \lambda_g}{\nu} K[\nabla P_c + (\rho_l - \rho_g)\vec{g}].$$

(Eq. 6-17)

The capillary pressure P_c is the pressure difference across the two-phase interface, determined by the (porosity, wetting) and transport properties and liquid saturation. An example of the capillary pressure correlation is given below:

$$P_g - P_l = P_c = \sigma(\cos\theta_c)\left(\frac{\varepsilon}{K}\right)^{1/2} J(s),$$

(Eq. 6-18)

where σ is the surface tension. The Leverett J function $J(s)$ is determined by the material wettability [17]:

$$J(s) = \begin{cases} 1.417(1-s) - 2.120(1-s)^2 + 1.263(1-s)^3 & \text{for } \theta_c < 90° \\ 1.417s - 2.120s^2 + 1.263s^3 & \text{for } \theta_c > 90° \end{cases}.$$

(Eq. 6-19)

It should be noted that the above Leverett function was originally developed for two-phase flow in soils; as such, it is not directly applicable to liquid transport in fibrous GDLs. Kumbur et al. [18] showed the above Leveret function exhibits significant deviation from their experimental data. Gostick et al. [19] investigated the capillary pressure for fibrous porous media, and their experimental set up is shown in Figure 6-9. Figure 6-10 presents the results obtained by Gostick et al. [19] for Toray 120 (left) and Toray 060 (right). Breakthrough saturations of 0.33 and 0.41

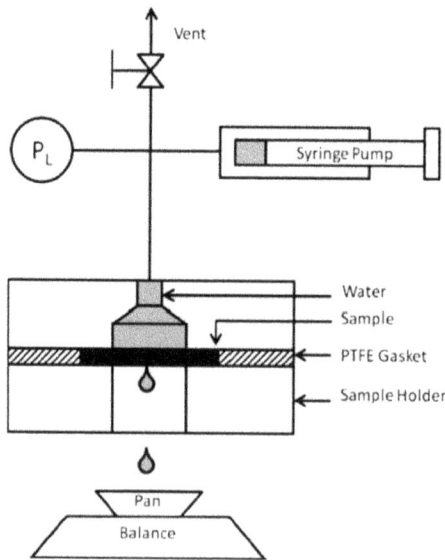

Figure 6-9. Schematic diagram of water injection experimental setup employed by Gostick et al. [19] (Courtesy of the Electrochemical Society).

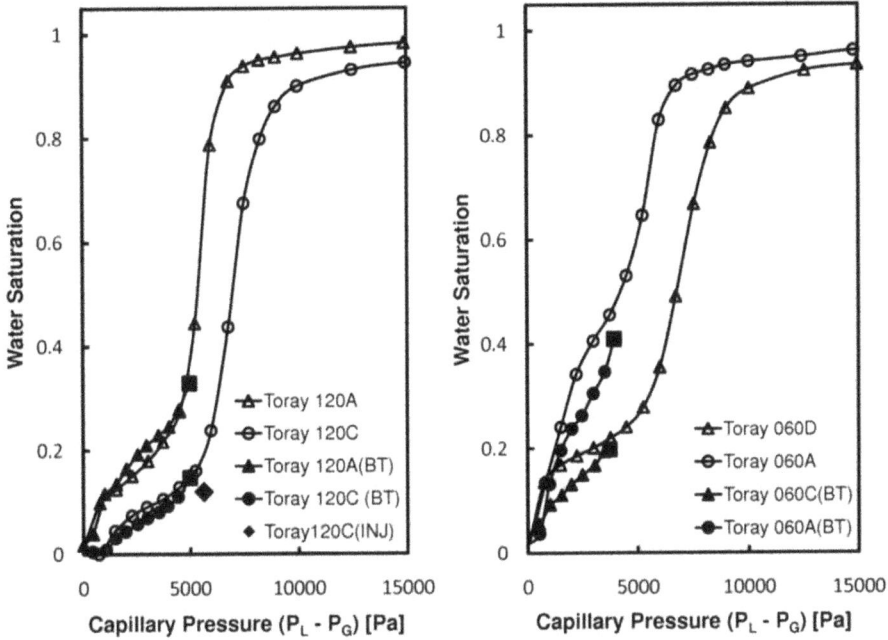

Figure 6-10. Results from breakthrough tests on Toray 120 (left) and Toray 060 (right) with (circle markers) and without (triangle markers) PTFE coating. Breakthrough points determined by the modified capillary pressure method (BT) are marked with a large black square at the termination of each breakthrough experiment. The large black diamond indicates the breakthrough point determined by the injection experiment [19] (Courtesy of the Electrochemical Society).

were obtained in Toray 120A and 060A without any PTFE contents, respectively, whereas the values of 0.14 and 0.20 were measured in Toray 120C and 060C with treated PTFE, respectively. The PTFE content was found to reduce the breakthrough saturation by about half in both the thin (Toray 060) and thick (Toray 120) materials.

6.2.2.3 Model Prediction

One-Dimensional (1D) Analysis

For 1D steady-state flow in the through-plane direction, the assumption of capillary action being the only driving force for liquid flow yields

$$\frac{1}{2F} M_w = \frac{k_{rl}}{\nu} K \nabla p_c. \tag{Eq. 6-20}$$

Substituting the Leveret function, Eq. 6-18, into the above equation leads to

$$\frac{1}{2F} M_w = -\frac{s^3}{\nu} K \sigma \cos \theta_c \left(\frac{\varepsilon}{K}\right)^{1/2} \nabla J(s). \tag{Eq. 6-21}$$

For hydrophobic and hydrophilic GDLs, the J function takes different forms. Substituting the J function yields

for hydrophobic GDL,

$$s^1(0.35425 - 0.8480s + 0.6135s^2) = \frac{1}{2F} M_w \frac{\nu}{\sigma \cos\theta_c (\varepsilon K)^{1/2}} x + C_1,$$

for hydrophilic GDL,

$$s^4(-0.2415 + 0.6676s - 0.6135s^2) = \frac{1}{2F} M_w \frac{\nu}{\sigma \cos\theta_c (\varepsilon K)^{1/2}} x + C_1. \quad \text{(Eq. 6-22)}$$

Figure 6-11 displays the liquid profile predicted by the 1D analytical model (Eq. 6-22). Since in this case study, the left side is adjacent to the catalyst layer where water is produced by the ORR, and thus shows a higher saturation. The maximum saturation is predicted to be around 10%. The two GDLs, despite their different wettabilities, exhibit the similar trend of decreasing liquid-saturation along the GDL.

Three-dimensional (3D) analysis

A PEM fuel cell has three physical dimensions. To probe on the 3D two-phase flow phenomena in all the dimensions, a PEM fuel cell with a single channel, 30-μm-thick Gore™ membrane, and carbon cloth was chosen for a case study. Figure 6-12 presents the predicted liquid-water

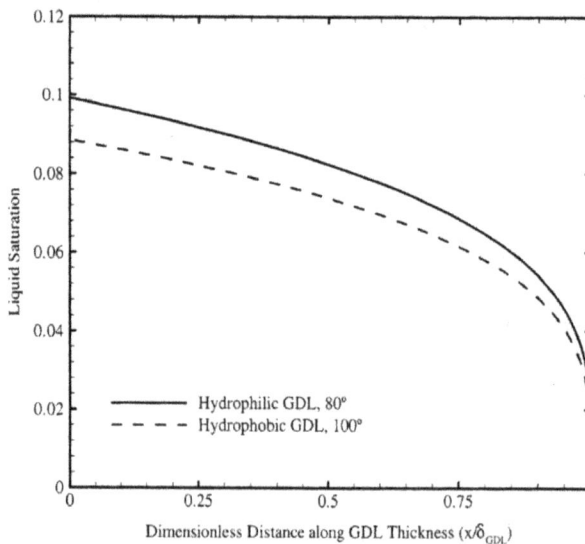

Figure 6-11. Liquid profiles in a GDL predicted by the 1D analytical model [20] (Courtesy of the Electrochemical Society).

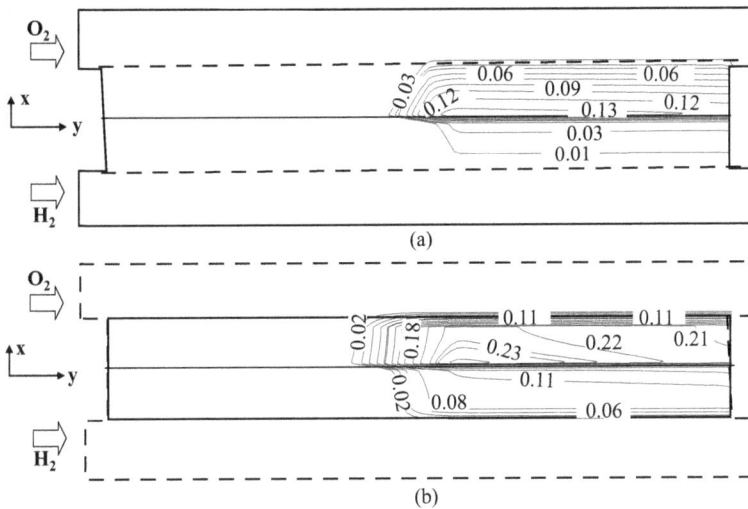

Figure 6-12. Water saturation contours in a GDL: (a) under the channel and (b) under the land (operating condition: 2 atm, 0.8 A/cm², 80°C, RHa/c = 66%/66%, and Stoich.a/c = 1.5/2) [21].

saturation contours in the GDL. Single- and multiphase flows coexist inside the PEM fuel cell under the low-humidity condition. The single-phase region is near the inlet, where dry reactant flows are fed in. Product water increases the water content in the gas flow, leading to two-phase flow downstream. Liquid water emerges first on the cathode side. Liquid saturation reaches as high as ~25%. Water back-diffusion rehydrates the anode side, leading to anode two-phase flow. Furthermore, increasing the stoichiometric flow ratio improves the capability of water vapor removal by channel streams, therefore reducing the two-phase region.

Figure 6-13 shows the reactant contours, indicating a decreasing trend down the gas flow channel as a result of hydrogen consumption due to the HOR and oxygen consumption due to the ORR, respectively. In contrast to the considerable reduction observed along the channel, the concentration change cross the GDLs is small. The land imposes a larger transport resistance, resulting in a larger drop in the reactant concentration under the land.

Figure 6-14 shows the relative humidity (RH) distributions. The anode RH decreases initially due to water loss by the electro-osmotic drag, followed by an increase when back diffusion takes effect. Here, the RH is defined by accounting for the water in both liquid and gaseous phases; thus the value of RH of over 1.0 (100%) means two-phase region. Figure 6-14(b) shows the RH distribution at the cross section of plane A–A where liquid water emerges. At this location, there emerges two-phase flow under the anode land, whereas no liquid exists under the anode channel. At the location of B-B, that is, upstream region, the RH is lower than 1.0 (100%) at the location of Figure 6-14 (c), that is, no liquid water emerges.

Figure 6-15 shows the velocity and saturation distributions in the cathode GDL at two different cross sections. At the location of half way down the channel, a slight liquid flow occurs under the channel, as shown by the small liquid velocity in Figure 6-15 (a). The liquid flows are directed from the catalyst layer toward the gas flow channel and from under-land region to that under the channel. The liquid flow is accompanied by a reverse gas flow, indicative of the occurrence of complex multiphase flow phenomena in the fuel cell.

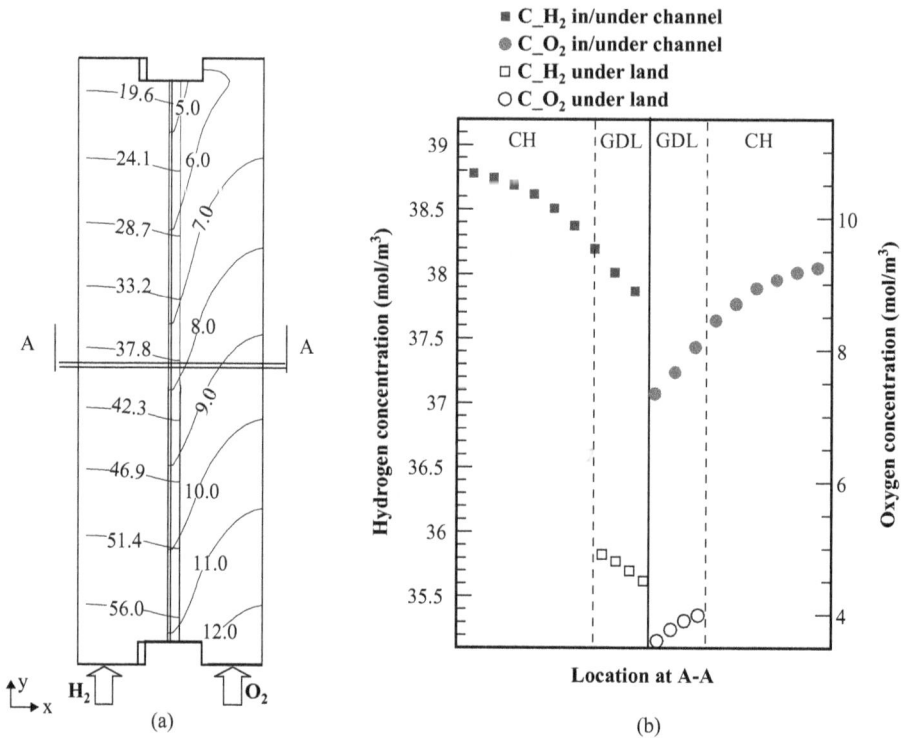

Figure 6-13. Hydrogen and oxygen concentration contours (a) at the middle section of the PEM fuel cell and (b) at the A–A location (operating condition: 2 atm, 0.8A/cm², 80°C, RHa/c = 66%/66%, and stoich.a/c = 1.5/2).

The above simulation results were computed for a PEM fuel cell with co-flow configuration. Alternative configuration is the counter flow, that is, hydrogen gas and air flows are arranged in the opposite directions. Figure 6-16 (a) shows the schematic of a counter-flow configuration. The counter-flow design promotes water recirculation, improving water management. Figure 6-16 (b) and (c) present the liquid contours under the channel and land, respectively. In contrast to what happen in the co-flow configuration, liquid water appears only in the middle of the PEM fuel cell. As the outlet is set to be on the same side of the dry inlet on the other side, water near the cathode outlet is transported to the anode via the membrane, drying out liquid water in the cathode. As the anode downstream humidifies the cathode inlet dry reactant, liquid water emerges earlier in the cathode than that in the co-flow configuration.

6.2.3 CARBON PAPER (CP) VERSUS CARBON CLOTH (CC)

Two types of GDLs are commonly used in PEM fuel cells: carbon paper and carbon cloth; both are commercially available. For convenience, carbon paper and carbon cloth are herein denoted as CP and CC, respectively. Both of them are carbon-fiber-based porous materials: carbon paper is non-woven, whereas carbon cloth is woven fabric; thus no binder is needed. Figure 6-2 shows the SEM pictures of these two GDL substrates. It has been experimentally observed that the

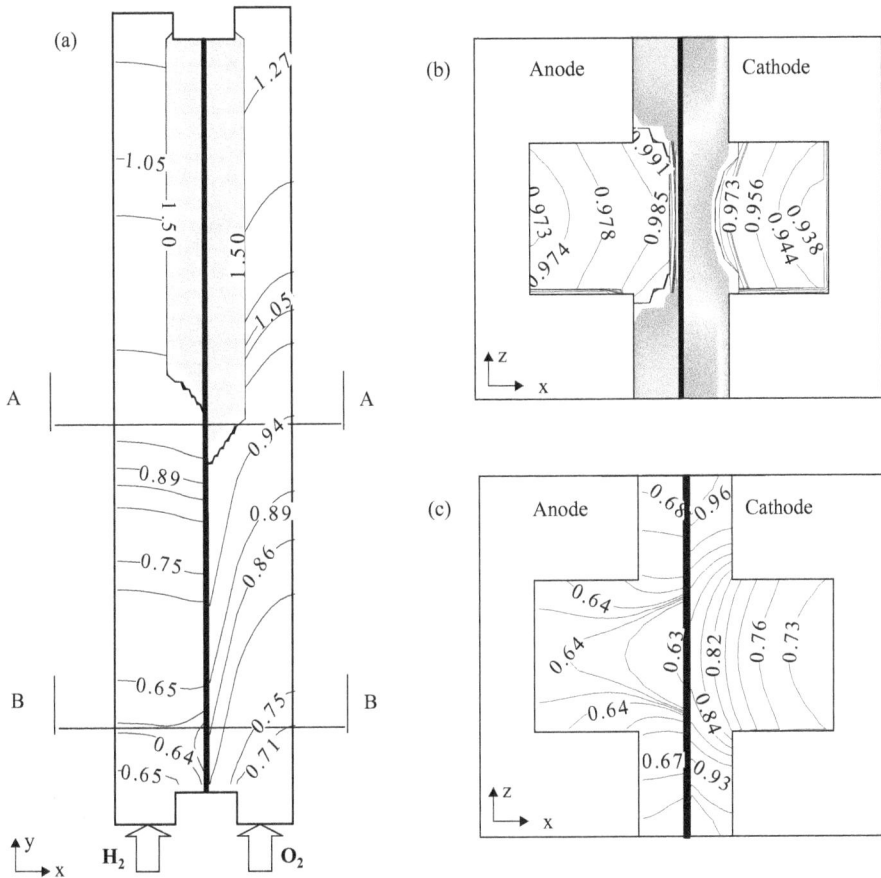

Figure 6-14. Relative humidity (RH) distributions: (a) along the gas flow channel; (b) at the plane A–A; and (c) at the plane B–B. The gray areas denote the multiphase region. (Operating condition: 2 atm, 0.8 A/cm², 80°C, RHa/c = 66%/66%, and Stoich.a/c = 1.5/2).

cell performance of PEM fuel cells employing CC GDLs is different from that with CP GDLs under low- or high-humidity operations, respectively. Figure 6-17 shows that under the low humidity the CP yields better performance, while under the fully humidified condition, the two materials give rise to similar performances at low current densities and the CC shows superior performance over 0.6 A/cm². The structural differences (non-woven vs. interwoven) is a major factor. For example, CC is more porous and less tortuous than CP. In addition, the rougher carbon cloth has less liquid water coverage as compared with carbon paper, which is clearly shown in Figure 6-18.

In addition, CC differs from CP in several other aspects that may result in the distinct performance observed. Table 6-1 lists the porosity and contact angle for both CC and CP at various PTFE loadings, which were obtained by Benziger *et al.* [23]. The schematic of the apparatus employed by Benziger *et al.* [23] is shown in Figure 6-19. In their experiments, a 5.0 cm diameter piece of the diffusion media was placed on a porous plate in a pressurized membrane filtration cell. The cell was 12 cm high. Clear tygon tubes (1.2 cm diameter) were attached to the top of the cell; one was used to fill the cell and the second tube allowed air to escape.

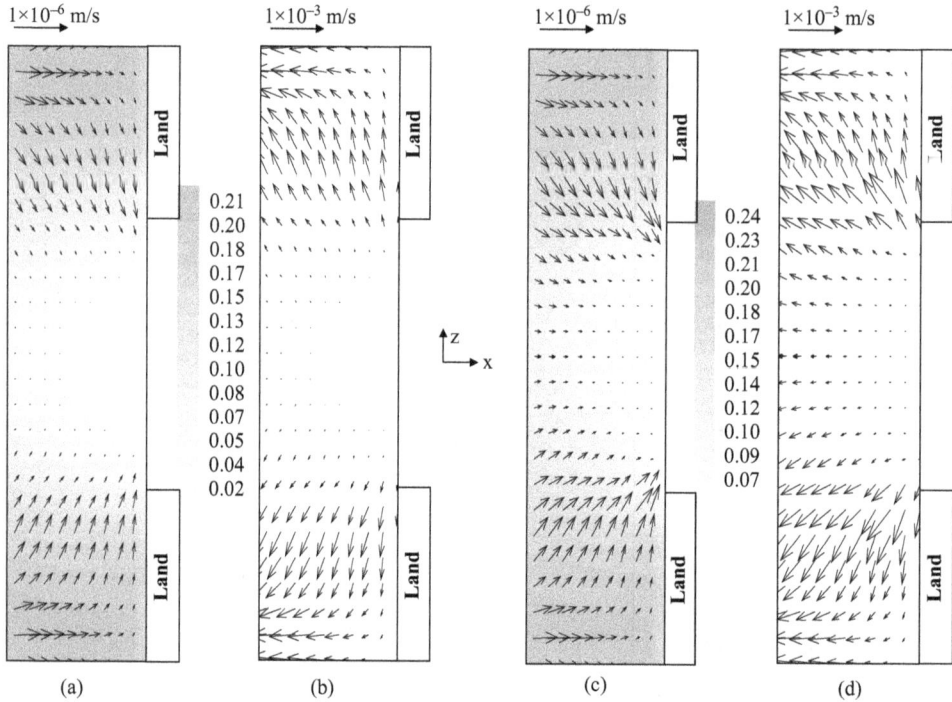

Figure 6-15. Velocity and saturation distributions in a cathode GDL: liquid velocity and saturation at (a) 50% and (c) 90% fraction distance from the inlet and gas velocity at (b) 50% and (d) 90% fraction distance from the inlet (operating condition: 2 atm, 0.8 A/cm^2, 80°C, RHa/c = 66%/66%, and Stoich.a/c = 1.5/2).

The results for flow through an initially wet GDL indicate there is a small drop in the necessary pressure head to cause water to enter the pores of the GDL, compared to an initially dry GDL (see Table 6-2). The results also show that there is a slight decrease in the necessary pressure head to cause flow between a virgin GDL and a previously used GDL. The results in Table 6-2 indicate that the woven cloth has the largest pores with their diameter $\sim$250 μm. The carbon paper has pores about five times smaller and the ELAT™ catalyst layer has pores that are only 20 μm in diameter. The pores in the woven carbon cloth can be imaged and the model of the parallel pores is close to reality.

Figure 6-20 shows a schematic water imbibition in a GDL. For simplicity, the pores are assumed to be cylindrical and run across the GDL. When liquid water flows through a cylindrical pore it must wet the walls of the pore. If the pore walls are hydrophilic, water will be drawn into the pores. For hydrophilic GDLs, there would be no restriction to liquid flow and imbibition would begin instantaneously without having to apply any pressure. On the other hand, if the pore walls are hydrophobic liquid water is excluded from the pores until sufficient work is done to overcome the surface energy. The carbon fibers of both the cloth and paper are slightly hydrophobic; work must be done to push the water into the hydrophobic pores. The larger the pores, the less the work is required to overcome the unfavorable surface energy that is inversely proportional to the pore radius. Coating the carbon fibers with Teflon makes the pores highly hydrophobic; thus a higher pressure is required to push the water into the pores. The pressure, ΔP, that is required to force water into the pores of radius R, is given by the Young–Laplace equation:

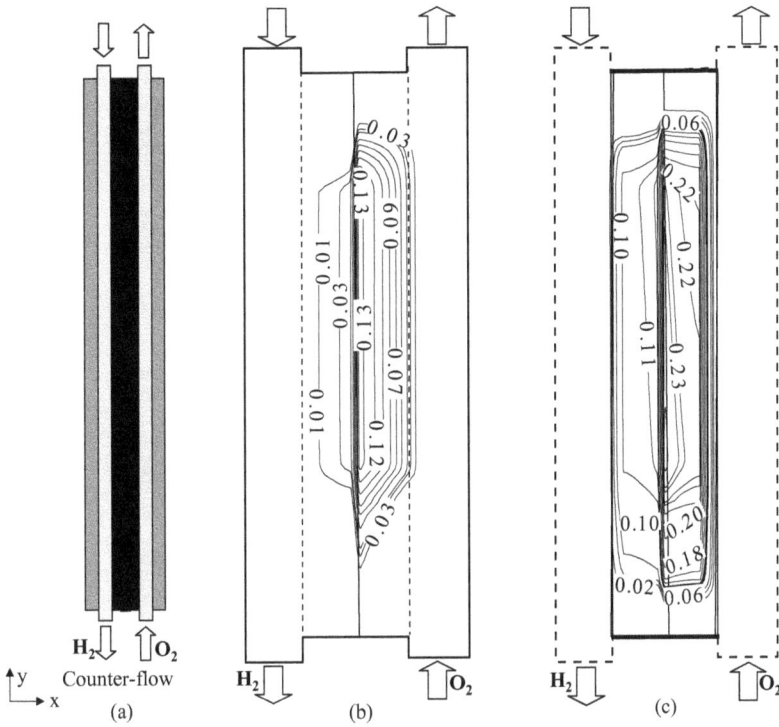

Figure 6-16. (a) Schematic of the counter-flow configuration and computed liquid water saturation contours **(b)** under the channel and **(c)** under the land (operating condition: 2 atm, 0.8 A/cm², 80°C, RHa/c = 66%/66%, and Stoich.a/c = 1.5/2).

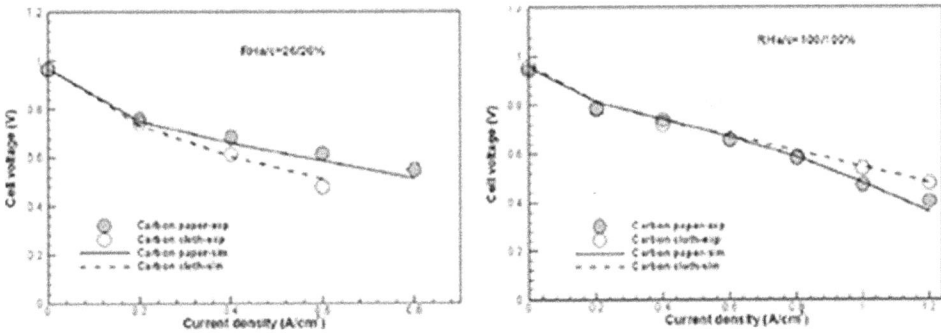

Figure 6-17. Performance of PEM fuel cells with carbon paper and carbon cloth GDLs, respectively [22].

$$\Delta P = \frac{2\sigma \cos \theta_c}{R}, \qquad \text{(Eq. 6-23)}$$

where σ is the surface tension and θ_c is the contact angle of water at the surface of the pore. If the applied pressure is less than the right-hand side of Eq. 6-23, the water is excluded from the pore. The experimental results of the minimum pressure for water flow through different GDL media yields a measurement of the largest pores in the GDL material because it is the

(a)

(b)

Figure 6-18. *In-situ* images of liquid water on the GDL surface in operating PEM fuel cells with: (a) carbon paper GDL and (b) carbon cloth GDL. The channels shown are 1 mm wide, with droplets on carbon paper being ~200 μm and those on carbon cloth ~10 μm (operating condition: 0.8 A/cm², 80°C, and 2 atm) [22].

largest pores that require the smallest pressure for liquid water to penetrate. Table 6-3 shows the permeability of different GDLs as a function of initially applied hydrostatic head. Larger hydrostatic pressure permits water to enter into smaller pores and thus increases the flow of liquid water through the GDL.

6.2.4   SPATIALLY VARYING PROPERTIES

GDLs are usually treated as uniform media with constant properties in all the three dimensions. In this treatment, the material properties such as permeability and contact angle are constant and thus taken out of spatial derivative operators in the governing equations.

6.2.4.1   Through-Plane Variation in the GDL Property

To examine the effect of spatial property variation, we use the through-plane liquid transport as an example. The capillary action can be treated as a "diffusive" term, using the following one-dimensional (1D) problem as an example:

Table 6-1. Several physical characterizations of carbon cloths and papers [23] (Courtesy of Elsevier)

Media	Dry areal mass (kg/m^2)	Areal mass after liquid water contact (kg/m^2)	Void fraction	Advancing / receding contact angle
Carbon paper (Toray)	0.259 ± 0.007	0.469 ± 0.052	0.72 ± 0.05	115°/30°
Carbon paper + 20%Teflon	0.374 ± 0.007	0.423 ± 0.035	0.69 ± 0.05	170°/120°
Carbon paper + 40% Teflon	0.456 ± 0.010	0.525 ± 0.047	0.59 ± 0.05	170°/120°
Carbon paper + 60% Teflon	0.476 ± 0.009	0.526 ± 0.044	0.50 ± 0.05	170°/120°
Carbon cloth	0.355 ± 0.009	0.484 ± 0.027	0.75 ± 0.05	95°/30°
Carton cloth+ 20% Teflon	0.476 ± 0.012	0.551 ± 0.047	0.73 ± 0.05	170°/120°
Carton cloth+ 40% Teflon	0.595 ± 0 011	0.648 ± 0.041	0.68 ± 0.05	170°/120°
Carton cloth+ 60% Teflon	0.697 ± 0.017	0.839 ± 0.055	0.52 ± 0.05	170°/120°
E-TEK/ELAT electrode	0.435 ± 0 011	0.620 ± 0.036	0.74 ± 0.05	170°/120°

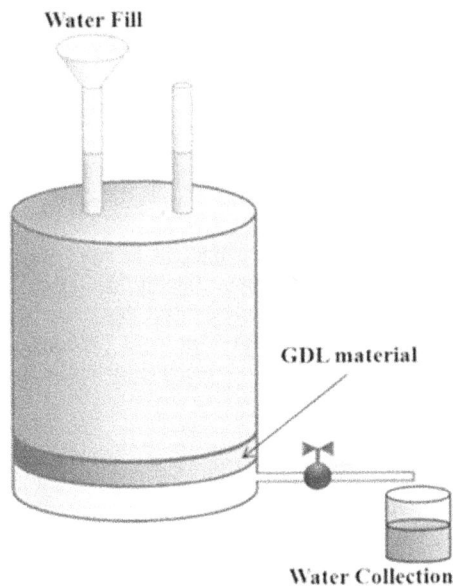

Figure 6-19. Schematic of the flow measurement. The hydrostatic pressure is measured from the height of the water above the GDL. A clamp on the outlet can stop the flow to allow greater hydrostatic pressure heads to be applied (redrawn from [23] with modification).

Table 6-2. Minimum pressure head required to promote water flow [23] (Courtesy of Elsevier)

Media	Minimum pressure for flow in an initially dry gdl first trial/ second trial (Pa)	Pore radius based on minimum pressure (μm)	Minimum pressure for flow in an initially wet GDL (Pa)
Carbon paper (Toray)	5300/3300	25	1200
Carbon paper + 20% Teflon	7100/3800	21	3200
Carbon paper + 40% Teflon	7200/4800	21	4200
Carbon paper + 60% Teflon	7400/4700	21	4000
Carbon cloth	400/200	140	200
Carbon cloth + 20% Teflon	1800/850	140	500
Carbon cloth + 40% Teflon	2200/1500	130	800
Carbon cloth + 60% Teflon	2400/1200	130	850
E-TEK/ELAT electrode	14.000	10	11,000

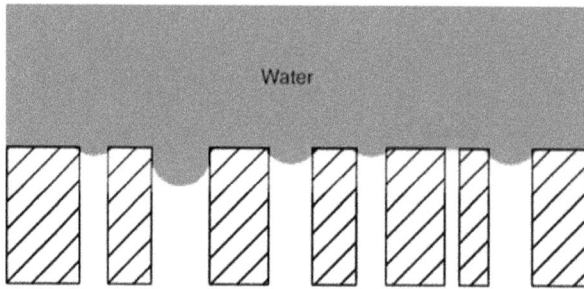

Figure 6-20. Water imbibition in an idealized model of a GDL that consists of parallel cylindrical pores of different diameters. By applying pressure, water is forced into the hydrophobic pores. The largest pores are penetrated first. The more hydrophobic the surface (or the larger the contact angle), the more pressure is required to force the water into the pores [23] (Courtesy of Elsevier).

$$\frac{\lambda_l \lambda_g}{\nu} K \frac{dP_c}{dx} = \frac{\lambda_l \lambda_g}{\nu} K \frac{dP_c}{ds} \frac{ds}{dx} = -D_c \frac{ds}{dx},$$ (Eq. 6-24)

where $D_c = -\frac{\sigma}{\nu} \cos\theta_c \left(K\varepsilon\right)^{1/2} k_{rl} \frac{dJ(s)}{ds}$. Thus, in the absence of any water source/sinks within a GDL and other driving mechanisms of liquid water besides the capillary pressure, no maximum or minimum of water saturation appears inside the GDL at steady state. Figure 6-21 presents 1D model prediction based on Eq. 6-24 under various conditions. The water saturation deceases monotonically from the left to right or from the catalyst-layer side to the channel due to the ORR's water production. However, the above is only valid for uniform GDL properties.

Table 6-3. Water permeability in various GDLs [23] (Courtesy of Elsevier)

Media	Initial hydrostatic pressure applied (kPa)	Water permeability at applied hydrostatic head of 1000 Pa, K (m^4 s/kg)
Carbon paper + 20% Teflon	7.5	0.5×10^{-10}
Carbon paper + 20% Teflon	10	2.8×10^{-10}
Carbon Paper + 20% Teflon	12.5	5.2×10^{-10}
Carbon paper + 60% Teflon	10	1.7×10^{-10}
Carbon paper + 60% Teflon	12.5	2.6×10^{-10}
Carbon cloth + 60% Teflon	3.6	3.5×10^{-10}
Carbon cloth + 60% Teflon	5.0	7.2×10^{-10}
Carbon cloth + 60% Teflon	10.0	11.1×10^{-10}

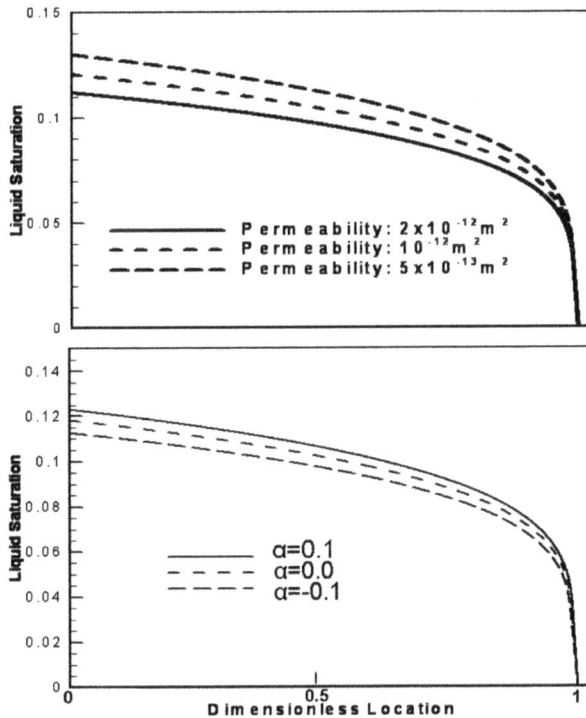

Figure 6-21. 1D liquid water profile across a cathode GDL with uniform property: (a) effect of permeabilities, (b) effect of the net water transfer coefficient α.

When property spatial variation is present, the above mathematic transform in Eq. 6-24 becomes invalid. In this case, extra terms must be added to account for this spatial variation based on the product rule of derivatives. For example, upon a spatial variation of permeability $K(x)$, the following term should be added to the right-hand side of Eq. 6-24 [24]:

$$\frac{\lambda_l \lambda_g \sigma \cos \theta_c \varepsilon^{1/2}}{\nu} J(s) K \frac{d}{dx}(K^{-1/2}). \qquad \text{(Eq. 6-25)}$$

This term may lead to local maximum/minimum in the water profile. It is also physically sound: for example, the media with larger pores trap more water. Such maximums and minimums were observed in both X-ray imaging and high-resolution neutron imaging experiments. Recently, Hinebaugh *et al.* [25] experimentally measured the local porosity of a carbon paper, showing a spatially varying porosity of GDLs as presented in Figure 6-22. The property variation is rooted in the GDL fabrication process: Carbon papers are common GDL materials based on carbon fibers. It is nonwoven with fibers tied by binders as shown in Figure 6-2, and the fiber dimension is around 10 μm at its cross-section. Both fibers and binders are randomly distributed, leading to local heterogeneity. In the carbon-paper GDL of a PEM fuel cell, the average pore diameter is around 10–30 μm, with large pores as large as 50 μm. As a result of porosity variation, the permeability K, a factor that has a great impact on two-phase flow, also varies spatially, as indicated by the Blake–Kozeny equation:

$$K = \frac{d^2}{150} \frac{\varepsilon^3}{(1 - \varepsilon)^2}. \qquad \text{(Eq. 6-26)}$$

where d is the nominal pore dimension. The calculated permeability profile is plotted in Figure 6-22 as well, given a uniform compression ratio (CR) of 0.72. By incorporating the permeability profile for both anode and cathode GDLs, Wang and Chen [24] obtained a liquid profile close to that probed by the X-ray imaging as displayed in Figure 6-23.

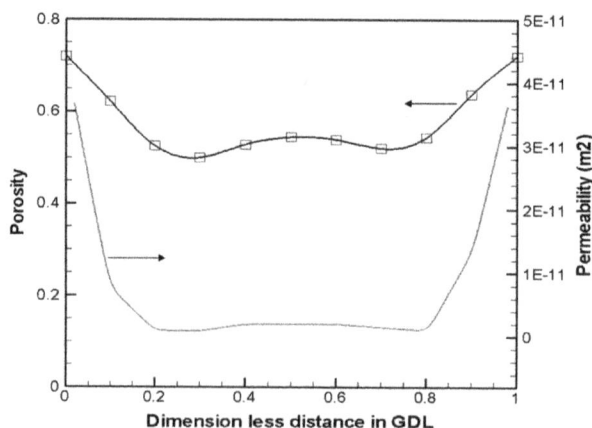

Figure 6-22. Spatial variation of porosity [25] and corresponding permeability in a GDL using the Blake–Kozeny equation (Courtesy of the Electrochemical Society).

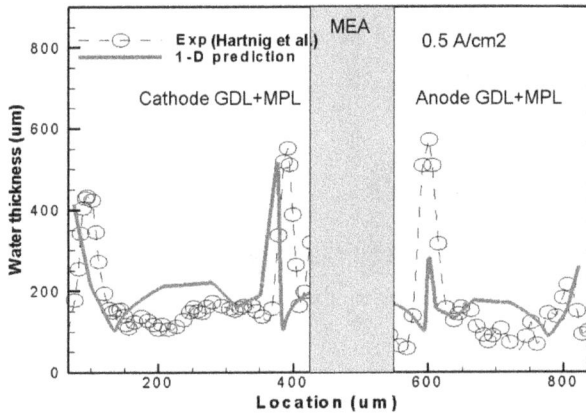

Figure 6-23. 1D prediction of liquid profile in the anode and cathode GDLs, and comparison with the X-ray imaging data [24] (Courtesy of the Electrochemical Society).

In addition to porosity and permeability, changes in other properties such as contact angle and surface tension result in similar effects, given that the capillary pressure P_c is a function of several parameters, that is, $P_c(\sigma, \theta_c, \varepsilon, K, s)$. A more general expression of the water flux driven by the capillary pressure accounting for spatially varying properties can be derived as follows [24]:

$$\frac{\lambda_l \lambda_g}{\nu} K \nabla P_c(\sigma, \theta_c, \varepsilon, K, s) = \frac{\lambda_l \lambda_g}{\nu} K \left(\frac{\partial P_c}{\partial \sigma} \nabla \sigma + \frac{\partial P_c}{\partial \theta_c} \nabla \theta_c + \frac{\partial P_c}{\partial \varepsilon} \nabla \varepsilon + \frac{\partial P_c}{\partial K} \nabla K + \frac{\partial P_c}{\partial s} \nabla s \right).$$

(Eq. 6-27)

6.2.4.2 In-Plane Property Variation and the Effect of Land Compression

The basic land-channel structure raises the in-plane gradient of water content. Tabuchi *et al.* [26] revealed that liquid saturation is higher under the lands through neutron radiography. Knowledge on the in-plane water distribution is essential to the understanding and optimization of PEM fuel cell structure.

In addition to the structural heterogeneity arising from the GDL fabrication, another cause of varying property stems from PEM fuel cell assembly–land compression. Upon compression, the GDL under lands reduces its thickness and hence porosity. The degree of compression or compression ratio (CR) can be defined as the ratios of the compressed and uncompressed GDL thicknesses. The resultant porosity is evaluated by the following expression:

$$\varepsilon = \frac{\varepsilon_0 - CR}{1 - CR},$$

(Eq. 6-28)

where ε_0 is the uncompressed porosity of a GDL. Equation 6-26 shows that local permeability deceases as a result of the reduced porosity. Figure 6-24 displays the liquid saturation

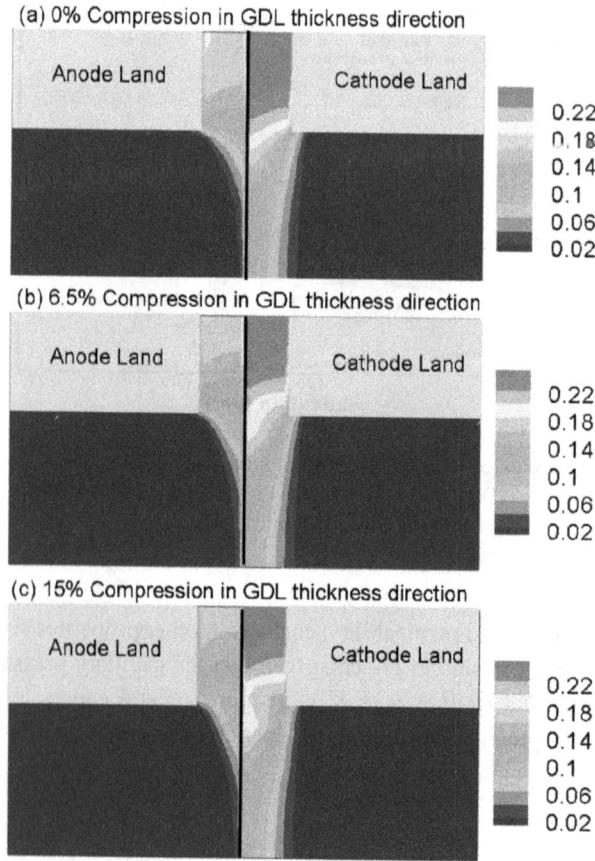

Figure 6-24. Liquid saturation contours at different GDL compression ratios under the operating condition of 0.8 A/cm^2, 80°C, and 2 atm [24] (Courtesy of the Electrochemical Society).

contours for three compression scenarios and shows that with compression the water saturation under the land decreases. For 15% compression (that is, the GDL thickness is reduced by ~15%), on the cathode side, a minimum in the liquid saturation is predicted near the interface between the land and channel. This can be explained by the GDL being more compact under the land, which contains lower liquid saturation in the pores to balance the liquid capillary pressure in the vicinity of the uncompressed portion of the GDL, that is, under the channel. The in-plane profiles of liquid water saturation plotted in Figure 6-25 for three land compressions show that land compression can greatly affect the liquid water profile under the land, particularly near the land-channel edge. For a GDL compression of 15%, discontinuity in liquid saturation near the edge occurs.

6.2.4.3 Microporous Layers (MPLs)

To improve the multiphase flow characteristics, a microporous layer or MPL is often added and placed between the GDL and CL. This layer is usually composed of carbon or graphite particles

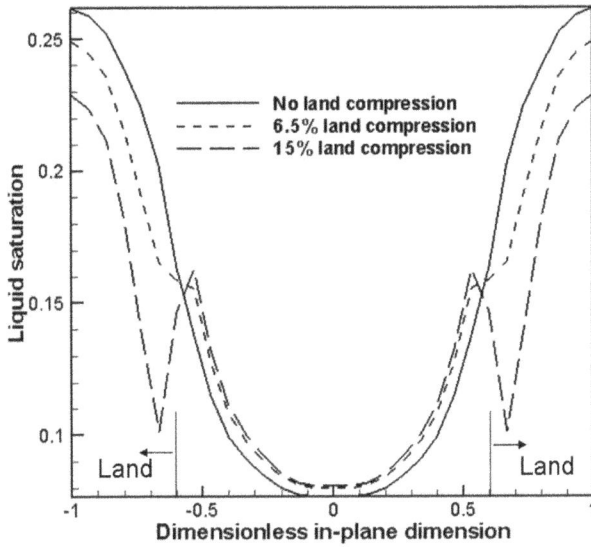

Figure 6-25. In-plane profiles of liquid saturation at different GDL compressions at 0.8 A/cm², 80°C, and 2 atm [24] (Courtesy of the Electrochemical Society).

mixed with polymeric binders, such as PTFE, with a fine pore structure. It has a pore dimension ranging from 0.1 to 1.0 μm, in comparison with carbon paper substrates that have mean pore size round 20 μm. MPLs can be fabricated by several methods, which follow that of applying catalyst pastes to substrates, or more specifically, using doctor blade, screen printing, spraying, and coating techniques. The solvent material, solid content, and substrate penetration can be altered to modify MPL properties. The solvent in the pastes must be evaporated slowly in order to prevent mud cracking. Heating is often applied to remove organics and to sinter the binder. Physical press is also used to push the paste into the substrate. MPLs are typically around 50 μm thick.

Many studies show that adding MPLs improves the water drainage characteristics and PEM fuel cell performance; (see Figure 6-26). Gostick *et al*. [28] reported that the liquid saturation in a GDL for water breakthrough is drastically reduced from ca. 25% to ca. 5% in the presence of a MPL. It was hypothesized that the MPL acts as a valve that keeps water away from electrodes to reduce electrode flooding. At the MPL–GDL interface, assumption of continuous pressure yields the following relation:

$$\sigma \cos(\theta_{c,\text{GDL}}) \left(\frac{\varepsilon_{\text{GDL}}}{K_{\text{GDL}}} \right)^{1/2} J(s) = P_{c,\text{GDL}} = P_{c,\text{MPL}} = \sigma \cos(\theta_{c,\text{MPL}}) \left(\frac{\varepsilon_{\text{MPL}}}{K_{\text{MPL}}} \right)^{1/2} J(s). \quad \text{(Eq. 6-29)}$$

The above adopts the Leverret relation. Generally, the MPL porosity and mean pore size are much smaller than those of the GDLs, respectively. The above equation is a key interfacial condition for the two-phase transport in GDLs and MPLs.

To shed light on the physical process, Fig. 6-27 shows the capillary pressure curves for two hydrophobic porous media. Porous medium I has a smaller characteristic capillary radius

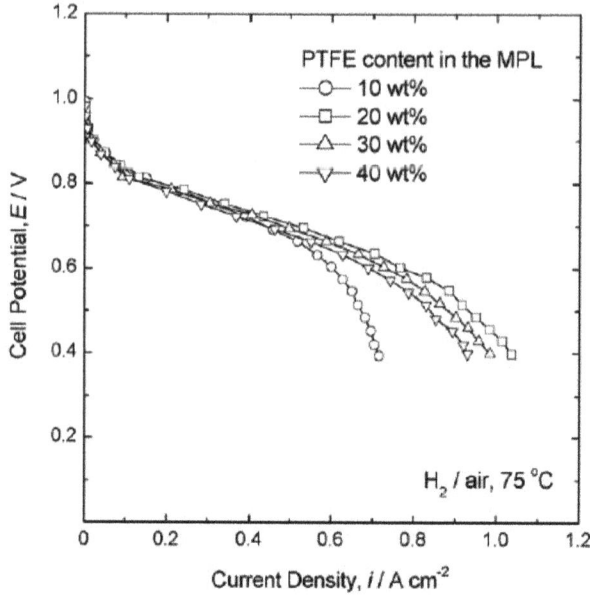

Figure 6-26. Polarization curves when using MPLs with different PTFE loadings [27] (Courtesy of Elsevier).

r, compared to medium II, which is achieved by changing the fiber diameter d or the porosity ε. When partially saturated (e.g., $s = 0.2$) porous media I and II are brought together, the saturation distribution in each porous medium evolves to make the capillary pressure P_c equal at the interface. This requires the transport of water from the medium I to II (fine to coarse medium), and the resultant saturation jump is shown in Figure 6-27 [5].

Figure 6-28 presents the P_c–s curves for GDL-MPLs under various arrangements. Regarding the plain GDL substrate material (SGL10BA), it is evident that the P_c–s relationship for water injection is significantly affected by the size of the GDL area directly in contact with water. Since the GDLs are thin (ca. 10–20 pores across), a significant fraction of the sample pore volume is invaded by non-percolating clusters before breakthrough if the entire inlet face of the sample is exposed to water (SGL10BA-F). In this case, water breakthrough occurs at $P_c \sim 2500$ Pa and a high saturation ($s \sim 0.27$). By comparison $P_c \sim 3800$ Pa and s ~ 0.05 when only a very small fraction of the GDL inlet face is exposed to water (SGL 10BA-P). This situation is illustrated in the left and middle panels of Figure 6-29 (shown on the following page).

The P_c–s relationship for water injection into SGL10BB is different from that for SGL10BA, and it depends on whether the MPL is placed on the inlet face (SGL10BB-I) or the outlet face (SGL10BB-O) of the sample. In the former case, the data are displaced towards higher capillary pressures, as expected since water invasion into the GDL is controlled by the smaller pore sizes of the intervening MPL. In the latter case, the P_c–s relationship in the low capillary pressure range is essentially the same as that in SGL10BA-F, as expected since water is first invading the substrate material in both cases. In either case (SGL10BB-I or SGL10BB-O), the maximum water saturation is about 75%, demonstrating that the MPL is not significantly filled with water at the maximum capillary pressure established. When the MPL is on the outlet face, water breakthrough in SGL10BB occurs at $P_c \sim 6700$ Pa and $s \sim 0.57$. Such high water saturation is easily understood since the GDL is filled substantially by water at the capillary pressure

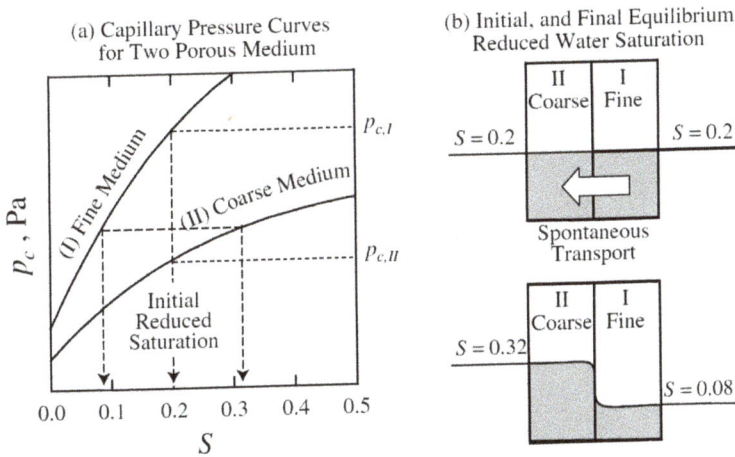

Figure 6-27. Saturation jump across the interface of porous media I and II, due to capillary action. Medium I has a finer pore structure than that of medium II [5] (Courtesy of Elsevier).

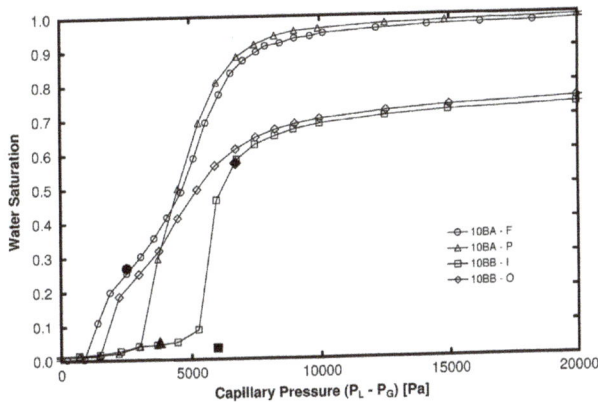

Figure 6-28. Capillary pressure curves for SGL10BA (plain GDL substrate) and SGL10BB (with MPL) with breakthrough points overlaid as filled markers with corresponding shapes. F: full face exposed to water, P: punctured mask between water and GDL, I: MPL facing water inlet, O: MPL facing away from water inlet [19] (Courtesy of the Electrochemical Society).

necessary for water to invade the smaller MPL pores. Water breakthrough in sample SGL10BB when the MPL is on the inlet face occurs at similarly high capillary pressure ($P_c \sim 6800$ Pa), but in this case $s \sim 0.03$ [19].

Water breaks through the MPL at a few isolated locations at the MPL–GDL interface; in the same way, water droplets emerge from the GDL at the GDL–channel interface, regardless of whether or not the MPL is cracked. Subsequent water percolation through the GDL starts from these locations, as opposed to the entire face of the GDL, as depicted in Figure 6-29 (right). This renders most pores on the GDL face inaccessible to water, dramatically decreasing the GDL

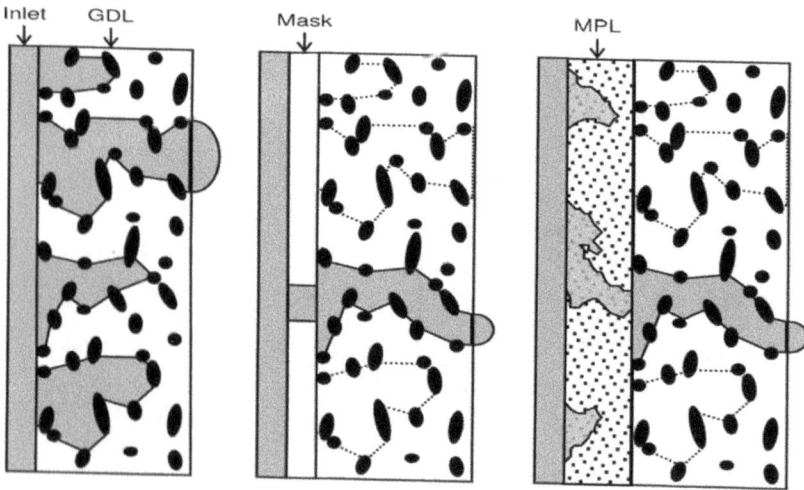

Figure 6-29. Proposed configuration of water in the GDL with different injection conditions. Left: full face injection leads to several dead-end clusters and a single breakthrough cluster. Middle: point source injection leads to a single breakthrough cluster. Right: percolation through the MPL creates conditions similar to point-like injection into the GDL. The gray area represents water and dashed line marks regions that would be filled if they were accessible [19] (Courtesy of the Electrochemical Society).

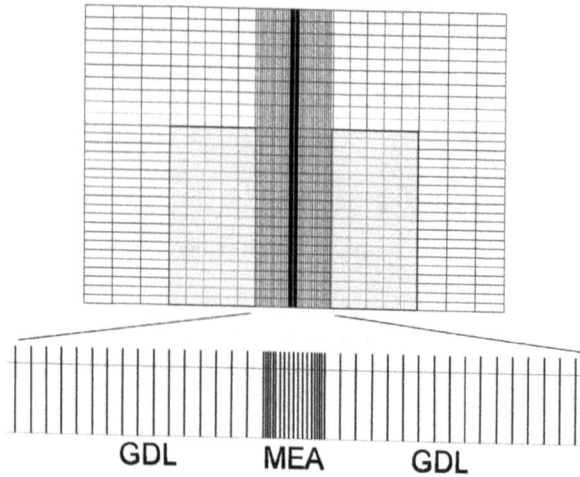

Figure 6-30. 2D computational domain of a PEM fuel cell. The MPL is part of the GDL.

saturation at breakthrough by reducing the number of dead end clusters (a finite-size effect). In this manner, reduced gas diffusivity is confined to a smaller part of the transport domain (the MPL), resulting in better fuel cell performance at high current densities.

In addition, one can follow the previous methodology to treat the MPL as part of the varying-property GDL; see the computational domain as shown in Figure 6-30. This scenario is closer to the real-world situation in which the MPL–GDL interface is not a 2D plane but a

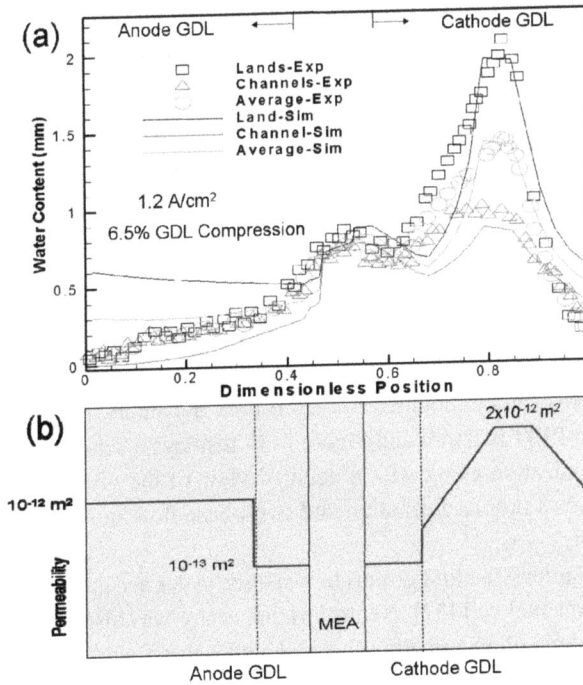

Figure 6-31. (a) Comparison of model prediction and neutron imaging data, and (b) spatially varying GDL property (including MPLs) used in the model prediction [24] (Courtesy of the Electrochemical Society).

transition zone in which the property changes sharply (not discontinuously) from one component to another. Figure 6-31 shows comparison between model prediction that accounts for spatial variation of the GDL (the MPL is part of the GDL) and neutron imaging data, along with a plot of the property variation used in the prediction.

Figure 6-31 (a) displays liquid water profiles for under a channel and land from both model prediction and neutron imaging. Inside the GDL, the curves exhibit maximum water contents under both land and channel locations in the neutron imaging data. In the model prediction, a 6.5% land-compression and varying GDL property in the through-plane direction are set as shown in Figure 6-31 (b) in order to achieve the best match with the experimental data. The location of the maximum water content occurs at the locally large permeability. The under-land area shows much higher water content than the under-channel region on the cathode side.

6.3   MULTIPHASE FLOW IN GAS FLOW CHANNELS (GFCS)

The multiphase flow in gas flow channels delivers gaseous reactants and removes product water. PEM fuel cell performance will be greatly reduced when considerable amount of liquid water is accumulated in the channels. Figure 6-32 shows the cell voltage variation over time (the blue or higher trace) for five air stoichiometric flow ratios (ξ) at the current density of 0.2 A/cm^2 in a PEM fuel cell having an active area of 14 cm^2. It can be seen that the cell voltage becomes oscillatory with a magnitude of ~120 mV at a stoichiometric flow ratio of 2. Thus, cathode flooding results in a performance loss (~120 mV) that completely negates any potential improvement from catalyst development: for instance, a 4-fold increase in catalytic activity yields only ~45 mV gain in cell voltage. Figure 6-32 also shows the variation of two-phase pressure drop as a function of time, which can be correlated with the cell voltage variation. Moreover, the voltage fluctuation induced by channel flooding may set up a voltage cycling at high potentials, which could result in serious durability issues.

6.3.1 EXPERIMENTAL VISUALIZATION

Liquid water in GFCs can be visualized optically by designing transparent fuel cells. This is done by replacing the bipolar plates with one transparent plate (side wall) plus a conductive channel plate. Cameras can be mounted to record flow dynamics. Figure 6-33 shows a photo of such a transparent PEM fuel cell and Figure 6-34 displays a schematic of the experimental setup for direct visualization along with a detailed view of the visualization device and flow filed. Figure 6-35 shows droplet formation and two-phase flow in an operating PEM fuel cell through visualization.

Tuber *et al.* was among the first groups to visualize water accumulation in a gas flow channel using a transparent fuel cell [32]. Neutron radiography was also employed to probe water retention in the flow field of an operating fuel cell, and it was found that higher water retention may occur at the U-bends of the serpentine channels [33].

6.3.2 TWO-PHASE FLOW PATTERNS

In PEM fuel cells, water is produced in the cathode through the oxygen reduction reaction (ORR). Liquid water emerges when the partial vapor pressure reaches its saturated value and eventually enters the channel via the GDL, resulting in air–water two-phase flow. Liquid water

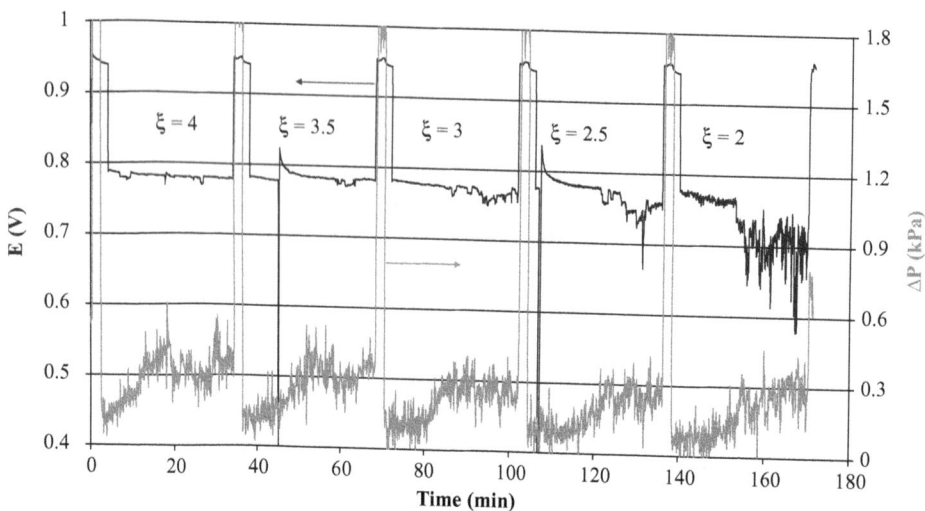

Figure 6-32. Cell voltage variation for different air stoichiometric ratios (0.2 A/cm², RHa/c = 70%/70%, 80°C, back pressure = 150 kPa) [29].

Figure 6-33. Specially designed, transparent, fuel cell direct-visualization apparatus [30].

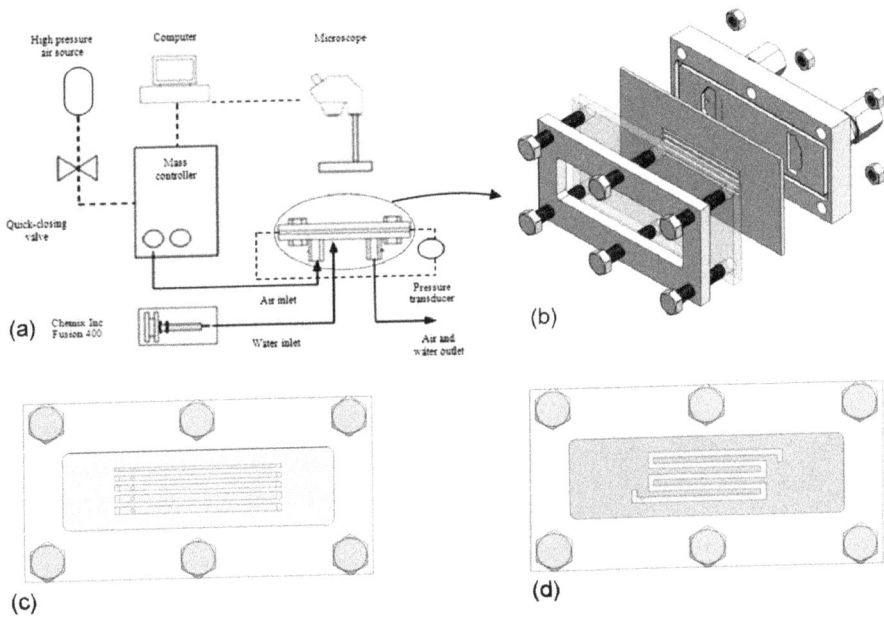

Figure 6-34. (a) Schematic of a transparent experimental setup for *ex situ* experiment of two-phase flow; (b) the experimental device; (c) the straight channel flow field; (d) the serpentine flow field [31].

Figure 6-35. Visualization of two-phase flow in PEM fuel cell channels [29].

can accumulate to a degree that greatly affects reactant supply. Through the net water transport coefficient α, the net water gain in the cathode is expressed by

$$S_w = (1+2\alpha)\frac{I}{2F} \quad \text{and} \quad \dot{m}_l = (1+2\alpha)\frac{IM_w A_{mem}}{2FA_{ch}}. \qquad \text{(Eq. 6-30)}$$

Assuming that all the water addition is in the liquid phase when entering the channels, integrating the above flux equation over a fuel cell yields the total liquid flow rate in the cathode channel. The inlet gas flow rate is determined by the stoichiometric flow ratio ξ_c:

$$u_g = u_{in,c}\Big|_{inlet} = \frac{\xi_c I A_{mem}}{4FC_{O_2} A_{ch}} \quad \text{and} \quad \dot{m}_g = \frac{\xi_c I A_{mem}\rho_g}{4FC_{O_2} A_{ch}}, \qquad \text{(Eq. 6-31)}$$

where A_{ch} and A_{mem} are the flow cross-sectional areas of the cathode GFCs and the membrane area, respectively. In the above derivation, it was assumed that the gas flow rate is constant along the channel. For typical conditions of 2 atm, 80°C, 1.5 stoichiometric flow ratio, and 1 A/cm^2 current density, the gas velocity is on the order of 1 m/s. In gas flow channels, flow regimes can be displayed using the superficial gas (U_G) and liquid (U_L) velocities as coordinates:

$$U_G = \frac{\dot{m}x_g}{\rho_g} \quad \text{and} \quad U_L = \frac{\dot{m}x_l}{\rho_l} \quad \text{where} \quad \dot{m} = \dot{m}_l + \dot{m}_g, \qquad \text{(Eq. 6-32)}$$

where x_g and x_l are the mass flow fraction of gas and liquid, respectively, and $\dot{m}$ the mass flux.

Figure 6-36 (a) displays a two-phase flow regime map for a GFC without GDLs (the GDL side is replaced by an aluminum plate). Figure 6-36 (b) shows several typical flow patterns in the experiment: as gas flow rate decreases, the two-phase flow experiences annulus, wavy annulus, wavy, and slug in the range of experiment conditions. The shear stress by the gas stream is a major force driving the liquid flow. In the occasion of insufficient gas flows, liquid accumulates, forming slug flow. When the gas flow increases, the two-phase stream becomes more stable with liquid spreading over the channel walls.

The frictional pressure drop (ΔP_f) is an important parameter characterizing channel two-phase flows, and its correlation can be developed using the concept of the two-phase flow multiplier ϕ_G^2, defined as follows:

$$\phi_G^2 = \frac{-\left(\dfrac{\partial P}{\partial z}\right)_{fr,2\phi}}{-\left(\dfrac{\partial P}{\partial z}\right)_{fr,1\phi}} \qquad \text{(Eq. 6-33)}$$

Chisholm [34] examined the Lockhart–Martinelli correlating procedure and developed the following pressure multiplier for pipe two-phase flows:

$$\phi_G^2 = 1 + CX + X^2 \qquad \text{where the Lockhart–Martinelli parameter}$$

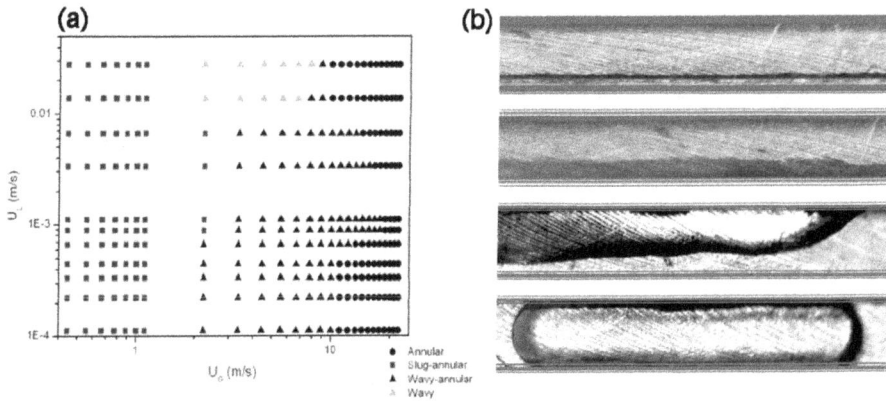

Figure 6-36. Flow patterns of two-phase flow in a straight channel with a smooth aluminum surface on the GDL side (a); from the top to bottom, the patterns are annular, wavy-annular, wavy, and slug-annular flows (b) [31].

$$X = \frac{\dot{m}_l}{\dot{m}_g} \sqrt{\frac{\rho_g}{\rho_l}}.$$ (Eq. 6-34)

Mishima *et al.* [35] indicated that the coefficient C depends on the microchannel diameter:

$$C = 21 \times (1 + e^{-0.319 D_h}),$$ (Eq. 6-35)

where the hydraulic diameter is defined as

$$D_h = \frac{4 \times \text{cross sectional area}}{\text{channel perimeter}}.$$ (Eq. 6-36)

Figure 6-37 displays the two-phase pressure amplifier and comparison between the Lockhart–Martinelli correlation and model prediction. Figure 6-38 (a) shows that the amplifier factor is slightly higher than one (or unity) at high gas flow rate and increases to around 10 at low gas flow rates. The liquid flow rate affects the amplifier as well. The model prediction using a two-fluid model (to be described in the next section) are also provided for higher gas flow rates. The model predictions are in line with the experimental data. For low gas flow rates, the model parameters must be changed, for example, as a function of different liquid flow rates, in order to fit the experimental data. This is likely due to the change in two-phase flow patterns. Figure 6-37 (b) compares the two-phase amplifier with the Lockhart–Martinelli correlation. The Mishima and Hibiki approach is adopted for the coefficient C in the Chisholm equation [34, 35]. The gas pressure drop, along with the flow rate, determines the pumping power for feeding the reactants. Another issue closely related to pressure drop is flow mal-distribution occurring among parallel channels. Assuming two parallel channels follow the same two-phase pressure amplifier, the two channels will have totally different reactant flows when subjecting to different flow regimes.

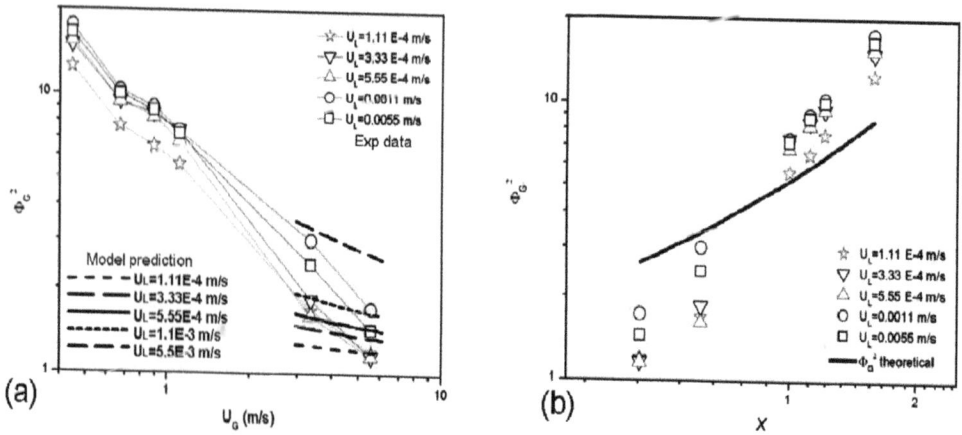

Figure 6-37. (a) ϕ_G^2 as a function of U_G and U_L and (b) comparison between the experimental and Lockhart–Martinelli values of ϕ_G^2 [31].

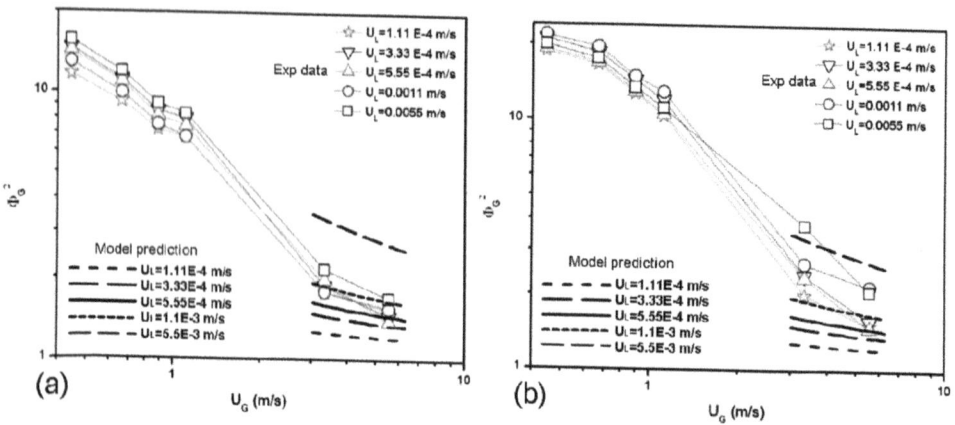

Figure 6-38. The two-phase pressure amplifiers for a straight channel with (a) carbon cloth and (b) carbon paper on the GDL side [31].

Figures 6-39 displays the experimental results of two-phase flow pattern for carbon paper and cloth, respectively, placed on the GDL side. Similar patterns are indicated as that in the aluminum surface.

6.3.3 MODELING TWO-PHASE FLOW

The reactant flow fields in an industry-size PEM fuel cell are featured by a number of parallel or serpentine channels with square, rectangular, semispherical, or trapezoid cross-sections and channel size around 0.5 mm; configurations, such as serpentine, parallel, and interdigitated flow fields. Figure 6-40 shows a typical serpentine flow field in industrial-size PEM fuel cells. Flow channels of similar structure are also encountered in miniature heat pipes and micro heat

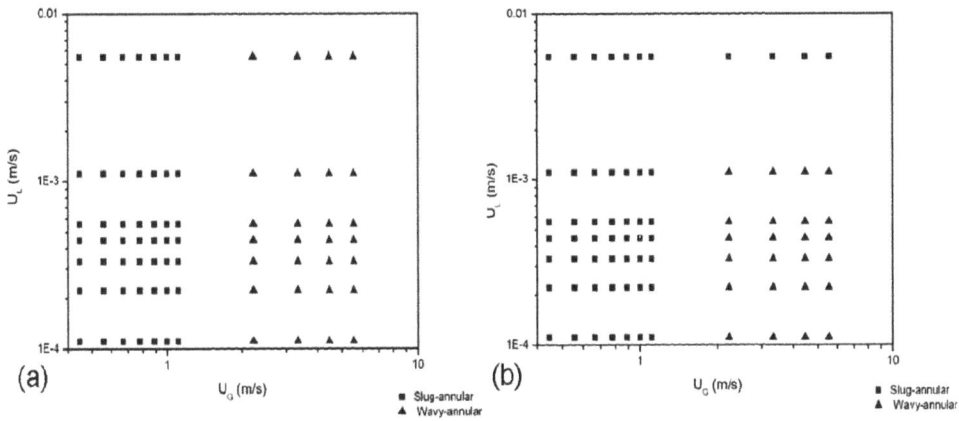

Figure 6-39. Flow patterns in the straight channel with carbon cloth (a) and paper (b) [31].

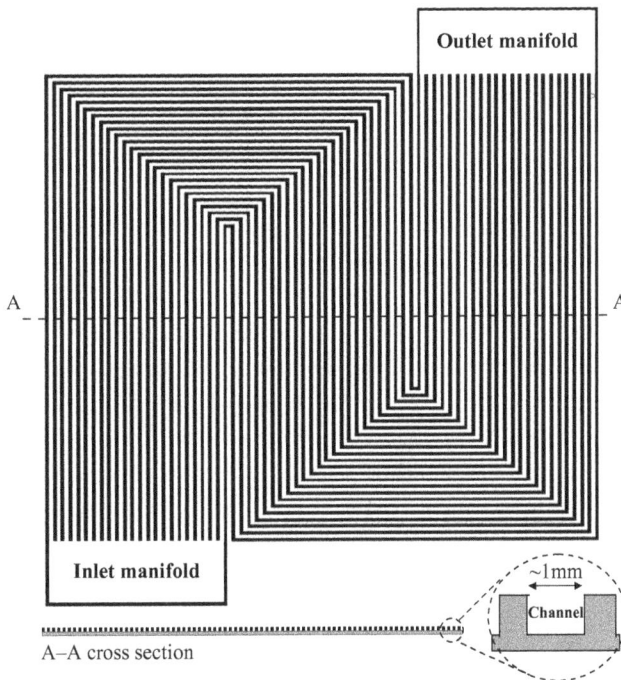

Figure 6-40. Flow field of an industrial PEM fuel cell.

exchangers. In these applications, evaporation and condensation occur to utilize the efficient latent heat transfer; therefore, two-phase flow is ubiquitous.

Another flow field worth mentioning is two-phase flow in geological media or petroleum reservoirs. Flow beds in these applications feature random solid particles with two-phase flow in pores. Figure 6-41 shows structures of several porous materials. Some porous formations feature a pore size range of ~1 mm, thus strongly resembling the PEM fuel cell channels. The structural similarity between the two, with the former being random porous media and the latter regular, ordered pores, was recognized by Chaouche *et al.* [36], Or and Tuller [37] and

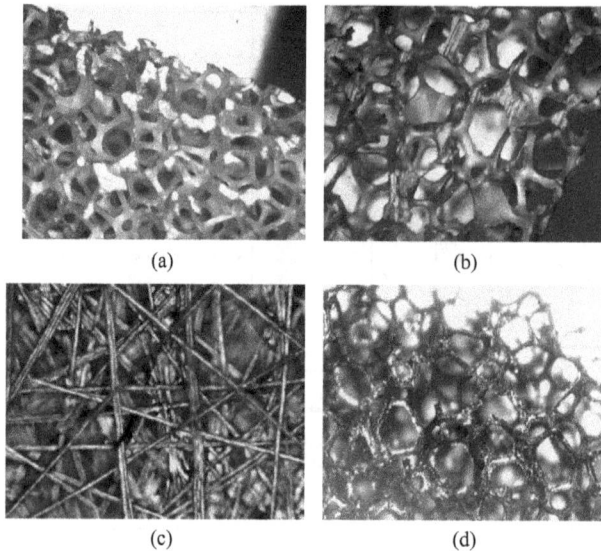

Figure 6-41. Structures of several porous media (a) Al foam; (b) Cu foam; (c) carbon paper; (d) carbon foam.

Li *et al*. [38]. In addition, there exists a flow analogy in that flow in PEM fuel cell channels is well within the laminar regime and exhibits a linear relationship between the pressure drop and velocity, the same relation as expressed by Darcy's law for flow in porous media. Indeed, the proportionality constant called hydraulic conductance in laminar flow through channels is physically equivalent to the permeability of a porous medium. Such a flow analogy was recognized by Chen [39], Wong *et al*. [40], Wang *et al*. [41], and Tio *et al*. [42], among others. In addition, the geometrical analogy between a porous medium and a capillary is traditionally employed to understand and develop flow theories in irregular porous media, in which disordered porous media with tortuous pore channels are simplified as a bundle of straight capillary tubes. Envisaging the geometrical and flow analogy between fuel cell channels and random porous media, a macroscopic model for two-phase flow in fuel cell channels can be developed using the two-phase Darcy's flow theory, as detailed in the next section. Similar approaches had been successfully implemented for miniature heat pipes and micro heat exchangers.

In this section, the focus is placed on a single straight channel in order to develop a channel two-phase flow model resolving axial variations in pressure, liquid saturation and other quantities. This is schematically shown in Figure 6-42 (a) for the cathode side. Partially or fully humidified air is fed in the inlet, and liquid water produced from ORR is injected into the channel from the sidewall facing the GDL. The channels can generally be treated as porous media with or without porous inserts. Without porous inserts, the pore size in the porous medium is exactly equal to the channel cross-section dimension with the pore torturosity and porosity equal to unity. This general porous approach is illustrated in Figure 6-42 (b) and Darcy's law is applied to describe the two-phase flow through the channels, provided that the Reynolds number based on the pore size is much smaller than 2,000 to ensure laminar flow. This condition is commonly met in typical PEM fuel cell operations. Appendix VI.B introduces the multiphase mixture model for multiphase flow in porous media.

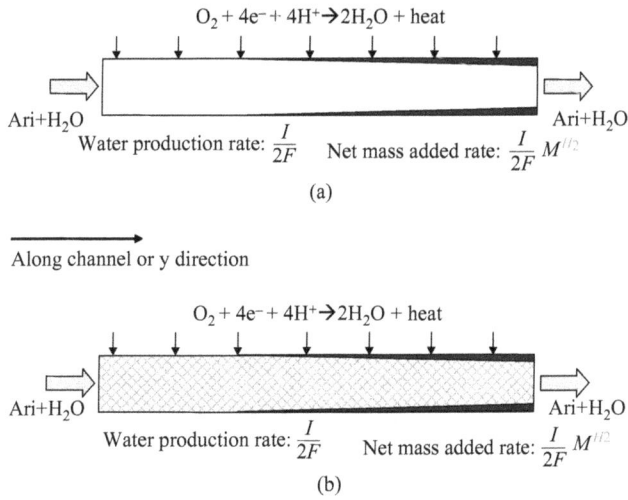

$O_2 + 4e^- + 4H^+ \rightarrow 2H_2O + heat$

Ari+H_2O

Ari+H_2O

Water production rate: $\dfrac{I}{2F}$ Net mass added rate: $\dfrac{I}{2F} M^{H_2}$

(a)

Along channel or y direction

$O_2 + 4e^- + 4H^+ \rightarrow 2H_2O + heat$

Ari+H_2O

Ari+H_2O

Water production rate: $\dfrac{I}{2F}$ Net mass added rate: $\dfrac{I}{2F} M^{H_2}$

(b)

Figure 6-42. Schematic of two-phase flow through a single channel of PEM fuel cells: (a) an open (hollow) channel; (b) a channel with porous inserts. The black shadowed regions on the upper or lower wall represent liquid water.

6.3.3.1 The Mixture Model

The primary species in GFCs are hydrogen, water, and nitrogen in the anode, and oxygen, nitrogen, and water in the cathode side. A general form of species transport equation for both single- and two-phase mixtures is given by

$$\varepsilon \frac{\partial C_k}{\partial t} + \nabla \cdot (\gamma_c \vec{u} C_k) = \nabla \cdot (D_k^{g,\text{eff}} \nabla C_k^g) - \nabla \cdot \left[\left(\frac{mf_l^k}{M_k^g} - \frac{C_k^g}{\rho_g} \right) \vec{j_l} \right], \quad \text{(Eq. 6-37)}$$

where γ_c is called the convection correction factor to correct the convective transport of the two-phase mixture due to difference between phase velocities. Assuming that other species, such as oxygen, hydrogen, and nitrogen, have negligible solubility in liquid water, γ_c can be expressed as

$$\gamma_c = \begin{cases} \dfrac{\rho}{C_w} (\dfrac{\lambda_l}{M_w} + \dfrac{\lambda_g}{\rho_g} C_{w,\text{sat}}) & \text{for water} \\[4mm] \dfrac{\rho \lambda_g}{\rho_g (1-s)} & \text{for other species.} \end{cases} \quad \text{(Eq. 6-38)}$$

The capillary action is a major force driving liquid water flow. Its flux $\vec{j_l}$ is placed in the second term on the right-hand side of Eq. 6-37. The capillary pressure p_c is defined as the pressure difference between gas and liquid phases, that is, $p_g - p_l$, determined by a number of factors such as surface tension σ, and contact angle, θ_c, pore dimension, and the liquid saturation, s. The capillary diffusion flux $\vec{j}_l$ and capillary pressure p_c can be calculated by

$$\vec{j}_l = \frac{\lambda_l \lambda_g}{\nu} K[\nabla p_c + (\rho_l - \rho_g)\vec{g}] \quad \text{where} \quad p_c = \sigma \cos(\theta_c)\left(\frac{\varepsilon}{K}\right)^{1/2} J(s), \quad \text{(Eq. 6-39)}$$

where $J(s)$ is the Leverett function, an empirical relation generally adopted for packed sand porous media. Here, we extend it to the structured porous medium representation of gas flow channels:

$$J(s) = \begin{cases} 1.417(1-s) - 2.120(1-s)^2 + 1.263(1-s)^3 & \text{for } \theta_c < 90° \\ 1.417s - 2.120s^2 + 1.263s^3 & \text{for } \theta_c > 90° \end{cases} \quad \text{(Eq. 6-40)}$$

GDLs are made hydrophobic by adding PTFE to avoid electrode flooding. Other channel walls are usually hydrophilic, enabling wicking of liquid water from the GDL surface. The following three issues are important to understanding the two-phase flow in PEM fuel cells: (a) water build up, (b) heterogeneous channel configuration, and (c) two-phase pressure drop.

(1) Two-Phase Transport Along GFCs

At steady state, integrating the 1D continuum equation along the channel yields

$$A_{xz}\rho u_y = \left(A_{xz}\rho u_y\right)\big|_{\text{in}} + \int_0^y S_m L_{z,\text{mem}} d\bar{y} = \left(A_{xz}\rho u_y\right)\big|_{\text{in}} + L_{z,\text{mem}} \frac{\int_0^y I(\bar{y})d\bar{y}}{2F} M_{H2}, \quad \text{(Eq. 6-41)}$$

where $(\ )\big|_{\text{in}}$ represents the inlet value of a quantity and S_m denotes the mass source due to the ORR. The net mass source accounts for water addition minus oxygen consumption. Since the capillary effect and diffusion along the channel are small relative to convection and thus negligible, integration of the 1D species water equation results in

$$A_{xz}\gamma_c u_y C_w = \left(A_{xz}\gamma_c u_y C_w\right)\big|_{\text{in}} + \int_0^y S_w L_{z,\text{mem}} d\bar{y}$$

$$= \left(A_{xz}\gamma_c u_y C_w\right)\big|_{\text{in}} + L_{z,\text{mem}} \frac{\int_0^y I(\bar{y})d\bar{y}}{2F}, \quad \text{(Eq. 6-42)}$$

where S_W denotes the ORR's water production rate. Substituting Eq. 6-41 into Eq. 6-42 to cancel the mixture velocity u_y on the left-hand side yields

$$\frac{\gamma_c C_w}{\rho} = \frac{\left(A_{xz}u_y C_w\right)\big|_{in} + L_{z,mem} \dfrac{\displaystyle\int_0^y I(\bar{y})d\bar{y}}{2F}}{\left(A_{xz}\rho u_y\right)\big|_{in} + L_{z,mem} \dfrac{\displaystyle\int_0^y I(\bar{y})d\bar{y}}{2F} M_{H_2}}. \qquad \text{(Eq. 6-43)}$$

Using the inlet condition, one can further simplify the above equation to

$$\frac{\gamma_c C_w}{\rho} = \frac{\left(\dfrac{\xi_c}{2C_{O_2}} C_w\right)\Big|_{in} + \dfrac{\displaystyle\int_0^Y I(\bar{Y})d\bar{Y}}{I_{av}}}{\left(\dfrac{\xi_c}{2C_{O_2}} \rho\right)\Big|_{in} + \dfrac{\displaystyle\int_0^Y I(\bar{Y})d\bar{Y}}{I_{av}} M_{H_2}}. \qquad \text{(Eq. 6-44)}$$

Here, Y represents the dimensionless distance from the inlet (see Fig. 6-42), namely:

$$Y = \frac{y}{L_y} = \frac{y}{A_{mem}} L_{z,mem}. \qquad \text{(Eq. 6-45)}$$

The average current density I_{av} is calculated by integrating local current density $I(Y)$:

$$I_{av} = \int_0^1 I(\bar{Y})d\bar{Y}. \qquad \text{(Eq. 6-46)}$$

Note that the left-hand side of Eq. 6-44 is solely determined by liquid water saturation whereas the right-hand side is a function of operating parameters, such as ξ_c and the inlet humidity, as well as the axial location Y. Given the expressions of the mixture density and advection correction factor, Eq. 6-38, the left-hand side can be further rearranged as

$$\frac{\gamma_c C_w}{\rho} = \begin{cases} \dfrac{C_w}{\rho_g} & Y < Y_o \\[2ex] \left(\dfrac{1}{M_w} - \dfrac{1}{\rho_g} C_{w,sat}\right)\lambda_l + \dfrac{1}{\rho_g} C_{w,sat} & Y_o < Y \end{cases}, \qquad \text{(Eq. 6-47)}$$

where Y_0 is the position of transition from the single- to two-phase flows, that is, when $C_w\big|_{Y_0} = C_{w,sat}$. That is, Y_0 can be determined by setting C_w on the left-hand side of Eq. 6-44 to be $C_{w,sat}$:

$$\frac{C_{w,\text{sat}}}{\rho_g} = \frac{\left.\left(\dfrac{\xi_c}{2C_w} C_w\right)\right|_{\text{in}} + \dfrac{\displaystyle\int_0^{Y_0} I(\bar{Y})d\bar{Y}}{I_{\text{av}}}}{\left.\left(\dfrac{\xi_c}{2C_w}\rho\right)\right|_{\text{in}} + \dfrac{\displaystyle\int_0^{Y_0} I(\bar{Y})d\bar{Y}}{I_{\text{av}}} M_{\text{H2}}}. \qquad \text{(Eq. 6-48)}$$

In the region of $Y < Y_0$, only single-phase flow is present. Note that ρ and ρ_g represent the mixture density and gaseous density, respectively. The water concentration is calculated through Eqs. 6-47 and 6-38. Other variables, such as the O_2 concentration and mixture velocity, can be readily obtained in a similar fashion.

Under a constant current density, Y_0 is explicitly expressed as

$$Y_0 = \frac{\xi_c}{2} \frac{\left.\left[\dfrac{1}{C_{O_2}}\left(\dfrac{C_{w,\text{sat}}}{\rho_g}\rho - C_w\right)\right]\right|_{\text{in}}}{1 - \dfrac{C_{w,\text{sat}}}{\rho_g} M_w}. \qquad \text{(Eq. 6-49)}$$

As $\xi_c \to \infty$, $Y_0 \to \infty$ unless $C_w|_{\text{in}} \geq C_{w,\text{sat}}$. The electrochemical reaction imposes little effect ($<5\%$) on the cathode gas density and flow velocity under common operation. Therefore, Eq. 6-49 can be further simplified to calculate Y_o by neglecting mass injection and gas density variation.

In the region of $Y > Y_0$, liquid emerges, resulting in two-phase flow in the channel. Substituting Eq. 6-38 into Eq. 6-47 yields

$$\lambda_l = \frac{\left.\dfrac{\xi_c}{2}\left(\dfrac{C_w}{C_{O_2}}\right)\right|_{\text{in}} + \dfrac{\displaystyle\int_0^Y I(\bar{Y})d\bar{Y}}{I_{\text{av}}} - \dfrac{1}{\rho_g}C_{w,\text{sat}}\left(\left.\dfrac{\xi_c}{2}\left(\dfrac{\rho}{C_{O_2}}\right)\right|_{\text{in}} + \dfrac{\displaystyle\int_0^Y I(\bar{Y})d\bar{Y}}{I_{\text{av}}} M_{\text{H2}}\right)}{\left(\dfrac{1}{M_w} - \dfrac{1}{\rho_g}C_{w,\text{sat}}\right)\left(\left.\dfrac{\xi_c}{2}\left(\dfrac{\rho}{C_{O_2}}\right)\right|_{\text{in}} + \dfrac{\displaystyle\int_0^Y I(\bar{Y})d\bar{Y}}{I_{\text{av}}} M_{\text{H2}}\right)}. \qquad \text{(Eq. 6-50)}$$

The liquid mobility λ_l is a function of liquid water saturation s, as shown in Figure 6-43. Through rearrangement from Eq. 6-47, one can express the liquid water saturation explicitly through the liquid mobility λ_l:

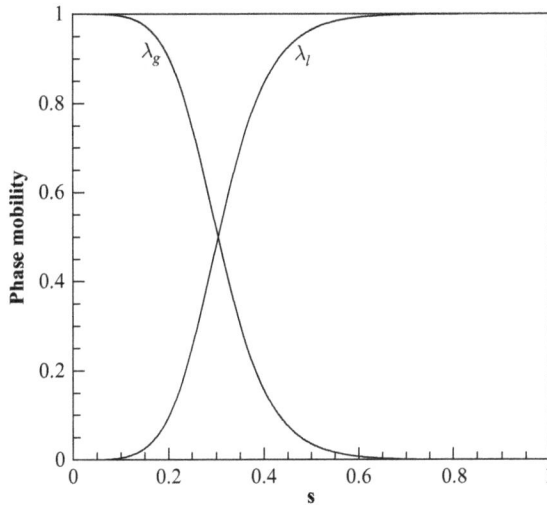

Figure 6-43. The phase mobility as a function of liquid water saturation ($n_k = 4$).

$$\frac{s - s_{ir}}{1 - s_{ir}} = \frac{1}{\left(\dfrac{1 - \lambda_l}{\lambda_l}(v_g / v_l)\right)^{1/n_k} + 1}. \quad \text{(Eq. 6-51)}$$

Substituting Eq. 6-50 into Eq. 6-51 yields an explicit solution of water saturation as a function of operating condition and axial location. Note that the effect of residual saturation S_{ir} is accounted for in Eq. 6-51.

Once the liquid saturation s is known, the mixture density can be determined and further be substituted to compute the mixture velocity:

$$u_y = \frac{(\rho u_y)\,|_{\text{in}} + \dfrac{\displaystyle\int_0^Y I(\bar{Y})d\bar{Y}}{2F} M_{H_2}}{\rho}. \quad \text{(Eq. 6-52)}$$

For constant current density, Eq. 6-50 changes to

$$\lambda_l = \frac{\dfrac{\xi_c}{2}\left.\left(\dfrac{C_w}{C_{O_2}}\right)\right|_{\text{in}} + Y - \dfrac{1}{\rho_g}C_{w,\text{sat}}\left(\left.\dfrac{\xi_c}{2}\left(\dfrac{\rho}{C_{O_2}}\right)\right|_{\text{in}} + YM_{H_2}\right)}{\left(\dfrac{1}{M_w} - \dfrac{1}{\rho_g}C_{w,\text{sat}}\right)\left(\left.\dfrac{\xi_c}{2}\left(\dfrac{\rho}{C_{O_2}}\right)\right|_{\text{in}} + YM_{H_2}\right)}. \quad \text{(Eq. 6-53)}$$

For fully humidified inlet reactant flows, Eq. 6-53 is further simplified to

$$\lambda_l = \frac{\left(1 - \dfrac{1}{\rho_g} C_{w,sat} M_{H_2}\right)}{\left(\dfrac{1}{M_{H_2}} - \dfrac{1}{\rho_g} C_{w,sat}\right)\left(\dfrac{\xi_c}{2}\left[\dfrac{\rho}{C_{O_2}}\right]\bigg|_{in} + Y M_{H_2}\right)} Y. \tag{Eq. 6-54}$$

In the above equation, $\xi_c \to \infty$ leads to $\lambda_l \to 0$ and hence $s \to s_{ir}$ as shown in Eq. 6-51. For the oxygen equation, following a similar integration, one reaches

$$A_{xz}\gamma_c u_y C_{O_2} = A_{xz}\left(\gamma_c u_y C_{O_2}\right)\big|_{in} + \int_0^y S_{O_2} L_{z,mem} dy = A_{xz}\left(\gamma_c u_y C_{O_2}\right)\big|_{in}$$

$$- L_{z,mem}\frac{\displaystyle\int_0^y I(\overline{y})d\overline{y}}{4F}. \tag{Eq. 6-55}$$

Thus,

$$C_{O_2} = \frac{\left(\gamma_c u_y C_{O_2}\right)\big|_{in} - \dfrac{\displaystyle\int_0^Y I(\overline{Y})d\overline{Y}}{4F}}{\gamma_c u_y}. \tag{Eq. 6-56}$$

Once the operating conditions and liquid saturation are known, the oxygen concentration can be determined using the above equation.

Liquid water saturation s is a key parameter characterizing two-phase flow. Figure 6-44 plots the liquid saturation profiles along a channel under fully humidified operation computed from Eq. 6-51 at 80°C and 100 kPa. The irreducible residual saturation S_{ir} is arbitrarily set to zero. Since the inlet reactant is already fully humidified, liquid water emerges at the beginning of the channel stream. The saturation rises drastically right after liquid emerges, followed by a gradual increase downstream. One reason for the trend is that the mobility of liquid water λ_l strongly depends on liquid saturation, as shown in Figure 6-43, with λ_l being approximately proportional to S^{nk}. The liquid mobility becomes vanishingly small at low saturations, and therefore liquid water builds up quickly as shown in Fig. 6-44. In addition, the larger the stoichiometric flow ratio ξ_c or the gas flow rate, the lower the saturation level. This can be explained by that higher air velocity is efficient in draining liquid water. At $\xi_c = 2.0$, the predicted saturation can reach as high as 20% near the outlet, in good agreement with many experimental observations. Furthermore, the cross-section-averaged saturation (which was computed from 3D numerical simulation by solving the partial differential equations) is plotted in the same figure; good agreement is seen between analytical solution and 3D model prediction, indicating validity of the assumptions made in deriving the analytical solutions.

Figure 6-45 (a) shows that liquid flow speeds up along the channel, as a result of water production by the ORR. As the inlet gas reactant is fully saturated, the flow channel totally relies

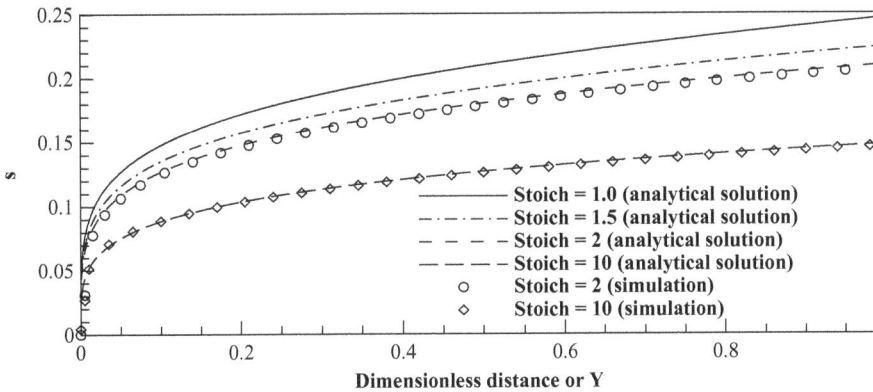

Figure 6-44. The liquid saturation along a GFC, computed from both the analytical solutions and a 3D model.

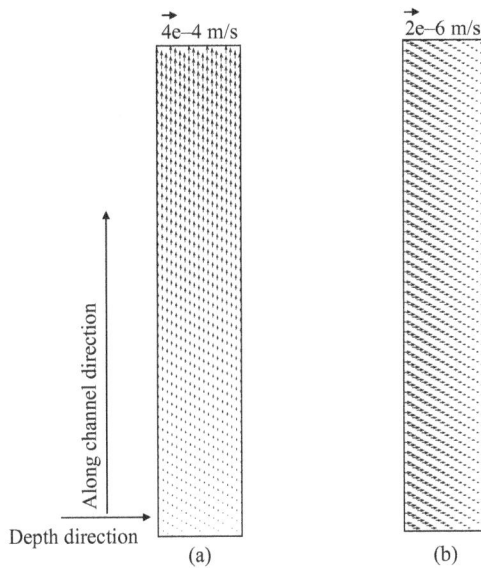

Figure 6-45. Simulation results of (a) liquid velocity and (b) liquid velocity component in the depth direction.

on the liquid phase to remove product water. In addition, the liquid axial velocity is small along the channel with a magnitude of 10^{-4} m/s. This velocity magnitude yields a time constant of ~1000 s for water drainage or liquid accumulation in gas flow channels, given a channel length of around 0.1 m. This residence time is consistent with experimental observation on fuel cell transients, and much larger than that in species diffusion, membrane hydration, GDL drying, and cold start. Figure 6-46 shows liquid saturation profiles under various distributions of local current density as computed from the analytical solution. As current density is proportional to water production, different current density profiles yield different two-phase flows as shown in Figure 6-46. Because the average current densities are the same, the liquid water saturations at the outlet are equal for all the cases.

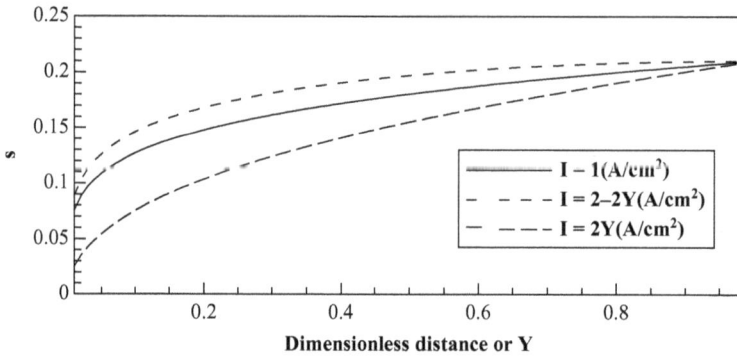

Figure 6-46. Analytical solution of liquid saturation profiles under different profiles of current density distributions.

(2) Heterogeneity in Flow Channels

The foregoing results and discussion are focused on flow channels with uniform cross-section and surface wettability. In practice, many PEM fuel cell designs have nonuniform channels which feature either geometrical heterogeneity or surface wettability variation along the flow, or both. The two types of heterogeneity are considered together and regarded generally as a spatial variation in channel permeability (or hydraulic conductance) and surface contact angle. For example, a branch-merge channel can be treated as a porous medium with differing permeability or hydraulic conductance, as shown in Figure 6-47. This branch-merge flow field is a popular design in PEM fuel cells due to its ability to mix and re-distribute reactants in channels, thereby alleviating the tendency of flow mal-distribution. However, as shown in Figure 6-47, this channel design has a problem of water trapping at the transition point. In addition, a branch of parallel channels in a PEM fuel cell usually share the same inlet or outlet, also called manifold as shown in Figure 6-40, which resembles either the merge or branched pattern.

For liquid water in homogeneous channels, the term that describes capillary action can be written, through substituting Eq. 6-39 into the second term on the right-hand side of Eq. 6-37, as

$$\nabla \cdot \left[\left(\frac{mf_l^k}{M_k} - \frac{C_k^g}{\rho_g} \right) \vec{j}_l \right] = \nabla \cdot \left[\left(\frac{mf_l^k}{M_k} - \frac{C_k^g}{\rho_g} \right) \frac{\lambda_l \lambda_g}{\nu} K \nabla P_c \right]. \qquad \text{(Eq. 6-57)}$$

The above term can be rearranged to a diffusion term as follows:

$$\nabla \cdot \left[\left(\frac{mf_l^k}{M_k} - \frac{C_k^g}{\rho_g} \right) \vec{j}_l \right] = \nabla \cdot \left[\left(\frac{mf_l^k}{M_k} - \frac{C_k^g}{\rho_g} \right) \frac{\lambda_l \lambda_g}{\nu} K \frac{dP_c}{ds} \nabla s \right]. \qquad \text{(Eq. 6-58)}$$

Thus, similar to the diffusion process, the capillarity force drives water from a higher liquid saturation to lower saturation regions. In heterogeneous porous media, however, the above arrangement needs to take into account the spatial derivation of porous properties as follows:

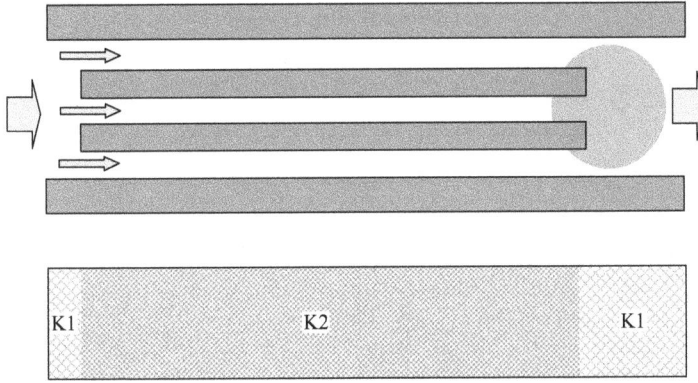

Figure 6-47. Schematic of a branch-merge flow field and the corresponding porous channel model ($K_1 > K_2$). The cycle shadow represents liquid water trapping at the heterogeneity.

$$\nabla \cdot \left[\left(\frac{mf_l^k}{M_k} - \frac{C_k^g}{\rho_g} \right) \frac{\lambda_l \lambda_g}{\nu} K \nabla P_c(\sigma, \theta_c, \varepsilon, K, s) \right] =$$

$$\nabla \cdot \left[\left(\frac{mf_l^k}{M_k} - \frac{C_k^g}{\rho_g} \right) \frac{\lambda_l \lambda_g}{\nu} K \left(\frac{\partial P_c}{\partial \sigma} \nabla \sigma + \frac{\partial P_c}{\partial \theta_c} \nabla \theta_c + \frac{\partial P_c}{\partial \varepsilon} \nabla \varepsilon + \frac{\partial P_c}{\partial K} \nabla K + \frac{\partial P_c}{\partial s} \nabla s \right) \right].$$

(Eq. 6-59)

It is seen that any spatial changes in surface tension σ, contact angle θ_c, porosity ε or permeability K add extra terms into capillary action in addition to the capillary diffusion under saturation gradient. For the sake of simplicity, we use heterogeneity in the permeability (or hydraulic conductance) $K(x, y, z)$ as an example. The additional term becomes

$$\nabla \cdot \left[\left(\frac{mf_l^k}{M_k} - \frac{C_k^g}{\rho_g} \right) \frac{\lambda_l \lambda_g}{\nu} K \left(\frac{\partial P_c}{\partial K} \nabla K \right) \right] =$$

$$\nabla \cdot \left[\left(\frac{mf_l^k}{M_k} - \frac{C_k^g}{\rho_g} \right) \frac{\lambda_l \lambda_g \sigma \cos(\theta_c) \varepsilon^{1/2}}{\nu} J(s) K \nabla(K^{-1/2}) \right].$$

(Eq. 6-60)

Assuming a linear profile of $K(y)$ along the channel, as shown in Figure 6-48:

$$K = k_1 y + k_2.$$

(Eq. 6-61)

One can reach

$$K \nabla(K^{-1/2}) = K \begin{pmatrix} 0 \\ -\frac{1}{2} k_1 K^{-\frac{3}{2}} \\ 0 \end{pmatrix} = -\frac{1}{2} k_1 \begin{pmatrix} 0 \\ K^{-\frac{1}{2}} \\ 0 \end{pmatrix}.$$

(Eq. 6-62)

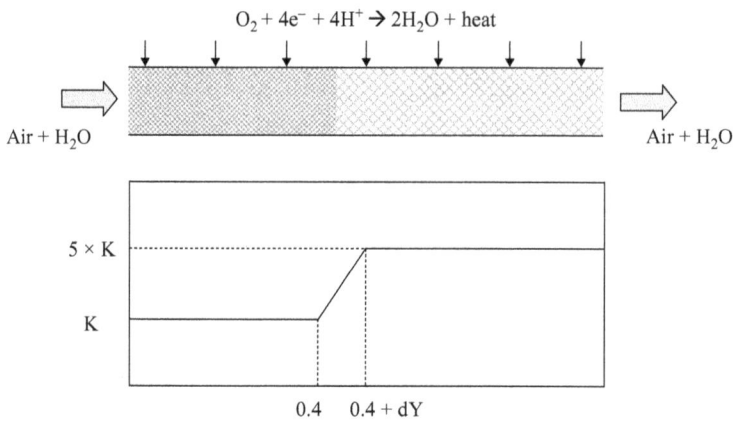

Figure 6-48. Schematic of the permeability heterogeneity in the analysis.

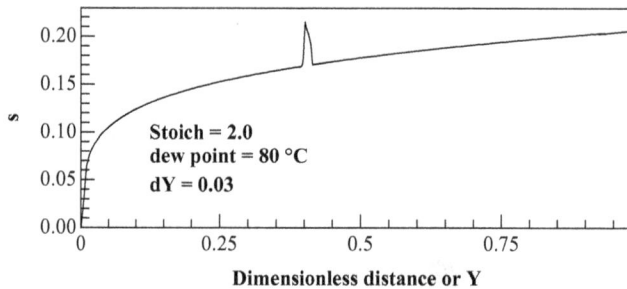

Figure 6-49. Liquid water saturation in a heterogeneous channel.

Figure 6-49 shows the liquid saturation along a heterogeneous channel with its permeability varying as given in Figure 6-48. The saturation profile remains almost the same as that in a homogeneous channel except of a spike occurring near the heterogeneity (i.e., the region with sharp increase in permeability). Thus, the heterogeneity affects liquid saturation locally where the property changes abruptly. This localized phenomenon can be readily explained by Eq. 6-59 in which the extra term acts like an additional capillarity force to hold up or repel water. On the global scale, since the extra terms in Eq. 6-59 are mathematically in conservative form, no extra water is actually added besides the water produced from ORR. Finally, it is noteworthy that under high liquid saturation, the localized spike may cause channel clogging and flow shutdown in a multiple, parallel channel configuration.

Note that though Eq. 6-59 is derived for isotropic porous media, similar analysis can be performed for anisotropic media. More discussion on two-phase flows in heterogeneous porous media can be found in the fields of petroleum and geological systems.

(3) Two-Phase Pressure Drop

Because the presence of liquid hinders gas flow, the gas pressure gradient in the two-phase regime is greater than that in the single-phase for given flow geometry. The gas pressure gradient is linked to the gas velocity through Darcy's law as follows:

$$u_g = -\frac{K(1-s)^{n_k}}{\mu_g L_y}\frac{dp_g}{dY}.$$ (Eq. 6-63)

In view of the nitrogen content of ~79% in dry air, the gas velocity can be taken to remain nearly constant along the channel and approximately equal to the inlet value, $u_g \approx u_{in,c}$. Thus, Eq. 6-63 can be rearranged to

$$\frac{dp_g}{dY} = -\frac{I_{av}A_{mem}}{4F}\frac{1}{C_{O_2}A_{xz}}\frac{\mu_g L_y}{K}\frac{\xi_c}{(1-s)^{n_k}}.$$ (Eq. 6-64)

If a dimensionless pressure is defined as

$$|\Delta P| = \frac{|\Delta p_g|}{\dfrac{I_{av}A_{mem}}{4F}\dfrac{1}{C_{O_2}A_{xz}}\dfrac{\mu_g L_y}{K}}.$$ (Eq. 6-65)

One obtains that

$$|\Delta P| = \xi_c \int_0^1 \frac{1}{(1-s_{(Y,\xi_c,RH_{in})})^{n_k}}dY = |\Delta P(\xi_c)|.$$ (Eq. 6-66)

Equation 6-66 shows that the dimensionless pressure drop is only a function of the air stoichiometric flow ratio (ξ_c) and inlet humidity (RH) based on the model. In the single-phase regime, Eq. 6-66 is reduced to

$$|\Delta P^{1\varphi}| = \xi_c.$$ (Eq. 6-67)

Thus, a two-phase pressure-drop factor ϕ_G^2 can be defined as

$$\phi_G^2 = \frac{|\Delta P|}{|\Delta P^{1\phi}|} = \int_0^1 \frac{1}{[1-s(Y,\xi_c,RH_{in})]^{n_k}}dY.$$ (Eq. 6-68)

Equation 6-68 indicates that the two-phase pressure amplifier ϕ_G^2 depends on the air stoichiometric flow ratio and inlet RH. Figure 6-50 shows the analytical result of the two-phase pressure factor as a function of stoichiometric flow ratio for different inlet dew points or RHs. At low stoichiometric flow ratios, that is, low channel flow rates, the two-phase pressure factor is greater than unity due to the presence of liquid water. As the gas flow rate increases, the onset site of two-phase flow shifts further downstream and hence the two-phase pressure factor is lower, approaching unity. Also, Eq. 6-68 clearly indicates that the exponent n_k is a key factor that determines the two-phase pressure drop.

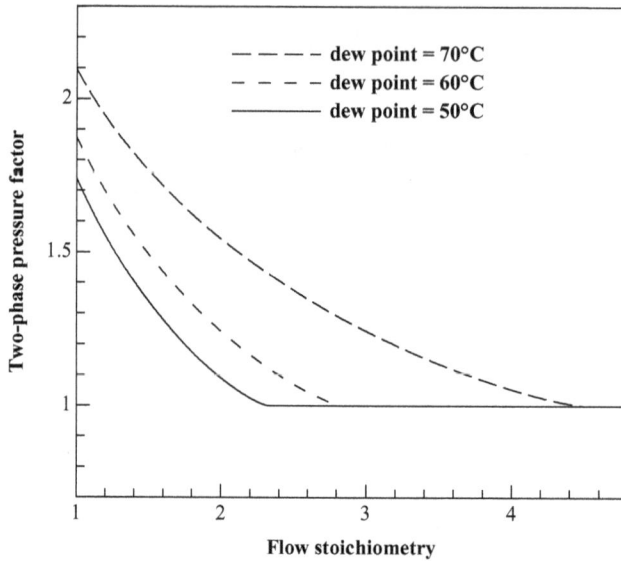

Figure 6-50. The two-phase pressure amplifier versus the stoichiometric flow ratio in a single flow channel ($n_k = 4$).

6.3.3.2 Two-Fluid Modeling

In the two-fluid modeling approach, flow of each fluid is formulated separately with the phase change rate explicitly appearing in the governing equations. A two-fluid model consists of two separate sets of flow equations, respectively, for the gas and liquid phases. Appendix VI.A summarizes the general governing equations of multiphase flow and heat transfer in porous media, along with examples for two-fluid equations in PEM fuel cells. The mass exchange occurs at the phase interface when water evaporation or condensation occurs, and the term describing this mass exchange explicitly appears in the governing equations, as shown in the mass conservation of each phase [43]:

gas-phase mass conservation: $\nabla \cdot \left(\rho_g \vec{u}_g \right) = S_m^{\text{lg}},$ (Eq. 6-69)

liquid-phase mass conservation: $\nabla \cdot \left(\rho_l \vec{u}_l \right) = -S_m^{\text{lg}},$ (Eq. 6-70)

where $\vec{u}_g$ and $\vec{u}_l$ are the superficial velocities of gas and liquid, respectively. The gas density can be calculated through the constituent species:

$$\rho_g = \sum_k M_k C_k,$$ (Eq. 6-71)

where k denotes the species index. The cathode stream contains nitrogen, water, and oxygen. By adopting Darcy's law, the momentum equations can be written as gas-phase momentum conservation: $\rho_g \vec{u}_g = -\dfrac{k_{rg} K}{\nu_g} \nabla P_g,$ (Eq. 6-72)

liquid-phase momentum conservation: $\rho_l \vec{u}_l = -\dfrac{k_{rl} K}{\nu_l} \nabla P_l.$ (Eq. 6-73)

The relative permeabilities k_{rg} and k_{rl} are defined as the ratio of the intrinsic permeability of liquid/gas phase to the total permeability of a porous medium. Physically, it describes the extent to which one fluid is hindered by others in pore space and can be approximated as a function of liquid saturation. The following set of correlations is widely used in PEM fuel cell modeling:

$$k_{rl} = s_e^{n_k} \text{ and } k_{rg} = (1 - s_e)^{n_k}, \text{ where the effective saturation } s_e = \frac{s_l - s_{ir}}{1 - s_{ir}}$$ (Eq. 6-74)

n_k of unity, that is, a linear relationship, is called the X-curve model. In addition, Corey [44] suggested a formula similar to Eq. 6-74 with k_{rl} having $n_k = 4$ whereas k_{rg} is given by

$$k_{rg} = (1 - s_e)^2 \left(1 - s_e^2\right).$$ (Eq. 6-75)

Fourar and Lenormand [45] proposed another expression accounting for the impact of the fluid viscosities:

$$k_{rl} = \frac{s_l^2}{2}(3 - s_l) \text{ and } k_{rg} = (1 - s_l)^3 + \frac{3}{2}\frac{\mu_g}{\mu_l} s_l (1 - s_l)(2 - s_l).$$ (Eq. 6-76)

Within gas flow channels, water transport occurs in both gas and liquid phases. Given that the liquid is essentially pure water, the water conservation equation is derived as follows:

Water conservation: $\nabla \cdot (\vec{u}_g C_w^g + \vec{u}_l \dfrac{\rho_l}{M_w}) = \nabla \cdot (\varepsilon s_g D_w^g \nabla C_w^g).$ (Eq. 6-77)

At 1D steady state, the boundary conditions for the mass conservation at the GDL–channel interface become a source term in the equation, rewritten as

$$L_z \frac{d\rho_g u_g}{dx} = S_m^{lg} + \beta S_m,$$ (Eq. 6-78)

$$L_z \frac{d\rho_l u_l}{dx} = -S_m^{lg} + (1 - \beta)S_m,$$ (Eq. 6-79)

where L_z denotes the channel depth, and u_g and u_l represent the gas and liquid velocities, respectively. Following the original work [43], we use x to denote the along-channel direction. Likewise, the 1D water equation reads

$$L_z \frac{d}{dx} \left(\vec{u}_g C_w^g + \vec{u}_l \frac{\rho_l}{M_w} \right) = L_z \frac{d}{dx} \left(\varepsilon s_g D_w^g \frac{d}{dx} C_w^g \right) + S_w. \qquad \text{(Eq. 6-80)}$$

Single-phase transport: When partially humidified air is injected, the channel stream is purely a gas-phase flow initially. In the absence of liquid, S_m^{lg} is zero while β is unity in Eq. 6-78. Integrating Eq. 6-78 from 0 to x yields

$$\rho_g u_g = \rho_{g,in} u_{g,in} + \frac{L_x}{L_z} \int_0^{\bar{x}} S_m(\hat{x}) d\hat{x}. \qquad \text{(Eq. 6-81)}$$

where $\bar{x} \left(= \dfrac{x}{L_x} \right)$ is the dimensionless distance from the inlet. ρ_g is determined by the gas composition. The dominant species in the air is nitrogen which is inactive in PEM fuel cells. Assuming a constant gas density, the gas velocity can be obtained directly from Eq. 6-81. One can estimate the magnitude of reactant gas flow acceleration under the condition that the entire channel is in the single-phase regime:

$$\frac{u_g - u_{g,in}}{u_{g,in}} = \frac{L_x \int_0^1 S_m d\hat{x}}{L_z \rho_g u_{g,in}}, \qquad \text{(Eq. 6-82)}$$

which is approximately 5% at 1.0 A/cm^2, 2 atm and $\alpha \sim 0.1$. For uniform current density I, the gas velocity can further be written as

$$\frac{u_g}{u_{g,in}} = 1 + \frac{4 C_{O_2,in}}{\rho_g \xi_c} \left(\frac{1}{2} M_{H_2} + \alpha M_w \right) \bar{x}. \qquad \text{(Eq. 6-83)}$$

Once the gas-phase velocity is obtained, the gas pressure can be determined directly from Darcy's law:

$$P_g = P_{g,in} - L_x \int_0^{\bar{x}} \frac{\mu_g u_g}{K} d\hat{x}. \qquad \text{(Eq. 6-84)}$$

The water vapor concentration is obtained by substituting u_g into Eq. 6-80. Assuming that convection dominates the channel water transport so as to eliminate the diffusion term, integration from 0 to $\bar{x}$ yields

$$C_w^g = \frac{u_{g,in} C_{w,in}^g}{u_g} + \frac{L_x \int_0^{\bar{x}} S_w d\hat{x}}{u_g L_z}. \qquad \text{(Eq. 6-85)}$$

Again for uniform current density, the above equations are written as

$$P_g = P_g(0) - \frac{\mu_g u_g(0) L_x}{K} \left[\bar{x} - 4 \left(\frac{1}{2} M_{H_2} + \alpha M_w \right) \frac{C^g_{O_2,in}}{\rho_g \xi_c} \bar{x}^2 \right], \quad \text{(Eq. 6-86)}$$

$$C^g_w = \frac{C^g_{w,in} + \dfrac{2(1+2\alpha)C^g_{O_2,in}\bar{x}}{\xi_c}}{1 + \dfrac{4C^g_{O_2,in}}{\rho_g \xi_c} \left(\dfrac{1}{2} M_{H_2} + \alpha M_w \right) \bar{x}}. \quad \text{(Eq. 6-87)}$$

The water vapor concentration increases downstream until it becomes saturated or when liquid water emerges. The transition point $\bar{x}^*$ is given by

$$\bar{x}^* = \left(\frac{\xi_c \rho_g}{2 C^g_{O_2,in}} \right) \frac{C_{w,sat} - C^g_{w,in}}{(1+2\alpha)\rho_g - 2 C_{w,sat} \left(\dfrac{1}{2} M_{H_2} + \alpha M_w \right)}. \quad \text{(Eq. 6-88)}$$

Two-phase transport: At $\bar{x}^*$, liquid water emerges, resulting in two-phase transport. Following the single-phase flow analysis, one can integrate both mass conservation equations from $\bar{x}^*$ to $\bar{x}$. yielding

$$\rho_g u_g = \rho_g(\bar{x}^*) u_g(\bar{x}^*) + \frac{L_x}{L_z} \int_{\bar{x}^*}^{\bar{x}} \left(S_m^{lg} + \beta S_m \right) d\hat{x}, \quad \text{(Eq. 6-89)}$$

$$\rho_l u_l = \frac{L_x}{L_z} \int_{\bar{x}^*}^{\bar{x}} \left(-S_m^{lg} + (1-\beta) S_m \right) d\hat{x} \quad \text{(Eq. 6-90)}$$

Applying the same integration for the water equation yields

$$u_g C^g_w + u_l \frac{\rho_l}{M_w} = u_g(\bar{x}^*) C_{w,sat} + \frac{L_x}{L_z} \int_{\bar{x}^*}^{\bar{x}} S_w d\hat{x}. \quad \text{(Eq. 6-91)}$$

In porous channels, liquid water attaches to the surface of the solid matrix; thus the interfacial area between the liquid and gas phases is large. One can assume liquid and gas phases are in equilibrium, that is, C^g_w equals to $C_{w,sat}$. In the above equation, we neglect the diffusion term because of this assumption. Again assuming a constant gas density ρ_g, one can solve for the superficial velocities to obtain

$$u_l = \frac{L_x M_w}{L_z \rho_l \left(C_{w,sat} M_w - \rho_g \right)} \left(C_{w,sat} \int_{\bar{x}^*}^{\bar{x}} S_m d\bar{x} - \rho_g \int_{\bar{x}^*}^{\bar{x}} S_w d\hat{x} \right). \quad \text{(Eq. 6-92)}$$

and

$$u_g = u_g(\overline{x}^*) - \frac{L_x}{L_z\left(C_{w,\text{sat}}M_w - \rho_g\right)}\left(\int_{\overline{x}^*}^{\overline{x}} S_m d\overline{x} - M_w \int_{\overline{x}^*}^{\overline{x}} S_w d\hat{x}\right). \qquad \text{(Eq. 6-93)}$$

Furthermore, the 1D Darcy's law for both phases are written, respectively, as

$$u_g = -\frac{k_{rg}K}{\mu_g}\frac{dP_g}{dx}, \qquad\qquad \text{(Eq. 6-94)}$$

$$u_l = -\frac{k_{rl}K}{\mu_l}\frac{dP_l}{dx} = -\frac{k_{rl}K}{\mu_l}\left(\frac{dP_g}{dx} - \frac{dP_c}{dx}\right). \qquad \text{(Eq. 6-95)}$$

Combining these two equations to eliminate $\dfrac{dP_g}{dx}$ and then substituting Eq. 6-39 yield

$$u_l = \frac{\mu_g}{\mu_l}\frac{k_{rl}}{k_{rg}}u_g + \frac{k_{rl}\left(K\varepsilon\right)^{1/2}}{\mu_l L_x}\sigma\cos\theta_c\frac{dJ}{ds}\frac{ds_l}{d\overline{x}}. \qquad \text{(Eq. 6-96)}$$

The above equation can be solved through numerical iteration. In addition, liquid transport along the channel is primarily driven by the two-phase interaction through the shear stress at the interface. Assuming that the capillary action is weak and negligible, one eliminates the derivative term on the right, arriving at an algebraic equation. Using the relative permeabilities (Eq. 6-74), one can further explicitly solve for the liquid saturation:

$$s_l = \frac{\left(\dfrac{u_l\mu_l}{\mu_g u_g}\right)^{\frac{1}{n_k}} + s_{ir}}{1 + \left(\dfrac{u_l\mu_l}{\mu_g u_g}\right)^{\frac{1}{n_k}}}. \qquad \text{(Eq. 6-97)}$$

With the saturation s_l given above, the relative permeabilities can be computed, which is further substituted into the pressure equations:

$$P_g = P_g(\overline{x}^*) - L_x\int_{\overline{x}^*}^{\overline{x}}\frac{\mu_g u_g}{k_{rg}K}d\hat{x}, \qquad \text{(Eq. 6-98)}$$

$$P_l = P_g(\overline{x}^*) - L_x\int_{\overline{x}^*}^{\overline{x}}\frac{\mu_l u_l}{k_{rl}K}d\hat{x}. \qquad \text{(Eq. 6-99)}$$

The gas-phase pressure drop along the channel is then given by

$$\Delta \overline{P}_g = \underbrace{\frac{\Delta P_g}{\mu_g u_{g,in} L_x}}_{K} = \int_0^{\overline{x}^*} \frac{u_g}{u_{g,in}} d\hat{x} + \int_{\overline{x}^*}^1 \frac{u_g}{u_{g,in}} \frac{1}{k_{rg}} d\hat{x}. \qquad \text{(Eq. 6-100)}$$

Assuming a constant gas velocity in flow channels, the scale of the pressure (i.e., the denominator in the middle term of the above equation) becomes the pressure drop for a single-phase channel stream. The above equation can further be rearranged as

$$\Delta \overline{P}_g = \overline{x}^* + \int_{\overline{x}^*}^1 \frac{1}{k_{rg}} d\hat{x} \qquad \text{(Eq. 6-101)}$$

Figure 6-51 plots the gas-phase velocity along a channel, computed from Eqs. 6-81 and 6-93. The apexes are the transition locations from the single- to two-phase flows. The gas velocity increases along channel at the initial single-phase stage, followed by a decrease in the two-phase region. The initial increasing trend arises from a mass source by the reaction and species transport added to the gas phase flow. The increment over the entire channel is, however, relatively small (<5%). After liquid water emerges, the mass addition is mostly into the liquid phase, as the gas phase is already saturated. The ORR, however, withdraws mass from the reactant gas, leading to a decreasing gas flow. The decelerating gas flow and oxygen consumption induce water phase change from vapor to liquid, further reducing the gas flow rate. The overall variation of the gas velocity is relatively small (<10%) in most cases.

Figure 6-52 presents the water vapor concentration along a channel. As the gas-phase velocity variation is small, the water concentration increases almost linearly along the channel in the single-phase region under the conditions of uniform current density and constant net water transport coefficient. In reality, these two factors vary spatially; one can then use Eq. 6-85

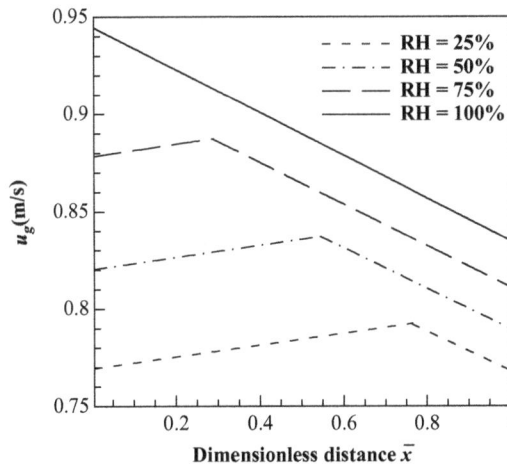

Figure 6-51. Gas-phase superficial velocity along a flow channel at four inlet relative humidities (RHs) [43] (Courtesy of the Electrochemical Society).

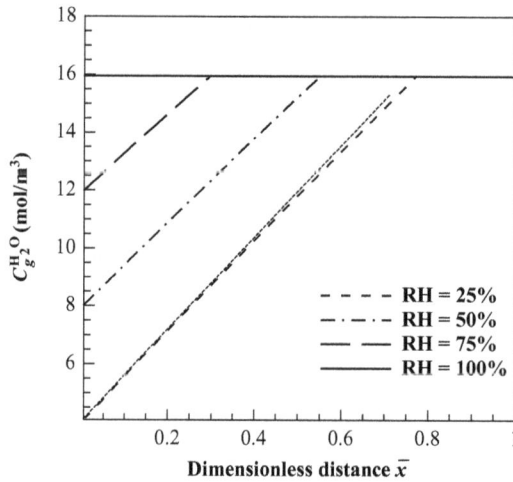

Figure 6-52. Vapor concentration along the channel at different relative humidities (RH). The thin dashed line near the curve of RH = 25% indicates the linear trend [43] (Courtesy of the Electrochemical Society).

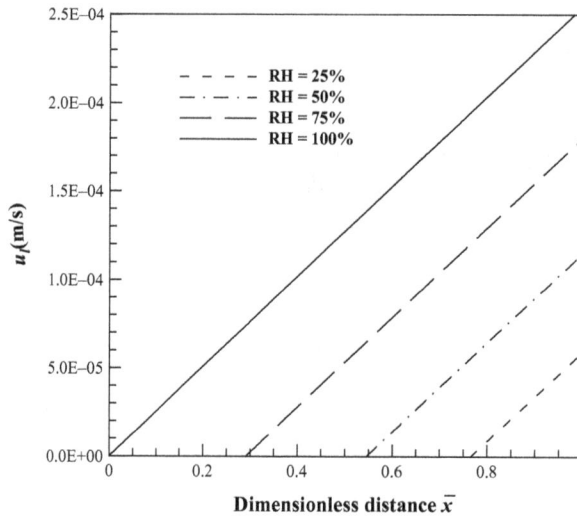

Figure 6-53. Liquid superficial velocities along a flow channel under various relative humidities (RHs) [43] (Courtesy of the Electrochemical Society).

to calculate the profile. In addition, the small deviation from the linear trend observed in this figure (e.g., see the case of RH = 25%) arises primarily from the gas flow change as shown in Figure 6-51. Once reaching the saturation level, the water vapor concentration remains constant as a result of the phase equilibrium.

Figure 6-53 displays the superficial liquid-phase velocity along the channel under four different RHs. In the single-phase region, the liquid velocity is zero, whereas in the two-phase region it increases almost linearly along the channel. The larger the RH, the earlier the liquid water appears in the channel and the higher the outlet liquid velocity. The magnitude of the superficial velocity is around 10^{-4} m/s.

Figure 6-54 shows the capillary effect, obtained by numerically solving the first-order ODE (ordinary differential equation), Eq. 6-96. Figure 6-54 also plots prediction by Eq. 6-97 for comparison: the two equations yield almost the same prediction. Only a minute difference is indicated, as shown in the close-up window, that is, the capillary mechanism can be neglected when analyzing the along-channel two-phase transport.

The relative permeabilities are important to two-phase flow. Figure 6-55 plots the saturation profiles at RH = 75% under various correlations of the relative permeabilities. A strong dependence of the two-phase flow prediction on the relative-permeability correlation is indicated. The outlet saturation varies from around 1% to 20%. In particular, the X-curve model gives the smallest fraction of liquid water in the two-phase region (<1%). Other correlations show similar trends but with different magnitudes. The correlation of Fourar & Lenormand [45] predicts a liquid saturation lower than 5%, whereas that of Corey [44] gives a profile higher than that of Eq. 6-74 with $n_k = 4$. Note that Corey has the same expression of the liquid relative permeability as Eq. 6-74 but differs in the gas relative permeability.

Figure 6-56 displays the gas pressure profiles computed by Eqs. 6-86 and 6-98 using the relative permeability of Eq. 6-74 with $n_k = 4$. Two stages of pressure change along the channel are shown: the first stage is in the single-phase region, and the other is in the two-phase one. Both trends are almost linear but different in slope. The two-phase region displays a more rapid decline in the gas pressure despite the decreasing gas velocity as shown in Figure 6-51 because the presence of liquid water narrows the gas passage. A slight difference is observed in the single-phase region among the three cases because the inlet flow rates differ under varying relative humidities.

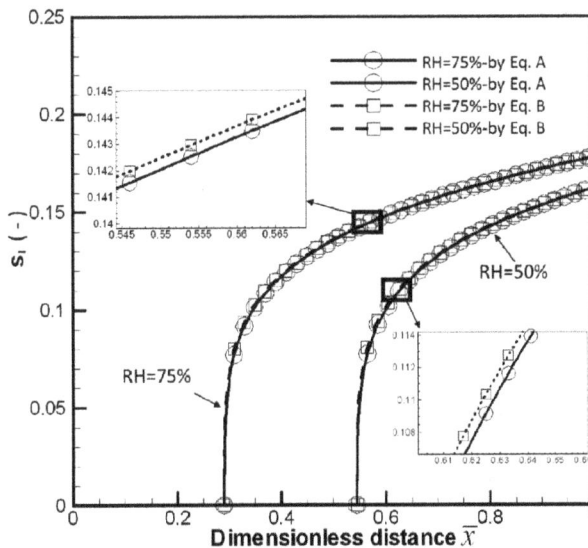

Figure 6-54. Liquid water profiles predicted by Eq. 6-96 (denoted as Eq. A in the figure) and Eq. 6-97 (denoted as Eq. B in the figure) using Eq. 6-74 for the relative permeability of $n_k = 4$ [43] (Courtesy of the Electrochemical Society).

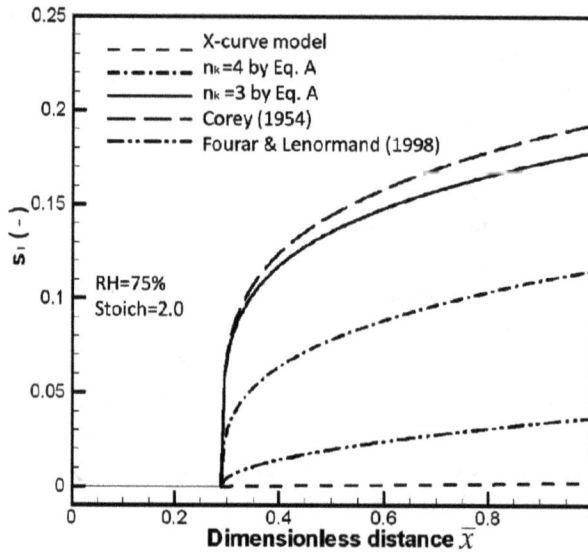

Figure 6-55. Liquid water profiles predicted using different relative permeability (RH) correlations. Eq. A in the figure refers to Eq. 6-74 [43] (Courtesy of the Electrochemical Society).

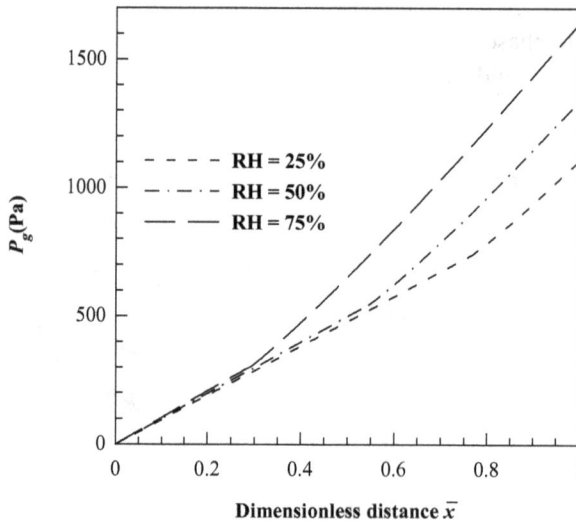

Figure 6-56. The gas pressure profiles under various inlet relative humidities (RH) [43] (Courtesy of the Electrochemical Society).

The gas pressure drops faster in the two-phase region. The relative humidity (RH) affects the onset site and saturation level; see Figure 6-54. The gas pressure drop through flow channels as a function of RH is quantified by Eq. 6-100 and approximated by Eq. 6-101, and this is plotted in Figure 6-57. This figure displays the dimensionless pressure drops at various RHs using Eq. 6-100. The magnitude of the pressure drop clearly increases with the RH. The pressure drop is dependent on the two-phase interaction. The relative permeability k_{rg} is a key parameter accounting for the interaction. Thus, the smaller the exponent in k_{rg} (e. g., $n_k = 3$), the lower the pressure drop. Furthermore, another dimensionless pressure drop is given by Eq. 6-101 under the assumption of constant gas velocity. The value computed from Eq. 6-101 is also plotted in Figure 6-57 for comparison, showing a similar trend as Eq. 6-100. Small deviations are observed: at low RHs, Eq. 6-101 underestimates the pressure drop, whereas under the higher RHs its prediction is higher. Predictions using both equations are close within 10% in difference.

The experimental data of the two-phase pressure drop are plotted in Figure 6-58 against the model prediction, showing a slight under-prediction using $n_k = 4$. The agreement between measured and computed pressure drops is good for both $n_k = 4.5$ and 5.0.

6.4   WATER DROPLET DYNAMICS AT THE GDL/GFC INTERFACE

At the cathode GDL/GFC interface, oxygen is transported toward the cathode catalyst layer to provide reactant for the ORR, whereas product water enters the GFC and is eventually removed out of the channel. The interfacial transport resistance can be significant when liquid water appears. Optical visualizations, see Figure 6-59, show that liquid water exists as droplets on the GDL surface; it is taken away by the gas flow or is attached to the channel walls. Liquid droplet dynamics at the GDL/GFC interface consists of three sub-processes: (1) liquid water transport from the catalyst layer to the GDL/GFC interface via capillary action; (2) droplet removal at the GDL/GFC interface via detachment or evaporation; and (3) liquid water transport through the GFC in form of films, droplets, and/or vapor. Using

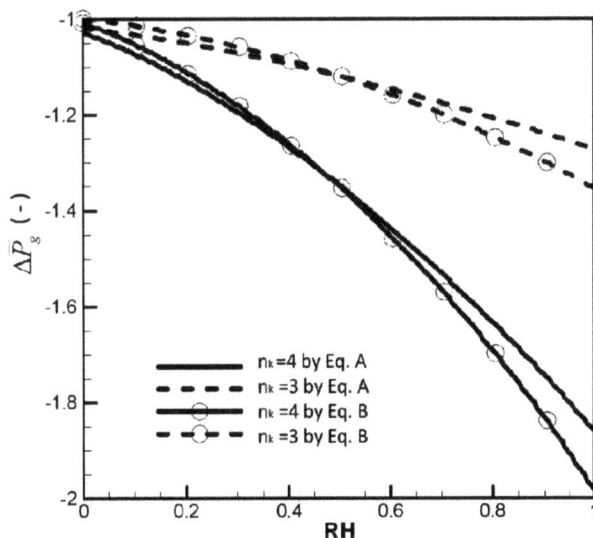

Figure 6-57. Dimensionless pressure drop in a GFC. Eq. A and Eq. B in the figure denote, respectively, Eq. 6-100 and Eq. 6-101 [43] (Courtesy of the Electrochemical Society).

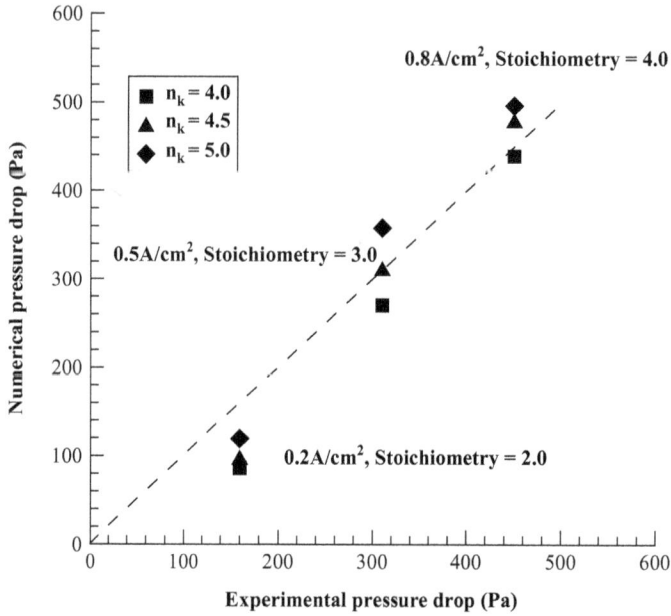

Figure 6-58. Experimental validation of the channel two-phase model [29].

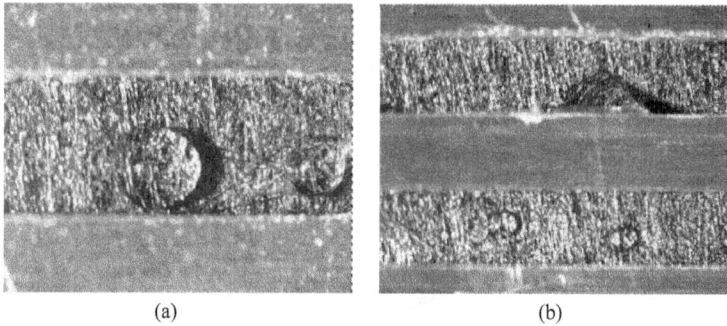

Figure 6-59. Visualization of droplet formation and dynamics at the GDL–GFC interface: (a) droplet formation; (b) droplets in the lower channel and droplet attached to the channel wall in the upper channel. The channel cross-section dimension is 1×1 mm [43] (Courtesy of the Electrochemical Society).

results computed by numerical simulation with the volume of fluid (VOF) method, Figure 6-60 shows the droplet dynamics on a hydrophobic surface.

6.4.1   FORCE BALANCE ON A SPHERICAL-SHAPE DROPLET

A water droplet in a gas flow channel experiences drag forces exerted by surrounding gas flow as shown in Figure 6-61. Because of the fluid viscosity, the gas velocity rapidly decreases to zero at the droplet surface if the no-slip condition applies. The drag force also arises from the

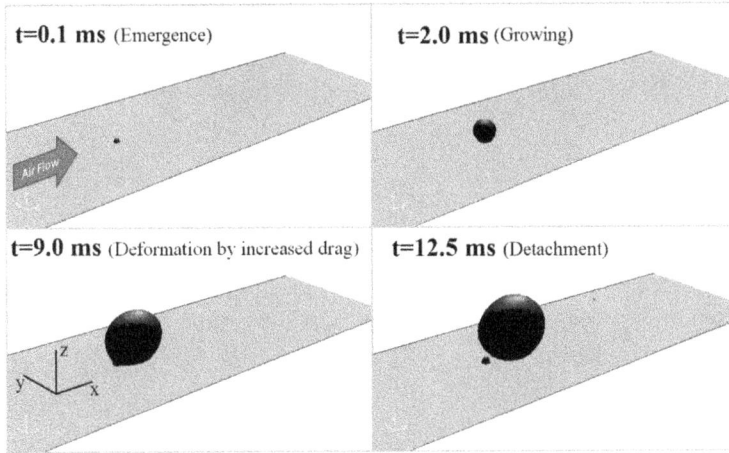

Figure 6-60. Water droplet dynamics predicted by a 3D VOF (volume of fluid) method (channel height = 1 mm, U_G = 2 m/s, θ = 150°) [46].

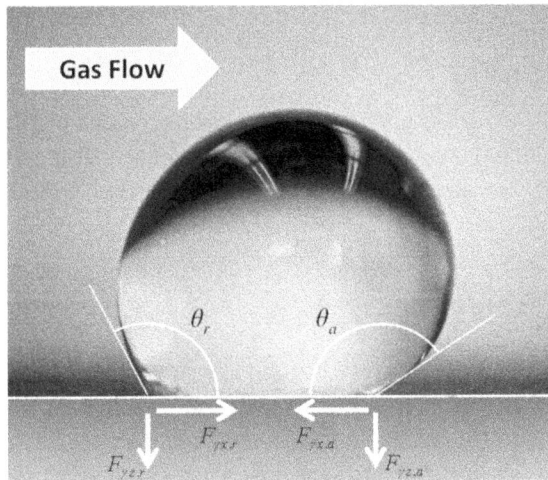

Figure 6-61. Wall adhesion on a droplet over a hydrophobic surface.

gas flow pressure–the upstream has a higher pressure than downstream. In GFCs, surface tension is another factor affecting droplet dynamics. The effect of the gravitational force is usually negligible in GFCs, as indicated by the Eötvos number:

$$E\ddot{o} = \frac{\Delta\rho g d^2}{\gamma}. \qquad \text{(Eq. 6-102)}$$

The Eötvos number ($E\ddot{o}$) is 0.0015–0.15 under the typical operating condition of PEM fuel cells.

At the onset of detachment (that is, immediately before the droplet is detached), all forces on a droplet, including the wall adhesion, are balanced. The drag, created by air flows near the droplet, is the sum of the pressure and viscous forces. The pressure force is obtained through the following equation:

$$F_p = \vec{i} \cdot \int_A p\vec{n} \, dA, \qquad \text{(Eq. 6-103)}$$

where $\vec{i}$ is the direction vector along the GFC, $\vec{n}$ is the normal vector of the infinitesimal surface dA, P is the local pressure, and A is the area of the dropletair interface or droplet surface. The viscous friction is calculated by

$$F_v = \vec{i} \cdot \int_A \mu_g \, \vec{t} \cdot \nabla \vec{u} \, dA, \qquad \text{(Eq. 6-104)}$$

where $\vec{t}$ is the tangential direction on the droplet surface.

The total drag force exerted on the droplet is the sum of both the pressure and viscous forces:

$$F_{\text{drag}} = F_p + F_v. \qquad \text{(Eq. 6-105)}$$

Wall adhesion (or force arising from surface tension) consists of the normal (z-direction) and tangential (x-direction) components. The maximum tangential force is determined by the static contact angle (θ_s), contact-angle hysteresis (θ_H), and surface tension (γ):

$$F_{\gamma x} = F_{\gamma x,r} + F_{\gamma x,a} = 2\gamma \pi d \cdot \sin^2 \theta_s \cdot \sin \frac{\theta_H}{2}. \qquad \text{(Eq. 6-106)}$$

The subscripts r and a denote the receding and advance contact angles, respectively. At the hydrophobic GDL surface, the effects of advancing and receding angles are opposite to each other and thus partly canceled out. Due to the droplet being dragged by gas flow, the advancing angle is greater than the receding angle, making the wall adhesion opposite to the gas flow direction. When the drag force overcomes the wall adhesion, the droplet is detached at the surface. Another parameter is the normal wall adhesion over the droplet contact line with the solid surface:

$$F_{\gamma z} = 2\gamma \cdot \left(\pi \frac{d}{2} \sin \theta_s \right) \cdot \int_0^\pi \sin\left(\theta_r + \frac{\theta_a - \theta_r}{\pi} \phi \right) d\phi, \qquad \text{(Eq. 6-107)}$$

where ϕ is the angle of the projected circle on the GDL surface. A simple relation is adopted between the static contact angle and contact-angle hysteresis:

$$\theta_r = \theta_s - \frac{\theta_a - \theta_r}{2}. \qquad \text{(Eq. 6-108)}$$

A local contact angle ξ can be defined along the contact line on the GDL surface as

$$\xi = \theta_r + \frac{\theta_a - \theta_r}{\pi}\phi.$$

(Eq. 6-109)

Eq. 6-107 can be re-arranged as

$$F_{\gamma z} = 2\gamma \cdot \left(\pi \frac{d}{2}\sin\theta_s\right) \cdot \left(\int_{\theta_r}^{\theta_a} \sin\xi \frac{\pi}{\theta_H}d\xi\right) \cdot \frac{1}{\pi}.$$

$$= 2\gamma \frac{\pi}{\theta_H}d \cdot \sin\theta_s \cdot \sin\frac{\theta_s}{2}\cdot \sin\frac{\theta_H}{2}$$

(Eq. 6-110)

This force component is the maximum that can balance the lift force caused by the surrounding flow. When the lift force overcomes this barrier, the droplet will be detached from the surface. Thus, droplet detachment occurs at

$$F_{\text{drag}} > F_{\gamma x} \text{ or } F_{\text{lift}} > F_{\gamma z}.$$

(Eq. 6-111)

In reality, the second criterion is more difficult to satisfy. Therefore, the first criterion is important to determine droplet detachment. The critical condition, that is, $F_{\text{drag}} = F_{\gamma x}$, is then expressed by

$$\hat{i} \cdot \int_A \left(\hat{n}\cdot p + \mu_g \nabla u\right)dA = 2\gamma\,\pi\,d \cdot \sin^2\theta_s \cdot \sin\frac{\theta_H}{2}.$$

(Eq. 6-112)

To evaluate the forces over a droplet, 3D numerical simulation based on CFD (computational fluid dynamics) is carried out with the droplet surface approximated by a solid spherical boundary as shown in Figure 6-62. The approximation is valid when the droplet has negligible deformation, thus follows a spherical-cap shape, and the liquid internal flow is also sufficiently slow and negligible.

In PEM fuel cells, a droplet will be removed when the air flow velocity exceeds the critical detachment velocity. For a parametric study, the gas velocity is allowed to exceed that limit and remains static. Figure 6-63 shows the drag forces exerted by the gas stream. Various droplet sizes are compared and results are presented for both the fully developed region and the entrance region. As expected, both drag forces increase rapidly with the gas velocity. For small droplets, the viscous force dominates whereas the pressure drag becomes more important for relatively large droplets. Figure 6-63 also compares the forces at two typical locations, that is, the entrance and fully developed regions. Over the entrance length, the boundary layer undergoes development, and it grows from the solid walls and eventually meets at the centerline. After that, the channel flow becomes fully developed with a parabolic velocity profile. In PEM fuel cells, the Reynolds number of channel flows is relatively low, falling in the laminar regime. The hydrodynamic entrance length L_E can be evaluated using that developed in tube flows: $L_E/d \sim 0.06Re$, where d is the tube diameter. A Re of 200 will lead to $L_E \sim 12\ d$, which is small but not negligible. For a 1.0-mm diameter tube, the entrance length is around 12 mm,

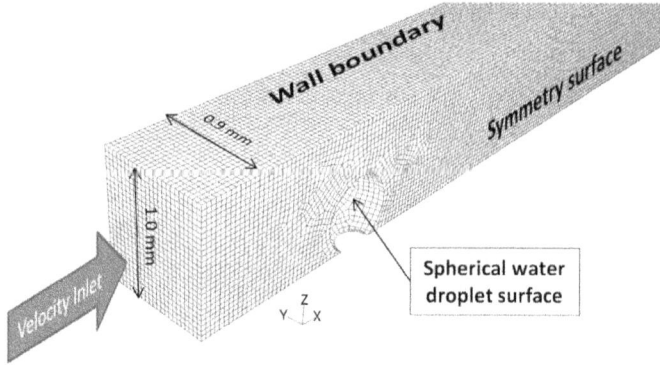

Figure 6-62. 3D computational domain for a case study of water droplet in a GFC.

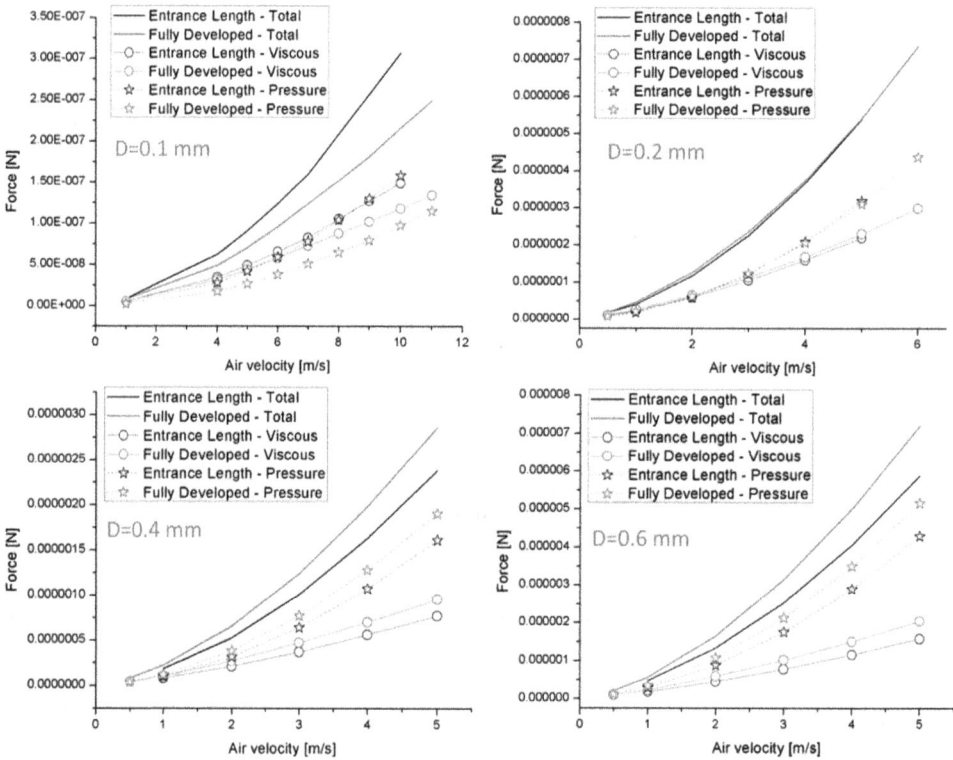

Figure 6-63. Drag forces in the entrance and fully developed regions.

which is about 10% of a 10-cm long channel. The velocity profiles near a water droplet in the entrance and fully developed stages are plotted in Figure 6-64. In the entrance region, a small droplet ($d < 0.2$ mm) experiences a stronger force than that in the fully developed region. This can be explained by that in the entrance region, the gas velocity is more uniform in the gas channel; therefore, the velocity around small droplets on the wall undergoes a stronger surrounding gas flow than that in the fully developed region. As a result, the detachment velocity

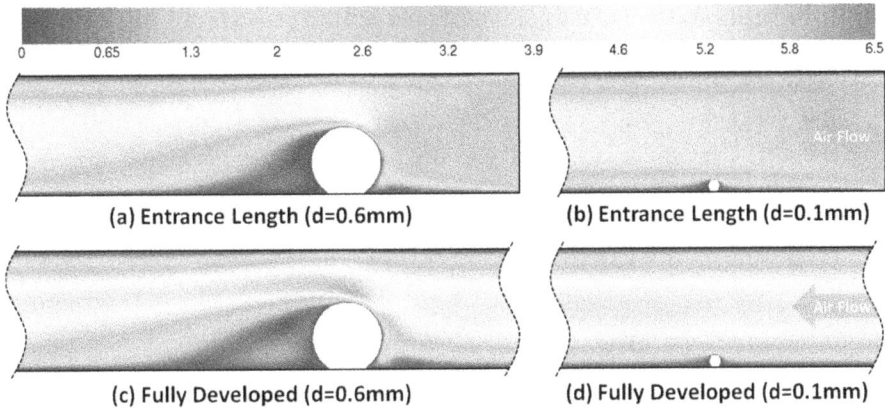

| 0 | 0.65 | 1.3 | 2 | 2.6 | 3.2 | 3.9 | 4.6 | 5.2 | 5.8 | 6.5 |

(a) Entrance Length (d=0.6mm)　　(b) Entrance Length (d=0.1mm)

(c) Fully Developed (d=0.6mm)　　(d) Fully Developed (d=0.1mm)

Figure 6-64. Velocity contours in the entrance (a & b) and fully developed (c & d) regions (u_{in} = 3.0 m/s).

of a small droplet is lower in the entrance region than that in the fully developed condition. As the droplet size increases, the pressure force becomes dominant. In the fully developed region, the local air velocity in the core of the channel is higher than the average value. A large droplet will physically extend to the core region, encountering higher air velocity in the fully developed region. Since the pressure force is approximately proportional to the square of the gas velocity magnitude, a larger droplet experiences a larger drag force, that is, a lower detachment velocity in the fully developed region.

6.4.2 DROPLET DEFORMATION

Droplets may deform upon forces exerted by surrounding air flow, deviating away from the spherical shape at the static state; see Figs. 6-61 and 6-65. Two major forces are exerted by air flow: one is the viscous force and the other is the pressure force. The former arises from the air viscosity, causing a sharp change of air velocity near the droplet surface, and is usually along the tangential direction of the surface. This force is determined by local air flow. The other arises from droplet obstruction, yielding pressure variation over the droplet surface. The pressure difference at two typical locations can be estimated: one is at the front surface of the droplet where air flow becomes stagnant; the other is at the side surface where the airflow cross-section reaches minimum. The air pressure changes as a result of air flow variation, which further alters the droplet's local curvature. The local curvature can be estimated using the Young–Laplace equation:

$$p_l - p_g = \frac{2\gamma}{R}.$$

(Eq. 6-113)

Assuming the liquid pressure within the droplet to be uniform, Eq. 6-113 can be differentiated to relate the gas pressure variation dp_g to the local curvature change [47]:

$$\frac{dR}{R} = \frac{R}{2\gamma} dp_g,$$

(Eq. 6-114)

where R is the local curvature of the gas–liquid interface. To evaluate dp_g, one can follow that in the orifice plates or Venturi tubes (both are the Bernoulli-type flow meters): Bernoulli's principle states that the gas-phase pressure drop is proportional to the square of the air velocity. To better elucidate the issue, we consider a 2D case, that is, the shape of liquid droplet is part of a cylinder. Then, the area ratio (ζ) between the upstream and minimum regions for 2D and 3D, respectively, is evaluated by

$$2D: \quad \zeta_{2D} = \frac{A_i}{A_2} = \frac{W \times H}{W \times \left[H - R_0(1 - \cos\theta_s)\right]} = \frac{H}{H - R_0(1 - \cos\theta_s)}, \quad \text{(Eq. 6-115)}$$

$$3D: \quad \zeta_{3D} = \frac{A_i}{A_3} = \frac{W \times H}{W \times H - R_0^2\left(\theta_s - \cos\theta_s \sin\theta_s\right)}, \quad \text{(Eq. 6-116)}$$

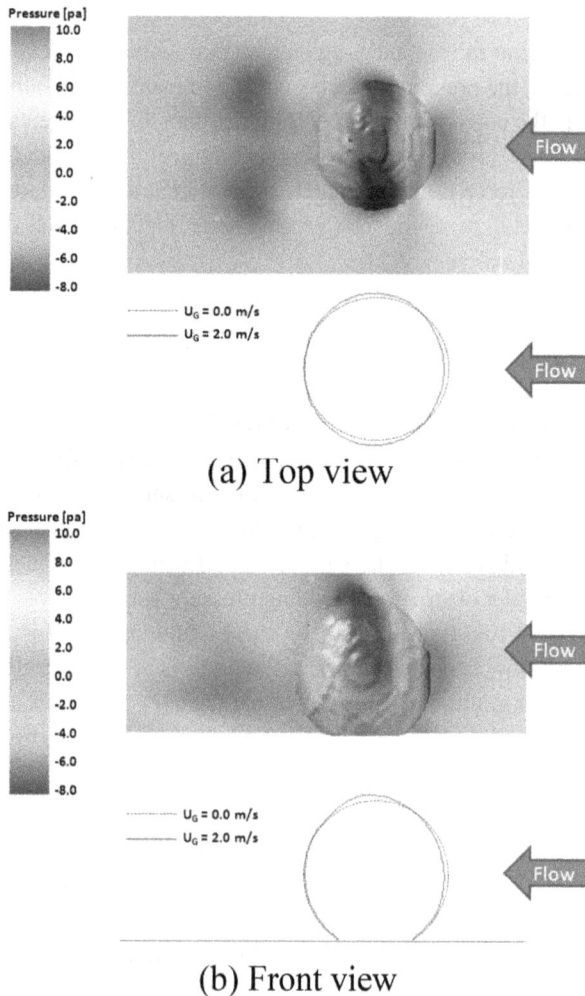

(a) Top view

(b) Front view

Figure 6-65. Air pressure distribution near a water droplet (top) and droplet shape change (bottom) [46].

where R_0 is the radius of curvature without deformation. Substitution of this area ratio into Bernoulli's equation yields

$$\text{2D: } dp_g = P_1 - P_2 = \frac{\rho u_{in}^2}{2} \zeta_{2D}^2, \qquad \text{(Eq. 6-117)}$$

$$\text{3D: } dp_g = P_1 - P_3 = \frac{\rho u_{in}^2}{2} \zeta_{3D}^2. \qquad \text{(Eq. 6-118)}$$

Substituting the pressure differences into Eq. 6-114 leads to

$$\text{2D: } \frac{dR}{R} = \frac{\rho u_{in}^2 R_0}{4\gamma} \frac{R}{R_0} \zeta_{2D}^2 = \frac{We_r}{4} \frac{R}{R_0} \zeta_{2D}^2, \qquad \text{(Eq. 6-119)}$$

$$\text{3D: } \frac{dR}{R} = \frac{\rho u_{in}^2 R_0}{4\gamma} \frac{R}{R_0} \zeta_{3D}^2 = \frac{We_r}{4} \frac{R}{R_0} \zeta_{3D}^2, \qquad \text{(Eq. 6-120)}$$

where We_r is the Weber number, defined as the ratio between the fluid inertia and surface tension:

$$We_r = \frac{\rho u^2 R_0}{\gamma}. \qquad \text{(Eq. 6-121)}$$

Figs. 6-66 and 6-67 present the analytical solutions along with the VOF prediction and experimental results.

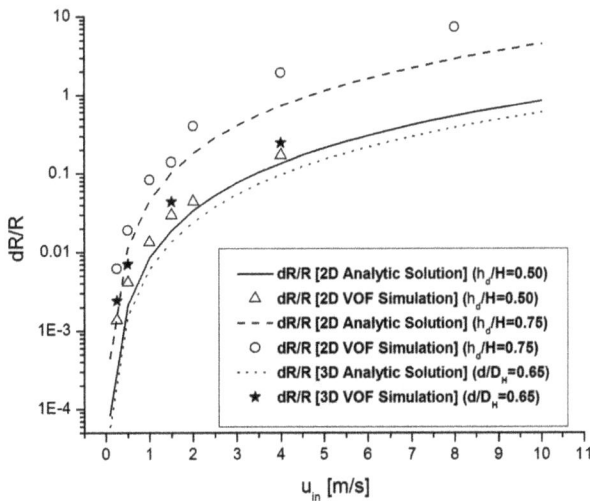

Figure 6-66. Comparison of analytical solution of droplet deformation with VOF numerical simulation results.

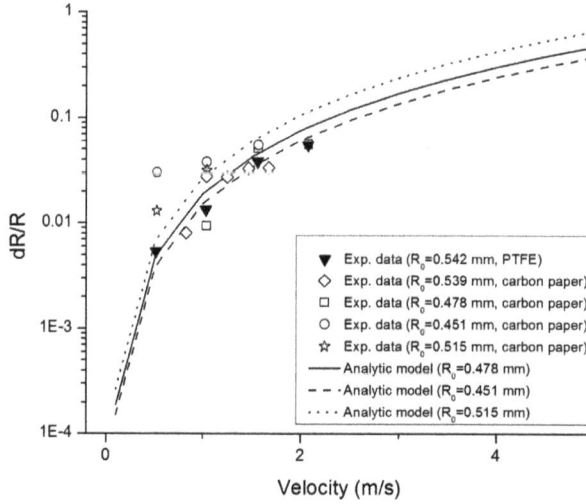

Figure 6-67. Comparison of analytical solution of droplet deformation with experimental data [46].

6.4.3 DROPLET DETACHMENT

Droplet's growth and detachment are affected by two factors: operating conditions (e.g., gas flow) and physical properties (e.g., surface roughness and wettability) of the GDL surface. Chen *et al.* [48] pioneered the analysis of droplet instability and detachment, indicating that the static contact angle (θ_s) and contact angle hysteresis (the difference between advancing and receding contact angles, i.e., $\theta_a - \theta_r$) are both important parameters determining the wall adhesion force.

6.4.3.1 Control Volume Method

Two fundamental shapes of droplets will be studied: they are (a) spherical shape; (b) cylindrical shape.

(1) A Spherical Droplet

Figure 6-68 shows a control volume (cross-section along the flow direction) enclosing a spherical water droplet situated on the GDL surface. The force balance along the channel-flow direction (i.e., z-direction) within the control volume yields

$$(p_0' - p_L')2Bl + l^2\tau_{xz,\text{wall}} + f_{\mu,d} = 0, \qquad \text{(Eq. 6-122)}$$

where p_0' and p_L' are the flowing gas pressure immediately upstream and downstream of the water droplet, respectively; $2B$ is the height of the flow channel; l is the maximum distance

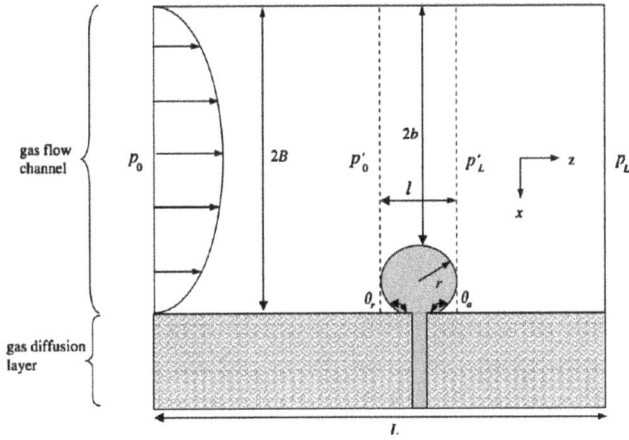

Figure 6-68. Control volume enclosing a liquid-water droplet.

between upstream droplet surface and downstream droplet surface; $l^2 \tau_{xz,\text{wall}}$ represents the shear stress exerted by the flowing gas on the top wall in an area of l^2 (which is the area projected by the droplet); $f_{u,d}$ is the viscous drag exerted on the water droplet by the flowing gas. By approximating the flow channel between the top wall surface and the droplet surface as that between two parallel plates with height $2b$, and taking the gas flow to be fully developed and laminar and the flowing fluid to be Newtonian, the top-wall shear stress per unit area, $\tau_{xz,\text{wall}}$, can be estimated as follows (see, e.g., Bird et al. 2002, p. 63 [49]):

$$\tau_{xz,\text{wall}} = \frac{(p_0' - p_L')}{l}(-b). \qquad \text{(Eq. 6-123)}$$

Substituting Eq. 6-123 into Eq. 6-122 yields an approximation for the viscous drag on the droplet surface:

$$-f_{\mu,d} = (p_0' - p_L')(2B - b)l. \qquad \text{(Eq. 6-124)}$$

It is interesting to check the limit when the droplet completely blocks the channel, that is, at $b = 0$ and $l = 2B$:

$$-f_{\mu,d,\max} = (p_0' - p_L')(2B)^2 = (p_0' - p_L')4B^2, \qquad \text{(Eq. 6-125)}$$

indicating that the shear stress on the top wall surface vanishes and the drag on the droplet achieves its maximum value, as expected. By taking the flowing gas to be a Newtonian fluid, the flow to be fully developed and laminar in a narrow slit (velocity profiles for such flows are well documented, see e.g., Bird et al. 2002 [49]), the pressure drop across the droplet, $(p_0' - p_L')$, can be related to the overall pressure drop, average velocity, viscosity, and channel height:

$$(p_0' - p_L') = (p_0 - p_L) - (L - l)\frac{3\mu U}{B^2} \qquad \text{(Eq. 6-126)}$$

and

$$(p_0' - p_L') = \frac{3\mu U'}{b^2}l, \qquad \text{(Eq. 6-127)}$$

where p_0 and p_L are the flowing gas pressure at the entrance and exit of the channel, respectively, L is the length of the flow channel, μ is the flowing-gas viscosity, U is the average velocity along flow direction in the upstream and downstream regions, and U' is the average velocity along flow direction in the region directly above the droplet. From mass balance,

$$U' = \frac{B}{b}U. \qquad \text{(Eq. 6-128)}$$

Substituting Eq. 6-128 in Eq. 6-127, solving for U' in terms of $(p_0' - p_L')$ and then substituting the resultant in Eq. 6-126 gives

$$(p_0' - p_L') = (p_0 - p_L)\frac{l}{l + (L - l)\left(\dfrac{b}{B}\right)^3}. \qquad \text{(Eq. 6-129)}$$

Combining Eqs. 6-125 and 6-129 yields an approximate viscous drag on the droplet surface in terms of the overall pressure drop and droplet size:

$$-f_{\mu,d} = (p_0 - p_L)\frac{l^2(2B - b)}{l + (L - l)\left(\dfrac{b}{B}\right)^3}. \qquad \text{(Eq. 6-130)}$$

From Figure 6-68, the height of the minimum flow passage between the top wall surface and the droplet surface is related to the channel height, droplet radius (r), and advancing contact angle (θ_a) by

$$2b = 2B - r\cos\theta_a \text{ or } b = B - \frac{r}{2}\cos\theta_a. \qquad \text{(Eq. 6-131)}$$

Upon substituting Eq. 6-131, Eq. 6-130 becomes

$$-f_{\mu,d} = \frac{(p_0 - p_L)B(\frac{l}{L})^2 L(1 + \widehat{H})}{(1 - \widehat{H})^3 + \frac{l}{L}[1 - (1 - \widehat{H})^3]}, \qquad \text{(Eq. 6-132)}$$

where

$$\widehat{H} = \frac{r(1 - \cos\theta_a)}{2B} \qquad \text{(Eq. 6-133)}$$

is the dimensionless droplet height based on the channel height with θ_a varying from $0°$ to $180°$. When the droplet grows to be sufficiently large and with $\theta_a \geq 90°$, $l = 2r$ as shown in Figure 6-68, Eq. 6-132 can be simplified to

$$-f_{\mu,d} = \frac{(p_0 - p_L)}{L} \frac{16B^3}{(1 - \cos\theta_a)} \frac{(1 + \widehat{H})\widehat{H}^2}{(1 - \cos\theta_a)(1 - \widehat{H})^3 + \frac{4B}{L}\widehat{H}[1 - (1 - \widehat{H})^3]} \qquad \text{(Eq. 6-134)}$$

or, in terms of average velocity in the upstream and downstream regions away from the droplet, the approximate viscous drag exerted on the droplet is given by

$$-f_{\mu,d} = \frac{48\mu UB}{(1 - \cos\theta_a)} \frac{(1 + \widehat{H})\widehat{H}^2}{(1 - \cos\theta_a)(1 - \widehat{H})^3 + \frac{4B}{L}\widehat{H}[1 - (1 - \widehat{H})^3]}. \qquad \text{(Eq. 6-135)}$$

The total surface tension force $f_{\gamma,d}$, which tends to hold the droplet in place, is given by

$$f_{\gamma,d} = \pi(r\sin\theta_a)\gamma\cos(180 - \theta_a) + \pi(r\sin\theta_a)\gamma\cos\theta_r, \qquad \text{(Eq. 6-136)}$$

where the first and second terms on the right-hand side represent the surface-tension forces acting on the advancing and receding contact lines, respectively; $r\sin\theta_a$ denotes the droplet's wetted radius; θ_a and θ_r refer to the advancing and receding contact angles, respectively; and γ is the surface tension of liquid water (more accurately, γ denotes the surface-tension force per unit length acting on the liquid–water/air interface). Making use of trigonometric relations and the fact that $r = 2B\widehat{H}/(1 - \cos\theta_a)$, Eq. 6-136 can be re-written as

$$f_{\gamma,d} = \frac{4\pi B\widehat{H}\sin\theta_a}{1 - \cos\theta_a}\gamma\sin\frac{1}{2}(\theta_a + \theta_r)\sin\frac{1}{2}(\theta_a - \theta_r)$$

$$= \frac{4\pi B\widehat{H}\sin\theta_a}{1 - \cos\theta_a}\gamma\left[(\sin\theta_a)y\sqrt{1 - y^2} - (\cos\theta_a)y^2\right], \qquad \text{(Eq. 6-137)}$$

where $y = \sin\frac{1}{2}(\theta_a - \theta_r)$. At the onset of droplet instability, the surface tension force and the viscous drag (which tends to shear the droplet along the GDL/GFC interface) balance each other:

$$f_{\gamma,d} + f_{\mu,d} = 0 \text{ or } -f_{\mu,d} = f_{\gamma,d}. \qquad \text{(Eq. 6-138)}$$

Substituting Eqs. 6-135 and 6-137 into Eq. 6-138 and approximating the advancing contact angle by the static contact angle (θ_s) result in a nonlinear equation relating the critical droplet height to the contact angle hysteresis, ($\theta_a - \theta_r$), at the onset of instability:

$$y\sqrt{1-y^2} - (\cot\theta_s)y^2 \frac{12\mu U}{\pi\gamma\sin^2\theta_s} \frac{\hat{H}(1+\hat{H})}{(1-\cos\theta_s)(1-\hat{H})^3 + \frac{4B}{L}\hat{H}[1-(1-\hat{H})^3]} = 0, \quad \text{(Eq. 6-139)}$$

where $0 < \theta_s < 180°$. Equation 6-139 can be conveniently solved for y using Newton's method with $\hat{H}$ varying from 0 to 1. Once y is computed, then the contact angle hysteresis can be determined from $(\theta_a - \theta_r) = 2\sin^{-1}(y)$. The above expression can be further rearranged as

$$y\sqrt{1-y^2} - (\cot\theta_s)y^2 = \frac{H}{r}\frac{12}{\pi\sin^2\theta_s} \frac{\text{Re}_\text{H}^{-1}We_r\hat{H}(1+\hat{H})}{(1-\cos\theta_s)(1-\hat{H})^3 + (4B/L)\hat{H}[1-(1-\hat{H})^3]}.$$

$$\text{(Eq. 6-140)}$$

where We_r and Re_H are defined using the droplet radius (r) and channel height (H), respectively, as length scales.

(2) A Cylindrical Droplet

The force balance over a control volume that encloses a cylindrical droplet having a length (normal to the flow direction or into the page) of unity yields an equation similar to Eq. 6-125 for a spherical droplet:

$$-f_{\mu,d,\text{cylindrical}} = (p_0' - p_L')(2B - b). \quad \text{(Eq. 6-141)}$$

Following the steps in the preceding section in deriving the viscous drag for a spherical droplet (that is, expressing the local pressure drop $p_0' - p_L'$ in terms of the overall pressure drop, $p_0 - p_L$ and the minimum flow passage height, b, in terms of the dimensionless droplet height, $\hat{H}$), the following equation relating viscous drag exerted on the cylindrical droplet to a length of unity by the flowing gas is obtained:

$$-f_{\mu,d,\text{cylindrical}} = \frac{(p_0 - p_L)}{L}4B^2 \frac{(1+\hat{H})\hat{H}}{(1-\cos\theta_a)(1-\hat{H})^3 + \frac{4B}{L}\hat{H}[1-(1-\hat{H})^3]}. \quad \text{(Eq. 6-142)}$$

In terms of average velocity in the upstream and downstream regions, this equation becomes

$$-f_{\mu,d,\text{cylindrical}} = 12\mu U \frac{(1+\hat{H})\hat{H}}{(1-\cos\theta_a)(1-\hat{H})^3 + \frac{4B}{L}\hat{H}[1-(1-\hat{H})^3]}. \quad \text{(Eq. 6-143)}$$

The total surface-tension force acting on the contact lines of a cylindrical droplet with a length (normal to flow direction or into the page) of unity is given by

$$f_{\gamma,d} = \gamma \cos(180 - \theta_a) + \gamma \cos\theta_r = 2\gamma \sin\frac{1}{2}(\theta_a + \theta_r)\sin\frac{1}{2}(\theta_a - \theta_r)$$

$$= 2\gamma[(\sin\theta_a)y\sqrt{1 - y^2} - (\cos\theta_a)y^2].$$

(Eq. 6-144)

Equating Eqs. 6-143 and 6-144 yields an equation relating the critical height of a cylindrical droplet to the contact angle hysteresis at the onset of instability:

$$y\sqrt{1 - y^2} - (\cot\theta_s)y^2 - \frac{6\mu U}{\gamma \sin\theta_s} \frac{\hat{H}(1 + \hat{H})}{(1 - \cos\theta_s)(1 - \hat{H})^3 + \frac{4B}{L}\hat{H}[1 - (1 - \hat{H})^3]} = 0, \quad \text{(Eq. 6-145)}$$

where $y = \sin\frac{1}{2}(\theta_a - \theta_r)$. In arriving at Eq. 6-145, the advancing contact angle θ_a was approximated by the static contact angle θ_s.

(3) Necessary Conditions for Preventing Droplets from Lodging in the Flow Channel

By considering the extreme scenarios in which droplets grow to the full channel height and using the simplified models developed in the preceding sections, necessary conditions for preventing droplets from lodging in the flow channel are developed and are presented in this section.

A spherical droplet: When the droplet grows to the full channel height, that is, when $\hat{H} = 1$, Eq. 6-135 with $\hat{H} = 1$ yields

$$-f_{\mu,d,\max} = \frac{24\mu UL}{1 - \cos\theta_a}.$$

(Eq. 6-146)

Similarly, Eq. 6-137 with $\hat{H} = 1$ yields

$$f_{\gamma,d,\max} = \frac{4\pi B\sin\theta_s}{1 - \cos\theta_s}\gamma\sin\frac{1}{2}(\theta_a + \theta_r)\sin\frac{1}{2}(\theta_a - \theta_r).$$

(Eq. 6-147)

To prevent droplets from lodging in the flow channel, it is necessary that the drag force exerted on the droplet be greater than the surface-tension force acting on the droplet's contact lines, that is,

$$-f_{\mu,d,\max} > f_{\gamma,d,\max}.$$

(Eq. 6-148)

Substituting Eqs. 6-146 and 6-147 into 6-148 yields a necessary condition for preventing spherical droplets from lodging in the flow channel:

$$\frac{L}{2B}\frac{\mu U}{\gamma} > \frac{\pi}{12}\sin\theta_s \sin\frac{1}{2}(\theta_a + \theta_r)\sin\frac{1}{2}(\theta_a - \theta_r).$$ (Eq. 6-149)

Since $\sin\theta_s \sin\frac{1}{2}(\theta_a + \theta_r)\sin\frac{1}{2}(\theta_a - \theta_r) < 1$ and $\sin\frac{1}{2}(\theta_a + \theta_r) < 1$, Eq. 6-149 can be further simplified to

$$\frac{L}{2B}\frac{\mu U}{\gamma} > \frac{\pi}{12} > \frac{\pi}{12}\sin\theta_s \sin\frac{1}{2}(\theta_a - \theta_r) > \frac{\pi}{12}\sin\theta_s \sin\frac{1}{2}(\theta_a + \theta_r)\sin\frac{1}{2}(\theta_a - \theta_r).$$ (Eq. 6-150)

Eq. 6-150 states simply that if the product of channel length-to-height aspect ratio by the capillary number is greater than $\pi/12$, then spherical droplets can be prevented from lodging in the flow channel. Some design rules can be readily developed from Eq. 6-150. Regardless wetting properties or characteristics at the GDL/GFC interface, water droplet removal can be enhanced by increasing the aspect ratio of channel length to channel height or by increasing the flowing-gas average velocity in the flow channel. The aspect ratio of channel length to channel height can be increased by either lengthening the channel or reducing channel height. Since the gas flow channel is usually a square duct, reducing the channel height implies reducing the channel width also. This design rule of increasing the channel aspect ratio may in part explain the serpentine flow channel design widely employed in practice and perhaps also the trend toward reducing channel width/height. It should be pointed out that the discussion presented above applies to the situation in which only a single droplet emerges from the GDL/GFC interface at any given time. In real-world operation, multiple water droplets may emerge simultaneously from the GDL/GFC interface along the flow-channel–this scenario may change the pressure gradients exerting on droplets downstream and thus affect their instabilities. Moreover, lengthening flow channel while maintaining a constant pressure gradient (or mean velocity) requires raising the pressure drop across the entire channel. With fixed channel geometry and flow conditions, droplet removal can be improved by making the GDL/GFC interface more hydrophobic or reducing contact angle hysteresis.

A *cylindrical droplet:* Similarly, a necessary condition for preventing cylindrical droplets from lodging in flow channels is developed from Eqs. 6-143 and 6-144 with $\hat{H} = 1$:

$$\frac{L}{2B}\frac{\mu U}{\gamma} > \frac{1}{6} > \frac{1}{6}\sin\frac{1}{2}(\theta_a + \theta_r)\sin\frac{1}{2}(\theta_a - \theta_r).$$ (Eq. 6-151)

Equation 6-151 states simply that if the product of channel length-to-height aspect ratio by the capillary number is greater than 1/6, then cylindrical droplets can be prevented from lodging in the flow channel. Since $\pi/12 > 1/6$, it is clear from comparing Eqs. 6-150 and 6-151 that it is easier to prevent cylindrical droplets from lodging in the flow channel as compared with spherical ones.

6.4.3.2 Derivation Using the Drag Coefficient (C_D)

When a solid object is emerged in a fluid flow, the fluid imposes drag forces over the object through pressure and sheer stress. In fluid mechanics, the drag coefficient C_D is defined to characterize the total drag force:

$$C_D = \frac{f_{\text{drag}}/A_p}{\frac{1}{2}\rho U^2}, \qquad \text{(Eq. 6-152)}$$

where A_p is the projected area of a spherical droplet which is estimated as $A_p = (\theta_s - \sin\theta_s \cos\theta_s)d^2/4$ The drag coefficient is determined by the Reynolds number. Chen [50] adopted the following formula:

$$C_D = \frac{30}{\sqrt{\text{Re}_H}}, \quad \text{Re}_H = \frac{\rho U H}{\mu}. \qquad \text{(Eq. 6-153)}$$

The surface tension force (Eq. 6-136) can be simplified as

$$F_{\gamma x} = \pi d \gamma \sin^2\theta_s \sin\frac{1}{2}(\theta_a - \theta_r) \qquad \text{(Eq. 6-154)}$$

Again, the critical gas velocity can be obtained by setting $F_{\text{drag}} = F_{\gamma x}$:

$$U_c = \left[\frac{H}{\rho\mu}\right]^{1/3} \left[\frac{4}{15} \frac{\pi\gamma\sin^2\theta_s \sin\frac{1}{2}(\theta_a - \theta_r)}{(\theta_s - \sin\theta_s \cos\theta_s)d}\right]^{2/3}. \qquad \text{(Eq. 6-155)}$$

A simplified relation between the Weber number and Reynolds number can be set up from the above equation at the critical gas velocity:

$$We_r \, \text{Re}_H^{-0.5} = \frac{2}{15} \frac{\pi\sin^2\theta_s \sin\frac{1}{2}(\theta_a - \theta_r)}{(\theta_s - \sin\theta_s \cos\theta_s)}. \qquad \text{(Eq. 6-156)}$$

Experiment can be carried out to obtain the droplet detachment velocity, as presented in Figure 6-69 for a 1.0×1.6 mm^2 channel with hydrophobic carbon paper (wet proofed with 30 wt% PTFE loading). The contact-angle hysteresis is set as 40° in the analytical solution. The operating condition is 1 atm and 25°C. It shows a reasonably good agreement with the new correlation of the drag coefficient C_D for the fully developed region. In the experiment, the droplets generated experimentally are relatively large with a diameter over 0.5 mm. The detachment velocity for droplets in the entrance length is higher than that in fully developed condition, consistent with the force-balance analysis.

The Re–We correlation is plotted in Figure 6-70, in comparison with the experimental data. The Re–We relation, based on a new correlation of the drag coefficient C_D, is shown for various H/r values. The ratio between channel height and droplet radius (H/r) of experimental data ranges from 1.9 to 2.9. Figure 6-71 presents other experimental data in larger channel dimension (7.0×2.7 mm^2) with similar GDL materials (hydrophobic carbon paper and cloth).

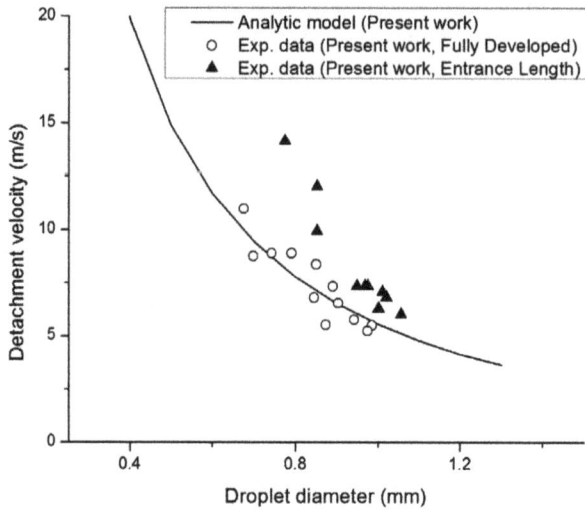

Figure 6-69. Detachment velocities at the fully developed and entrance regions [46].

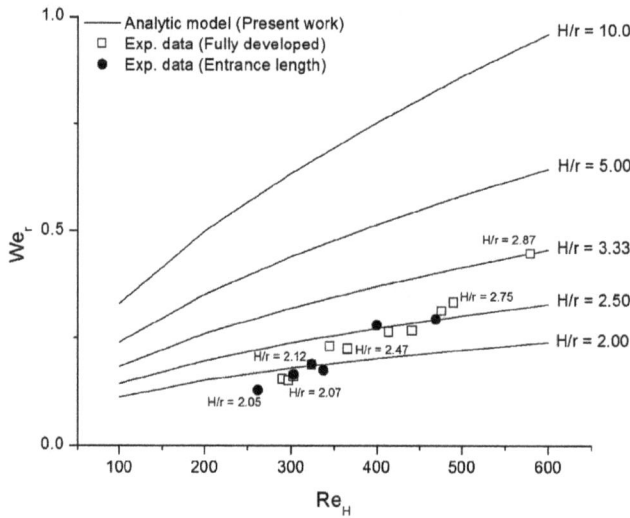

Figure 6-70. The Weber–Reynolds number relation at the detachment velocity [46].

The average static contact angle in the experiment is 130° for carbon paper and 145° for carbon cloth. The carbon paper has a higher detachment velocity than the carbon cloth and the analytical model also follows the same trend. As expected, the detachment velocity decreases as droplet diameter increases. Furthermore, in PEM fuel cells various reactants and flow conditions are encountered, such as different temperatures and reactants (hydrogen gas, air, and reformate). For example, a PEM fuel cell usually works in a range of temperature such as between 20°C and 80°C. The surface tension will change from around 0.0728 at 20°C to 0.0626 N/m at 80°C, about 15% difference. Gas composition changes with temperature as well. The above analytical solution provides a quick method to examine these different conditions.

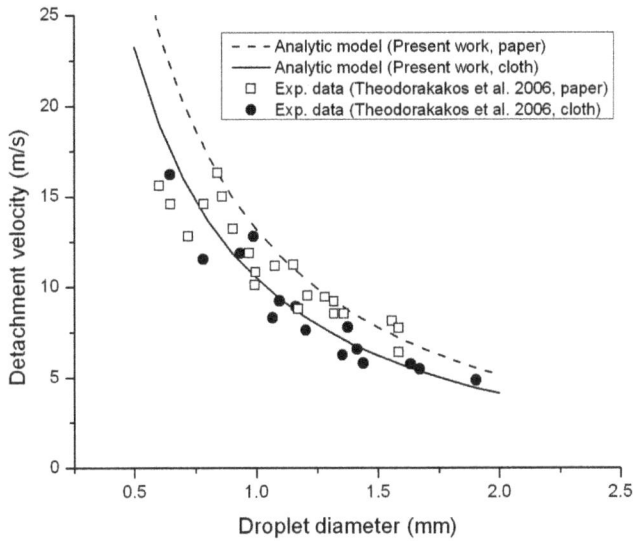

Figure 6-71. Comparison of the detachment velocity between the analytic solutions and experimental data [46]. Experimental data are from Ref. [51].

Note: Present work here refers to Eq. 6-155.

6.5 CHAPTER SUMMARY

This chapter delineates the critical two-phase flow phenomena in PEM fuel cells, which occur in three major components, namely gas diffusion layers (GDLs)/microporous layers (MPLs), catalyst layers (CLs), and gas flow channels (GFCs). Several key points of this chapter are summarized as follows:

- In GDLs/MPLs and CLs, the presence of liquid water hinders the transport of gaseous reactants, and it additionally reduces the active surface area for the HOR and ORR reactions in CLs; the multiphase mixture formulation is widely adopted to model the two-phase flow in PEM fuel cells. In homogeneous GDLs, the capillary action can be treated as "diffusion" (which results in liquid saturation varying monotonically within GDL), whereas in heterogeneous GDLs, local maximum/minimum may occur, as a result of spatial property variation, and a jump in liquid saturation at the GDL/MPL interface occurs, which arises from the condition of the capillary pressure being continuous at the interface.

- In gas flow channels (GFCs), liquid water accumulates, and consequently two-phase flow occurs and the pressure drop along the channel is increased dramatically (by as much as a factor of 10 or more). The inlet RH and stoichiometric flow ratio affect the onset and liquid saturation of two-phase flow; in the heterogeneous configuration of flow channels, liquid water may trap at the heterogeneities (which results in spikes in liquid saturation). Two-phase flow in GFCs experiences several flow regimes such as wave and annulus; both the multiphase mixture and two-fluid formulations are adopted to model the two-phase flow phenomena in PEM fuel cells, and analytical solutions are obtained, respectively, for the pressure drop and the 1D two-phase transport along GFCs.

- Liquid water dynamics at the GDL/GFC interface is important because it bridges the two-phase flows in GDLs and GFCs. Droplet formation increases the interfacial transport resistance; air flow exerts drag forces over droplets through the pressure variation and shear stress; liquid droplet deforms when subject to the viscous and pressure forces from air flow; a formula is developed to quantify the degree of deformation, and relate droplet deformation to the Weber number; and the force balance over a droplet is developed and the critical detachment velocity (in the Reynolds number) is obtained as a function of the Weber number.

REFERENCES

1. Nam, J.H., Lee, K.J., Hwang, G.S., Kim, C.J., and Kaviany, M. Microporous layer for water morphology control in PEMFC. *Int. J. Heat Mass Transfer*, *52* (2009): 2779–91. doi: http://dx.doi.org/10.1016/j.ijheatmasstransfer.2009.01.002

2. Crowe, C.T. *Multiphase Flow Handbook*, 1st Ed., CRC Press, (2005). doi: http://dx.doi.org/10.1201/9781420040470

3. Brennen, C.E. *Fundamentals of Multiphase Flow*. Cambridge University Press, (2005).

4. Whalley, P.B. *Two-Phase Flow and Heat Transfer*. Oxford, (1996).

5. Nam, J.H., and Kaviany, M. Effective diffusivity and water-saturation distribution in single- and two-layer PEMFC diffusion medium. *Int. J. Heat Mass Transf.* *46* (2003): 4595–611. doi: http://dx.doi.org/10.1016/S0017-9310(03)00305-3

6. Hussaini, I.S., and Wang, C.Y. Measurement of relative permeability of fuel cell diffusion media. *J. Power Sources 195* (2010): 3830–40. doi: http://dx.doi.org/10.1016/j.jpowsour.2009.12.105

7. Lenormand, R., Touboul, E., and Zarcone, C. Numerical models and experiments on immiscible displacements in porous media. *J. Fluid Mech. 189* (1988): 165–87. doi: http://dx.doi.org/10.1017/S0022112088000953

8. Aker, E., Jørgen Måløy, K., Hansen, A., and Batrouni, G.G. A two-dimensional network simulator for two-phase flow in porous media. *Transport Porous Med. 32* (1998): 163–86. doi: http://dx.doi.org/10.1023/A:1006510106194

9. Sinha, P.K., Halleck, P., and Wang, C.Y. Quantification of liquid water saturation in a PEM fuel cell diffusion medium using X-ray microtomography. *Electrochem. Solid-State Lett. 9* (2006): A344. doi: http://dx.doi.org/10.1149/1.2203307

10. Hartnig, C., Manke, I., Kuhn, R., Kardjilov, N., Banhart, J., and Lehnert, W. Cross-sectional insight in the water evolution and transport in polymer electrolyte fuel cells. *Appl. Phys. Lett. 92* (2008): 134106. doi: http://dx.doi.org/10.1063/1.2907485

11. Hussey, D.S., Jacobson, D.L., Arif, M., Owejan, J.P., Gagliardo, J.J., and Trabold, T.A. Neutron images of the through-plane water distribution of an operating PEM fuel cell. *J. Power Sources* 172 (2007): 225. doi: http://dx.doi.org/10.1016/j.jpowsour.2007.07.036

12. Hickner, M.A., Siegel, N.P., Chen, K.S., Hussey, D.S., Jacobson, D.L., and Arif, M. *J. Electrochem. Soc.* 155 (2008): B427. doi: http://dx.doi.org/10.1149/1.2826287

13. Mukundan, R., and Borup, R.L. Visualizing liquid water in PEM fuel cells using neutron imaging. *Fuel Cells* 9 (2009): 499. doi: http://dx.doi.org/10.1002/fuce.200800050

14. Wang, Y., and Chen, K.S. Through-plane water distribution in a polymer electrolyte fuel cell: comparison of numerical prediction with neutron radiography data. *J. Electrochem. Soc. 157* (2010): B1878–86. doi: http://dx.doi.org/10.1149/1.3498997

15. Wang, C.Y., and Cheng, P. In Hartnett, J.P. Jr. et al. (Eds.) *Advances in Heat Transfer*, vol. 30. Academic Press, (1997): 93–196.

16. Koido, T., Furusawa, T., Moriyama, K., and Takato, K. Two-phase transport properties and transport simulation of the gas diffusion layer of a PEFC. *ECS Trans. 3* (2006): 425–34. doi: http://dx.doi .org/10.1149/1.2356163

17. Leverett, M.C. Capillary behavior in porous solids. *Trans. AIME 142* (1941): 151–69.

18. Kumbur, E.C., Sharp, K.V., and Mench, M.M. A validated Leverett approach to multiphase flow in polymer electrolyte fuel cell diffusion media: Part 1. Hydrophobicity effect. *J. Electrochem. Soc. 154* (2007): B1295–304. doi: http://dx.doi.org/10.1149/1.2784283

19. Gostick, J.T., Ioannidis, M.A., Pritzker, M.D., and Fowler, M.W. Impact of liquid water on reactant mass transfer in PEM fuel cell electrodes. *J. Electrochem. Soc. 157* (2010): B563–71. doi: http://dx.doi .org/10.1149/1.3291977

20. Pasaogullari, U., and Wang, C.Y. Liquid water transport in gas diffusion layer of polymer electrolyte fuel cells. *J. Electrochem. Soc. 151* (2004): 399.

21. Wang, Y. Modeling of two-phase transport in the diffusion media of polymer electrolyte fuel cells. *J. Power Sources 185* (2008): 261–71. doi: http://dx.doi.org/10.1016/j.jpowsour.2008.07.007

22. Wang, Y., Wang C.Y., and Chen, K.S. Elucidating differences between carbon paper and carbon cloth in polymer electrolyte fuel cells. *Electrochimica Acta 52* (2007): 3965–75. doi: http://dx.doi.org/10.1016/j .electacta.2006.11.012

23. Benziger, J., Nehlsen, J., Blackwell, D., Brennan, T., and Itescu, J. Water flow in the gas diffusion layer of PEM fuel cells. *J. Membr. Sci. 261* (2005): 98. doi: http://dx.doi.org/10.1016/j.mem- sci.2005.03.049

24. Wang, Y., and Chen, K.S. Effect of spatially-varying GDL properties and land compression on water distribution in PEM fuel cells. *J. Electrochem. Soc. 158* (2011): B1292–99. doi: http://dx.doi .org/10.1149/2.015111jes

25. Hinebaugh, J., Fishman, Z., and Bazylak, A. Proceedings of the ASME 2010 Eighth International Fuel Cell Science, Engineering and Technology Conference, (2010) FuelCell2010-33097.

26. Tabuchi, Y., Shiomi, T., Aoki, O., Kubo, N., and Shinohara, K. Effects of heat and water transport on the performance of polymer electrolyte membrane fuel cell under high current density operation. *Electrochim. Acta 56* (2010): 352–60. doi: http://dx.doi.org/10.1016/j.electacta.2010.08.070

27. Park, S., Lee, J.W., and Popov, B.N. Effect of PTFE content in microporous layer on water manage- ment in PEM fuel cells. *J. Power Sources* 177 (2008): 457–63. doi: http://dx.doi.org/10.1016/j.jpow- sour.2007.11.055

28. Gostick, J.T., Ioannidis, M.A., Fowler, M.W., and Pritzker, M.D. On the role of the microporous layer in PEMFC operation. *Electrochem. Comm.* 11 (2009): 576–79.

29. Wang, Y., Basu, S., and Wang, C.Y. Modeling two-phase flow in PEM fuel cell channels. *J. Power Sources 179* (2008): 603–617. doi: http://dx.doi.org/10.1016/j.jpowsour.2008.01.047

30. Chen, K.S., Hickner, M.A., Siegel, N.P., Noble, D.R., Pasaogullari, U., and Wang, C.Y. Final Report on LDRD Project: Elucidating Performance of Proton-Exchange-Membrane Fuel Cells via Computational Modeling with Experimental Discovery and Validation, Sandia Technical Report, Sandia National Laboratories, SAND2006-6964 (2006).

31. Adroher, X.C., and Wang, Y. *Ex-situ* and modeling study of two-phase flow in a single channel of polymer electrolyte membrane fuel cells. *J. Power Sources 196* (2011): 9544–51. doi: http://dx.doi.org/10.1016/j .jpowsour.2011.07.076

32. Tuber, K., Pocza, D., and Hebling, C. Visualization of water buildup in the cathode of a transpar- ent PEM fuel cell. *J. Power Sources 124* (2004): 403–14. doi: http://dx.doi.org/10.1016/S0378 -7753(03)00797-3

33. Trabold, T.A., Owejan, J.P., Jacobson, D.L., Arif, M., and Huffman, P.R. *In situ* investigation of water transport in an operating PEM fuel cell using Neutron radiography: Part I – experimental method and ser- pentine flow field results. *Int. J. Heat Mass Transf. 49* (2006): 4712–20. doi: http://dx.doi.org/10.1016/j .ijheatmasstransfer.2006.07.003

34. Chisholm, D.A. A theoretical basis for the Lockhart–Martinelli correlation for two-phase flow. *Int. J. Heat Mass Transf. 10* (1967): 1767–78. doi: http://dx.doi.org/10.1016/0017-9310(67)90047-6

35. Mishima, K., and Hibiki, T. Some characteristics of air–water two-phase flow in small diameter vertical tubes. *Int. J. Multiphase Flow 22* (1996): 703–12. doi: http://dx.doi.org/10.1016/0301-9322(96)00010-9

36. Chaouche, M., Rakotomalala, N., Salin, D. and Yortsos, Y.C. Capillary effects in immiscible flows in heterogeneous porous media. *Europhys. Lett. 21* (1993): 19–24. doi: http://dx.doi.org/10.1209/0295-5075/21/1/004

37. Or, D., and Tuller, M. Hydraulic conductivity of unsaturated fractured porous media: flow in a cross-section. *Adv. Water Resour. 26* (2003): 883–98. doi: http://dx.doi.org/10.1016/S0309-1708(03)00051-4

38. Li, W., Vigil, R.D., Beresnev, I.A., Iassonov, P., and Ewing, R. Vibration-induced mobilization of trapped oil ganglia in porous media: experimental validation of a capillary-physics mechanism. *J. Colloid Interf. Sci. 289* (2005): 193–99. doi: http://dx.doi.org/10.1016/j.jcis.2005.03.067

39. Chen, J.D. Measuring the film thickness surrounding a bubble inside a capillary. *J. Colloid Interf. Sci. 109* (1986): 341–49. doi: http://dx.doi.org/10.1016/0021-9797(86)90313-9

40. Wong, H., Morris, S., and Radke, C.J. Three-dimensional menisci in polygonal *capillaries. J. Colloid Interf. Sci. 148* (1992): 317–336. doi: http://dx.doi.org/10.1016/0021-9797(92)90171-H

41. Wang, C.Y., and Cheng, P. A multiphase mixture model for multiphase, multicomponent transport in capillary porous media—I. Model development. *Int. J. Heat Mass Transf. 39* (1996): 3607–3618. doi: http://dx.doi.org/10.1016/0017-9310(96)00036-1

42. Benning, T., and Djilali, N. A 3D, multi-phase, multicomponent model of the cathode and anode of a PEM fuel cell. *J. Electrochem. Soc. 150* (2003): A1589–98. doi: http://dx.doi.org/10.1149/1.1621412

43. Wang, Y. Porous-media flow fields for polymer electrolyte fuel cells II. Analysis of channel two-phase flow. *J. Electrochem. Soc. 156* (2009): B1134–1141. doi: http://dx.doi.org/10.1149/1.3183785

44. Corey, A.T. *Mechanics of Immiscible Fluids in Porous Media*, 2nd Ed. Water Resources, (1986).

45. Fourar, M., and Lenormand, R. A viscous coupling model for relative permeabilities in fractures, Paper SPE 49006 presented at the SPE Annual Technical Conference and Exhibition, (1998).

46. Cho, S.C., Wang, Y., and Chen, K.S. Droplet dynamics in a polymer electrolyte fuel cell gas flow channel: forces, deformation and detachment: II. Comparisons of analytical solution with numerical and experimental results. *J. Power Sources 210* (2012): 191–97. doi: http://dx.doi.org/10.1016/j.jpowsour.2012.03.033

47. Cho, S.C., Wang, Y., and Chen, K.S. Droplet dynamics in a polymer electrolyte fuel cell gas flow channel: forces, deformation, and detachment: I. Theoretical and numerical analyses. *J. Power Sources 206* (2012): 119–28. doi: http://dx.doi.org/10.1016/j.jpowsour.2012.01.057

48. Chen, K.S., Hickner, M.A., and Noble, D.R. Simplified models for predicting the onset of liquid water droplet instability at the gas diffusion layer/gas flow channel interface. *Int. J. Energy Res. 29* (2005): 1113–32. doi: http://dx.doi.org/10.1002/er.1143

49. Bird, R.B., Stewart, W.E., and Lightfoot, E.N. *Transport Phenomena*, 2nd Ed. Wiley, (2001).

50. Chen, K.S. Modeling water-droplet detachment from GDL/channel interface in PEM fuel cells, Sixth International Conference on Fuel Cell Science Engineering and Technology, (2008) FUEL-CELL2008-65137.

51. Theodorakakos, A., Ous, T., Gavaises, M., Nouri, J.M., Nikolopoulos, N., and Yanagihara, H. Dynamics of water droplets detached from porous surfaces of relevance to PEM fuel cells. *J. Colloid Interface Sci., 300* (2006): 673–87. doi: http://dx.doi.org/10.1016/j.jcis.2006.04.021

CHAPTER 7

ICE DYNAMICS AND REMOVAL

PEM fuel cells must have the ability to survive and startup from subfreezing temperatures for them to be successfully deployed in automobiles. Under freezing environmental conditions, water produced in the cathode has a tendency to freeze in open pores in the catalyst layers and GDLs, rather than be removed from the fuel cell, thus creating mass transport limitations that eventually shutdown the operation. Therefore, starting up a PEM fuel cell from subfreezing temperatures or simply "cold start" is of paramount importance to automobile applications. In this chapter, we provide an overview on important issues and fundamentals in subfreezing operation, such as stages of cold start, oxygen starvation, the cell voltage loss due to ice formation, and key factors that determine and control the cold-start capability for PEM fuel cells.

7.1 SUBFREEZING OPERATION–OVERVIEW

During startup from subzero temperatures or cold start, water produced in the electrodes freezes instantaneously on the reaction sites as illustrated in Figure 7-1, covering and hence reducing the electrochemical active surface and plugging the open pores that enable reactant passage. To ensure a successful startup below freezing point, the PEM fuel cell needs to be heated, by either external heat source or self-heating, to 0°C or higher to prevent the ice produced in the catalyst layer from causing severe reactant starvation and considerable reduction in the ECA (electrochemically active area) and consequent substantial voltage loss. In automobile applications, such a startup for a fuel cell must be short (on the order of minutes or less) in order to be able to compete with the traditional internal combustion engines (ICE).

Many studies have been reported in the open literature to investigate the steady-state and transient operations with temperature above the freezing point; however, the physics of cold start remains poorly understood. Under a subfreezing environment, the oxygen reduction reaction (ORR) is still active and able to produce water. Product water can be visualized directly using neutron radiography–Figure 7-2 shows the evolution of the water contents under the channel and land, respectively, obtained from a neutron radiography imaging experiment that was conducted at the National Institute of Standard and Technology's Center for Neutron Research (NCNR). The measurements were conducted using Beam #1 and Aperture #4

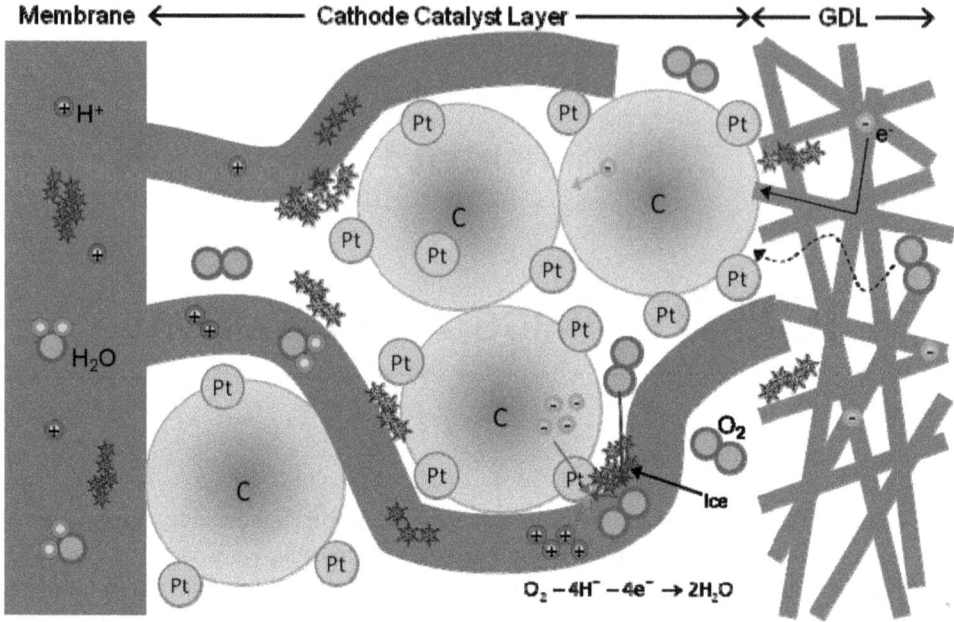

Figure 7-1. Schematic of ice formation in a PEM fuel cell during subfreezing operation [1] (Courtesy of the Electrochemical Society).

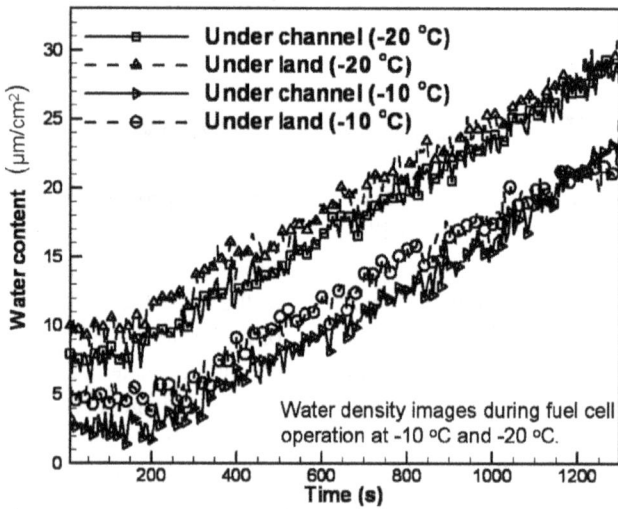

Figure 7-2. Water content evolution probed by neutron radiography during PEM fuel cell operation [2].

with a fluence rate of 2×10^7 neutrons/(cm^2·s). The fuel cell operates at subzero temperature from $-10°C$ and $-20°C$, respectively; consequently, water exists in its solid state, that is, as ice. The water content is shown to increase steadily with time as a result of the ORR's water production. At subzero temperatures, water freezing in the pores of the catalyst layer blocks the open pores required for reactant transport and eventually lead to voltage loss (Fig. 7-3).

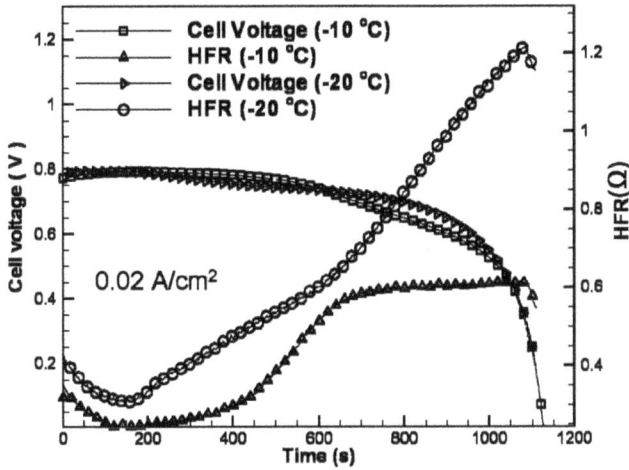

Figure 7-3. HFR (high-frequency resistance) and cell voltage evolutions during PEM fuel cell operation.

7.2 ICE FORMATION

7.2.1 WATER TRANSPORT AND CONSERVATION

In the cathode catalyst layer, the net water addition is the ORR's production plus that transported inward through boundaries. Assuming that no super-cooled liquid is present, water exists in the ionomer, gaseous, and ice phases under subfreezing conditions. Phase equilibrium usually holds true due to the small dimension of pores (~0.1 μm). Ice starts to emerge when the vapor phase reaches saturation. Cold start consists of three major stages: the first stage of cold start is water accumulation in the ionomer and gas phases until the vapor-phase pressure reaches the saturation pressure limit. In the second stage, the ionomer and gaseous phases are already saturated, and hence water is added to the solid phase; therefore, this stage is characterized by ice volume increase within the catalyst layer (Fig. 7-4). The third stage starts when the fuel cell reaches 0°C or so, and thus the residual ice in the catalyst layer melts due to fuel cell heat generation or external heating. The three stages are schematically illustrated in Figure 7-5. In what follows, the conditions of constant current density, net water transfer coefficient α and heat generation during cold start are considered and discussed.

For the open system of a cathode catalyst layer, the net water storage rate is equal to the production rate plus that transported across the catalyst layer's boundaries, as expressed in the following equation:

$$\iiint_{\text{CL}} S_w dv = \iiint_{\text{CL}} S_{\text{prod}} dv + \iiint_{\text{CL}} S_{\text{EOD}} dv - \iint_{\text{MEM-CL}} \vec{G}_{\text{MEM}} \cdot d\vec{s} - \iint_{\text{GDL-CL}} \vec{G}_{\text{GDL}} \cdot d\vec{s}. \quad \text{(Eq. 7-1)}$$

In the above, $S_{\text{prod}} = -\dfrac{j_c}{2F}$ and $S_{\text{EOD}} = -\nabla \cdot (\dfrac{n_d}{F}\vec{i}) = -\bar{n}_d \dfrac{j_c}{F}$ where j_c is the transfer current density. The electro-osmotic drag (EOD) coefficient n_d is a function of membrane water content, and for analysis purpose, a constant value of $\bar{n}_d$ is used. $\vec{G}_{\text{MEM}}$ denotes the water flux

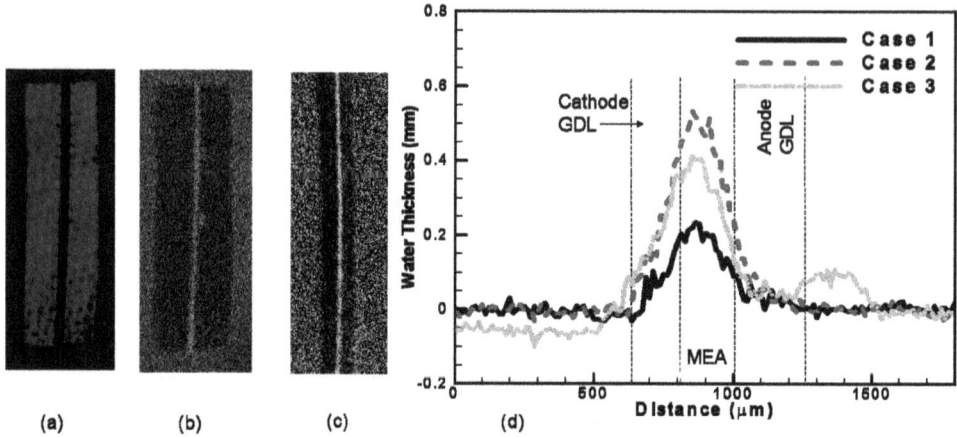

Figure 7-4. Neutron imaging of an operating PEM fuel cell: (a) grayscale; (b) in color; (c) close-up image; and (d) water thickness detected by high-resolution neutron imaging. Case 1: cooling from 100% RH, 0.4 V, 80°C and cold start at 0.04 A/cm^2; Case 2: cooling from 50% RH, 0.8 V, 80°C, and cold start at 0.04 A/cm^2; Case 3: cooling from 50% RH, 0.8 V, 80°C and cold start at 0.094 A/cm^2.

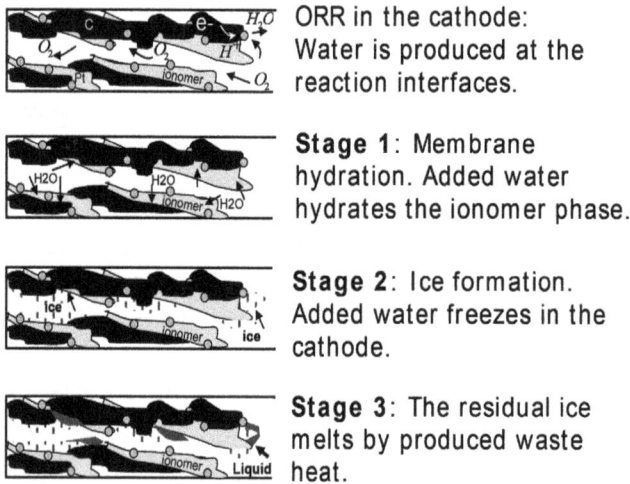

ORR in the cathode: Water is produced at the reaction interfaces.

Stage 1: Membrane hydration. Added water hydrates the ionomer phase.

Stage 2: Ice formation. Added water freezes in the cathode.

Stage 3: The residual ice melts by produced waste heat.

Figure 7-5. The three stages of PEM fuel cell cold-start.

back to the membrane. The net water transfer coefficient α is conventionally used to quantify the water exchange between the anode and cathode sides. Herein, this term is extended to cold start to show the net water gain in the cathode in addition to water production. Thus, the coefficient α also accounts for the water absorbed/released by the ionomer membrane. Assuming no spatial variations, that is, $\iiint_{CL} []dv = []V_{CL}$, the first two terms on the right-hand side of Eq. 7-1 become

$$-(1+2\alpha)\frac{j_c}{2F}V_{CL} = (1+2\alpha)\frac{I}{2F}A_m. \qquad \text{(Eq. 7-2)}$$

The water flux at the catalyst layer-GDL interface can be estimated by

$$\vec{G}_{GDL} = \left(-D_w^{g,\text{eff}}\nabla C_w^g + \vec{u}C_w^g\right)_{\text{GDL-CL}}. \qquad \text{(Eq. 7-3)}$$

In most flow-field designs, diffusion dominates reactant transport in GDLs. Assuming no ice or liquid water formation in GDLs, water conservation in the cathode channel and GDL yields

$$\iint\limits_{\text{GDL-CL}} \vec{G}_{GDL}\cdot d\vec{s} = \iint\limits_{\text{outlet}} \vec{u}C_w^g\cdot d\vec{s} + \iint\limits_{\text{inlet}} \vec{u}C_w^g\cdot d\vec{s} = \frac{IA_m}{F}\frac{\xi_c(\bar{C}_{w,c,\text{out}}^g - \bar{C}_{w,c,\text{in}}^g)}{4C_{O2,c,\text{in}}}, \qquad \text{(Eq. 7-4)}$$

where $\bar{C}$ is the average value, defined by $\iint \vec{u}C\cdot d\vec{s} \Big/ \left|\iint\limits_{\text{inlet}} \vec{u}\cdot d\vec{s}\right|$. If the flux in Eq. 7-4 was larger than the water gain rate in the cathode, no ice would form in the cathode. A dimensionless parameter $\beta_{1,c}$ can then be defined as

$$\beta_{1,c} = \frac{\dfrac{IA_m}{F}\dfrac{\xi_c(\bar{C}_{w,c,\text{out}}^g - \bar{C}_{w,c,\text{in}}^g)}{4C_{O2,c,\text{in}}}}{(1+2\alpha)\dfrac{I}{2F}A_m} = \frac{\xi_c(\bar{C}_{w,c,\text{out}}^g - \bar{C}_{w,c,\text{in}}^g)}{2(1+2\alpha)C_{O2,c,\text{in}}}, \qquad \text{(Eq. 7-5)}$$

where $\beta_{1,c}$ compares the water removal by the channel stream with that gained in the cathode. Under the condition of $\beta_{1,c}$ less than unity, water accumulates either in the solid state or the ionomer phase. $\beta_{1,c}$ may be a function of time during transient; at steady state beyond cold-start operation $\beta_{1,c}$ equals to 1. Under cold start or low humidity, water in the gas flow channel is totally taken away via the gas phase (assuming no liquid water/ice formation in the gas stream). Using the relative humidity (RH), Eq. 7-5 changes to

$$\beta_{1,c,g} = \frac{\xi_c(RH_{c,\text{out}}C_{w,\text{sat,out}} - RH_{c,\text{in}}C_{w,\text{sat,in}})}{2(1+2\alpha)C_{O2,c,\text{in}}}. \qquad \text{(Eq. 7-6)}$$

The maximum value of $\beta_{1,c,g}$, that is, the capability of vapor removal by channel flow, is at $RH_{c,\text{out}} = 1$. The outlet and inlet temperatures can be different, which is determined by the cooling unit configuration. When the inlet and outlet temperatures are the same, Figure 7-6 plots the maximum $\beta_{1,c,g}$ at different temperatures and pressures (related to $C_{O2,c,\text{in}}$). For sub-freezing operation, $\beta_{1,c,g}$ is mostly small (<2%). Thus, the amount of water taken away by the channel gas flow is neglected in the discussion presented in the following section.

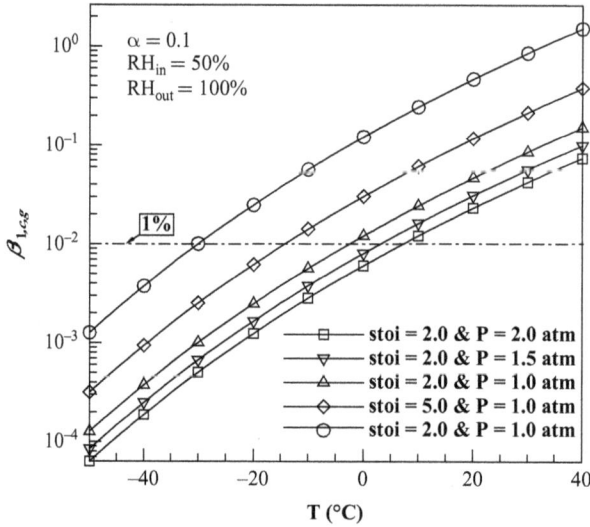

Figure 7-6. Functional dependence of $\beta_{1,c,g}$ on the stoichiometric ratio and pressure [3] (Courtesy of the Electrochemical Society).

7.2.2 THREE COLD-START STAGES

7.2.2.1 First Stage: Membrane Hydration

Below the freezing point, water exists in gas, solid, or ionomer phases. The ionomer, typically Nafion, holds much water: the amount of water stored in the membrane can be hundreds of times more than that as vapor in the channel space. Thus, the small amount of water in the gaseous phase in the catalyst layer is neglected. Cold start experiences two major stages before reaching the freezing point [3]: one is the hydration of the ionomer phase in the catalyst layer; the other is ice formation in the void space. Assuming that thermodynamic equilibrium among the gas, ionomer, and ice phases prevail due to the presence of a large interfacial area, no ice is produced in the first stage. The time constant $\tau_{ice,1}$ of this stage is estimated by

$$\int_0^{\tau_{ice,1}} S_w dt = \int_{\lambda_0}^{14} \varepsilon_m \frac{\rho_m}{EW} d\lambda \qquad \text{(Eq. 7-7)}$$

where ε_m is the ionomer volume fraction in the catalyst layer. For a constant rate of water addition S_w, $\tau_{ice,1}$ is then rewritten as

$$\tau_{ice,1} = \frac{\varepsilon_m \delta_{CL} \frac{\rho_m}{EW}(14-\lambda_0)}{(1+2\alpha)\dfrac{I}{2F}} = \frac{2F\rho_m \varepsilon_m \delta_{CL}(14-\lambda_0)}{EW(1+2\alpha)I}. \qquad \text{(Eq. 7-8)}$$

The ionomer volumetric content of 0.13–0.4 yields the range of $\tau_{ice,1}$ from 0 to 30 s for δ_{CL} of 10 μm, α of 0.1, and I of 0.1 A/cm². In this stage, product water is added to the ionomer phase, increasing the cathode hydration level and hence the ionic conductivity. Under subfreezing temperatures, membrane water diffusivity becomes relatively small, likely resulting in a net water flow to the cathode or a positive value of α. This can be qualitatively shown through an isothermal cold-start experiment, presented in Figure 7-3, which plots the HFR (high-frequency resistance) evolution during cold start from −10°C and −20°C, respectively. The initial HFR decrease is possibly due to the net water addition in the ionomer phase from water production. After the cathode-side ionomer is saturated (i.e., the second stage), water is added to the ice phase instead of hydrating ionomer and the anode-side membrane keeps losing water, thus increasing the HFR as indicated by Figure 7-3.

7.2.2.2 Second Stage: Ice Formation

Assuming that no liquid water is formed at super cooled state, ice will be produced upon water addition at $t > \tau_{ice,1}$. The ice volume fraction in the void space s_{ice} can be expressed as follows:

$$s_{ice}(t) = \frac{M_w}{\varepsilon_{CL}\rho_{ice}}\left(\int_{\tau_{ice,1}}^{t} S_w(t)dt - \int_{14}^{\lambda(t)} \varepsilon_m \frac{\rho_m}{EW}d\lambda\right). \qquad \text{(Eq. 7-9)}$$

For constant S_w, neglecting the ionomer absorption because of no liquid water formation yields

$$s_{ice}(t) = \frac{(1+2\alpha)M_w I}{2F\varepsilon_{CL}\rho_{ice}\delta_{CL}}(t-\tau_{ice,1}) = \frac{(t-\tau_{ice,1})}{\tau_{ice,2}} = \frac{t}{\tau_{ice,2}} - k_\tau \qquad \tau_{ice,1} < t \leq \tau_{ice,2}, \text{ (Eq. 7-10)}$$

where the time constant for the second stage $\tau_{ice,2}$ and ratio of the two time constants k_τ are defined as

$$\tau_{ice,2} = \frac{2F\varepsilon_{CL}\rho_{ice}\delta_{CL}}{(1+2\alpha)M_w I} \text{ and } k_\tau = \frac{\tau_{ice,1}}{\tau_{ice,2}} = \frac{\rho_m\varepsilon_m(14-\lambda_0)M_w}{\rho_{ice}\varepsilon_{CL}EW}. \qquad \text{(Eq. 7-11)}$$

In Eq. 7-11, k_τ is determined by the geometrical and material properties as well as the initial membrane conditions. Adding the two time constants yields

$$\tau_{s_{ice}} = \tau_{ice,1} + \tau_{ice,2} = \frac{2F\delta_{CL}}{(1+2\alpha)I}\left(\frac{\rho_m\varepsilon_m(14-\lambda_0)}{EW} + \frac{\varepsilon_{CL}\rho_{ice}}{M_w}\right). \qquad \text{(Eq. 7-12)}$$

This time constant $\tau_{s_{ice}}$ represents the duration it takes ice to occupy the entire void space (i.e., $s_{ice} = 1$). Assuming that the fuel cell shuts down at $s_{ice} = 1$ (this might not hold true when the tortuosity of the catalyst layer is large or under high-current operation; in that case, $\tau_{s_{ice}}$ can be defined as the time period for ice volume fraction reaching the critical value), $\tau_{T,1} > \tau_{s_{ice}}$ means startup failure due to ice occupation, while $\tau_{T,1} < \tau_{s_{ice}}$ means that the fuel cell is able to startup from subzero environment without any external aid. Therefore, one can further define a dimensionless parameter β_2 as

$$\beta_2 = \frac{\tau_{T,1}}{\tau_{s_{ice}}}. \qquad \text{(Eq. 7-13)}$$

Under the condition of $\beta_2 < 1$, a PEM fuel cell is able to startup. Once the cell temperature reaches 0°C, the ice volume starts decreasing as a result of fusion of the residual ice; thus, there exists a maximum ice volume fraction, given by

$$s_{ice}^{max} = \frac{\tau_{T,1}}{\tau_{ice,2}} - k_\tau. \qquad \text{(Eq. 7-14)}$$

Though physically s_{ice}^{max} cannot be over one (or unity), the above expression can be mathematically extended to the case of $\beta_2 > 1$. Similar to β_2, s_{ice}^{max} can be used as a criterion for cold-start analysis. A cold start with $s_{ice}^{max} > 1$ will fail.

Figure 7-7 displays the maximum ice volume fraction s_{ice}^{max}. The regime in which s_{ice}^{max} reaches unity corresponds to cold-start failure. Reducing λ_0 has limited benefits to cold-start operation, relative to the option of improving bipolar plate materials. For these bipolar-plate materials, self-cold-start from above -10°C is likely successful because s_{ice}^{max} is less than unity.

Figure 7-8 displays β_2 as a function of cold-start temperature, initial membrane water content (λ_0), and catalyst layer thicknesses. In the gray region in which $\beta_2 < 1$, cold start may succeed, whereas it fails in the above in which $\beta_2 > 1$. Similar to s_{ice}^{max}, β_2 increases with decreasing initial cold-start temperature. Similarly, reducing λ_0 shows a light improvement on cold-start characteristics as opposed to the option of increasing the catalyst layer thickness.

It is worthy to note that assuming a fuel cell shuts down at $s_{ice}=1$ might not hold true when the tortuosity of the catalyst layer is large or under high-current operation. Figure 7-9 (a) shows the cathode electrode cross-section following the operation at 0.01 A/cm^2, whereas Figure 7-9 (b) is an SEM image of the same electrode after ice sublimation and drying at high temperatures to remove the ice. Comparison of these two images shows that nearly all of the electrode porosity, across the entire cross section, is filled with ice generated by low-current subfreezing operation. In Figure 7-9 (c), it is clearly seen that the electrode void volume is largely free of ice from product water at 0.2 A/cm^2.

7.2.2.3 Third Stage: Ice Melting

For $\beta_2 < 1$, a PEM fuel cell will encounter the third stage of cold start, that is, ice melting. This stage is isothermal at 0°C due to phase change. The time constant for ice melting $\tau_{T,2}$ can be obtained as follows:

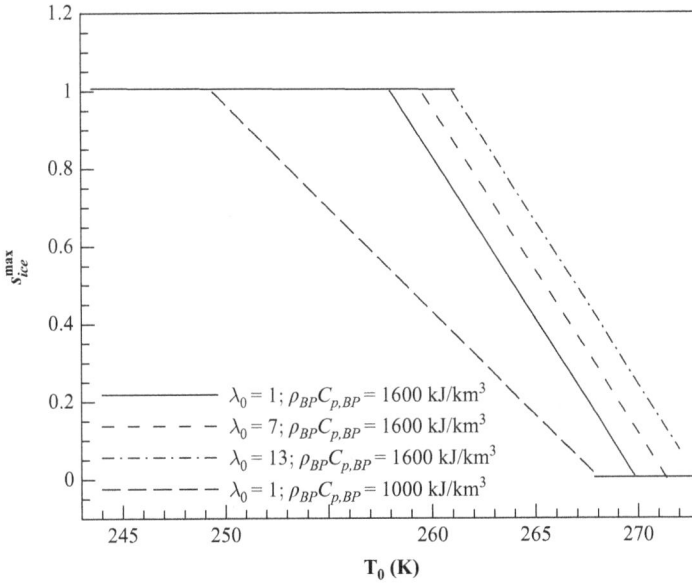

Figure 7-7. The maximum ice volume fraction in the cathode catalyst layer as a function of cold-start temperature and bipolar plate thermal properties [2].

$$\int_{\tau_{T,1}}^{\tau_{T,1}+\tau_{T,2}} I(E'_o - V_{\text{cell}})dt = \rho_{\text{ice}}h_{sl}\delta_{\text{CL}}\varepsilon_{\text{CL}}s_{\text{ice}}^{\max} = \rho_{\text{ice}}h_{sl}\delta_{\text{CL}}\varepsilon_{\text{CL}}\left(\frac{\tau_{T,1}}{\tau_{\text{ice},2}} - k_\tau\right), \quad \text{(Eq. 7-15)}$$

where E'_o is defined as $-\dfrac{\Delta h}{2F}$, representing the EMF (electromotive force) that all the energy from hydrogen/oxygen, the 'calorific value', heating value, or enthalpy of formation, were transformed into electrical energy with the ice as the reaction product. The EMF takes into account the latent heats of fusion and condensation as well as the sensible heat. h_{sl} is the latent heat during the phase change of liquid water and ice. Given the typical values of the parameters for PEM fuel cells, the maximum ice content, that is, $s_{\text{ice}}^{\max} = 1$, yields $\tau_{T,2} \sim 2$ s at 0.1 A/cm^2, indicating a fast ice-melting process. s_{ice} can then be expressed by

$$s_{\text{ice}} = s_{\text{ice}}^{\max} - \frac{\displaystyle\int_{\tau_{T,1}}^{t} I(E' - V_{\text{cell}})dt}{\rho_{\text{ice}}h_{sl}\delta_{\text{CL}}\varepsilon_{\text{CL}}} \quad \tau_{T,1} < t \leq \tau_{T,1} + \tau_{T,2} \qquad \text{(Eq. 7-16)}$$

For a constant rate of heat generation, the above expression changes to

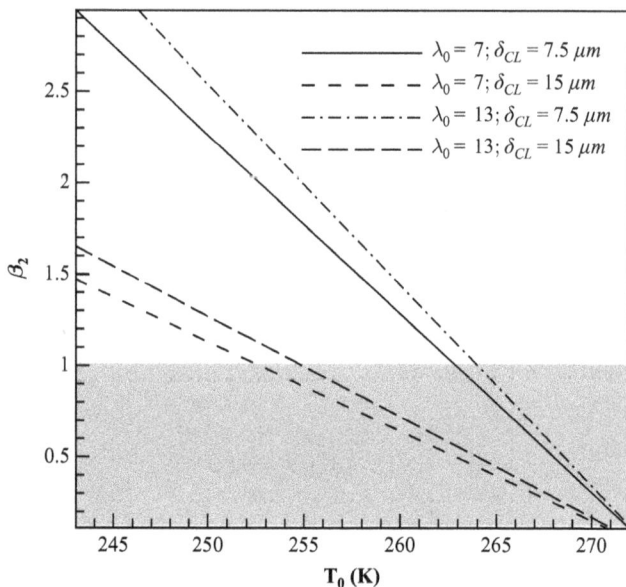

Figure 7-8. β_2 as a function of cold-start temperatures, initial membrane water content (λ_0), and catalyst-layer thickness [2].

$$s_{\text{ice}} = s_{\text{ice}}^{\max} \frac{\tau_{T,2} + \tau_{T,1} - t}{\tau_{T,2}} = \left(\frac{\tau_{T,1}}{\tau_{\text{ice},2}} - k_\tau \right) \frac{\tau_{T,2} + \tau_{T,1} - t}{\tau_{T,2}} \quad \tau_{T,1} < t \leq \tau_{T,1} + \tau_{T,2}. \quad \text{(Eq. 7-17)}$$

Of course, when all the ice is melted:

$$s_{\text{ice}} = 0 \quad t > \tau_{T,1} + \tau_{T,2} \quad \text{(Eq. 7-18)}$$

At $\tau_{T,1} < t \leq \tau_{T,1} + \tau_{T,2}$, ice and liquid water coexist. When $t > \tau_{T,1} + \tau_{T,2}$, all the ice changes to liquid water, which can be removed from the catalyst layer by capillary action. Because this stage is essentially a phase-change process, the above discussion will also be repeated in Chapter 9.

In addition, freezing-point depression may occur providing that the catalyst layer has small pore dimensions with a mean pore size ~50 nm. Ge and Wang [5] experimentally investigated the phenomenon of freezing-point depression and defined it as the difference between the freezing point in small pores and the normal equilibrium temperature of water (i.e., 0°C). Freezing-point depression is inversely proportional to the pore radius according to the well-known Gibbs–Thomson relation. That is, the freezing point of water becomes lower in a confined space as its size decreases. One way to distinguish water droplets from ice/frost particles on the catalyst layer surface during visualization is the fact that droplets are clearly growing. Ge and Wang [5] carried out a "quenching" experiment by initiating the cold start at −1°C and immediately turning off the fuel cell and then rapidly cooling it down to a lower temperature using a coolant bath. Under this lower temperature, the liquid water on the catalyst layer

(a)

(b)

(c)

Figure 7-9. Cryo-SEM images of a cross section of a cathode electrode under the initial membrane water content of 3.5 and $-20°C$: (a) following voltage failure at 0.01 A/cm^2, (b) following sublimation of ice to remove water, and (c) following voltage failure at 0.2 A/cm^2 [4] (Courtesy of the Electrochemical Society).

surface froze to form ice or frost. Figure 7-10 (a) displays the water droplet formation during cold start at $-1°C$. When temperature was lowered to $-10°C$ and maintained sufficiently long, ice was found on the surface, as shown in Figure 7-10 (b). Note that water droplets and ice particles differ in shape; the water droplets appear as spheres whereas the ice crystals have irregular shapes. Another quenching experiment was carried out after cold start from $-3°C$

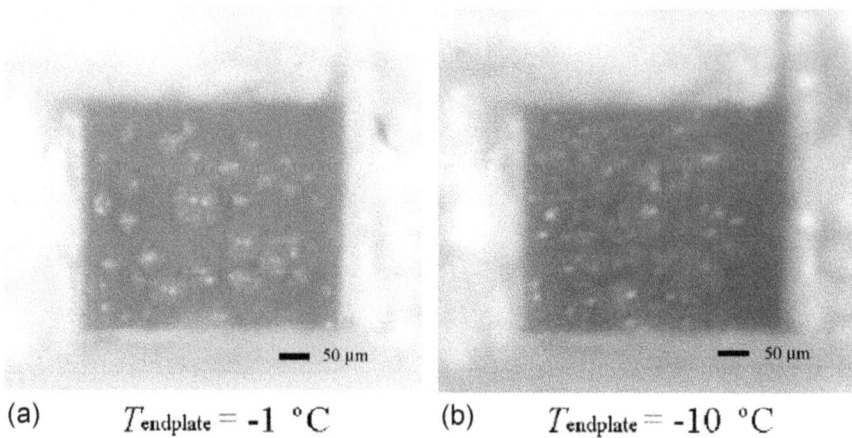

(a) $T_{\text{endplate}} = -1\ ^{\circ}\text{C}$ (b) $T_{\text{endplate}} = -10\ ^{\circ}\text{C}$

Figure 7-10. Formation of water droplets (a) and ice particles (b) on the catalyst layer surface. Droplet formation was observed in cold start at −1°C on the catalyst layer surface. When the surface was cooled to −10°C, water droplets froze [5] (Courtesy of the Electrochemical Society).

by maintaining cell temperature at −3°C for 2 h. Water droplets froze at −3°C when there was no electrochemical reaction and thus no waste heat generation inside the cell. Results from this second quenching experiment indicate that the freezing-point depression in the catalyst layer is small, and its effect on cold-start characteristics is negligible.

7.3   VOLTAGE LOSS DUE TO ICE FORMATION

7.3.1   SPATIAL VARIATION OF THE OXYGEN REDUCTION REACTION (ORR)

Though catalyst layers are thin (around 10 μm), the reaction rate changes spatially across the layer. A dimensionless parameter $\hbar$ can be defined as follows to quantify the degree of the non-uniformity [6]:

$$\hbar = \frac{|j_0 - j_\delta|}{\int_0^1 j(\bar{x})d\bar{x}} = \frac{|j_0 - j_\delta|}{\dfrac{I}{\delta_{\text{CL}}}}, \qquad \text{(Eq. 7-19)}$$

where j is the transfer current density, and j_0 and j_δ represent the transfer current densities at the two surfaces of the catalyst layer, respectively; $\bar{x}$ is the through-plane dimensionless distance, and δ is the layer's thickness. Through the rigorous derivation as shown in Chapter 4, Eq. 7-19 reduces to

$$\hbar = \frac{\Delta U}{2\dfrac{R_g T}{\alpha_c F}} \quad \text{where } \Delta U = \frac{I\delta}{\kappa^{\text{eff}}} = IR_\delta. \qquad \text{(Eq. 7-20)}$$

For a small $\hbar$ (e.g., at low currents), the ORR can be treated as uniform; therefore, product water is uniformly added to the CL. For subfreezing operation, though the current density is usually low, a large value of $\hbar$ can occur due to the low ionic conductivity at subfreezing temperature.

Figure 7-11 plots $\hbar$ as a function of temperature, showing small differences among the temperatures. It is seen that the reaction rate can be treated as uniform for $\Delta U < 0.05$ V ($\hbar < 1$). Note that ΔU is also a function of temperature because of the ionic conductivity dependence on temperature. The empirical correlation of ionic conductivity by Springer et al. was developed using measurements at 30°C for the Nafion® 117 membrane. A recent measurement made by Wang et al. [2] indicated that the ionic conductivity at subfreezing temperature is lower than that of Springer et al. [7]:

$$\kappa = (0.004320\lambda - 0.006620)\exp\left[4029\left(\frac{1}{273} - \frac{1}{T}\right)\right] \text{ for } \lambda \leq 7.22$$

$$\kappa = \kappa(\lambda = 7.22) \text{ for } \lambda > 7.22 \qquad \text{(Eq. 7-21)}$$

Figure 7-12 compares the values of ΔU at different current densities using the two ionic conductivity correlations (at $\lambda = 14$), respectively, by Springer et al. and Wang et al. At the lowest current both correlations yields a small value of $\Delta U < 0.02$ V at −30°C, whereas at 0.1 A/cm² ΔU is over 0.1 V at −30°C, and ΔU of 0.1 V yields $\hbar$ over 2, based on the correlation of Wang et al.; however, Springer et al. yields $\Delta U < 0.03$ V and

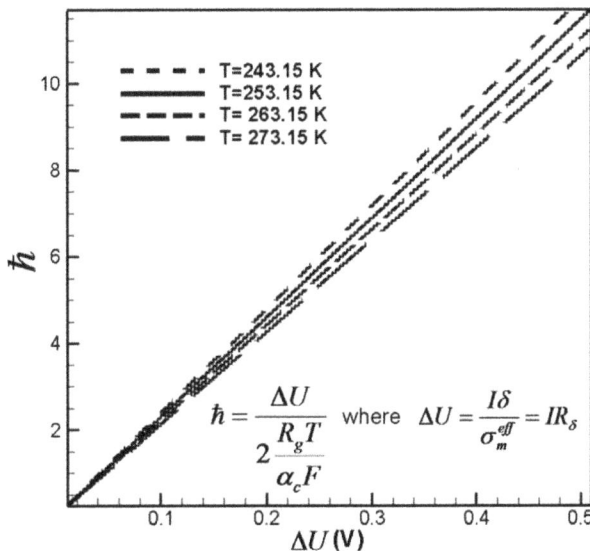

Figure 7-11. $\hbar$ as a function of temperature and ΔU for several typical subfreezing operating temperatures (σ_m^{eff} in the figure represents the ionic conductivity κ^{eff}).

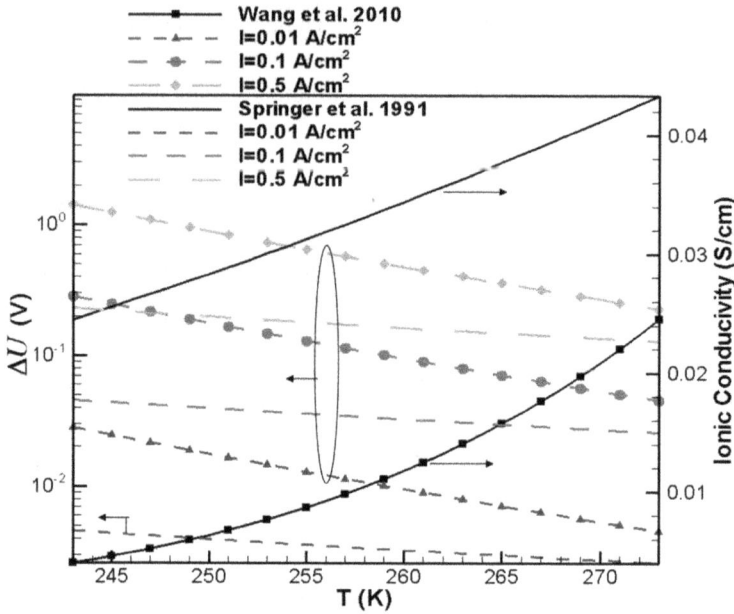

Figure 7-12. ΔU as a function of temperatur e and current density.

>1 at $-30°C$ under 0.1 and 0.5 A/cm^2, respectively. Thus, under the current of around 0.01 A/cm^2 or lower, a uniform current density and hence water production can be assumed across the cathode catalyst layer. For $-10°C$, both correlations yield a value of ΔU around 0.008 V at 0.01 A/cm^2, resulting in a very small $\hbar$ (<0.2). However, at 0.1 A/cm^2 and $-10°C$, the correlation of Wang *et al.* yields a ΔU around 0.08 V and consequently $\hbar$ around 2.

7.3.2 THE ORR RATE UNDER SUBFREEZING TEMPERATURE

Ice formation covers catalyst particles, reducing the total area of electrochemical active surfaces. The rate of the ORR electrochemical reaction is generally described by the Butler–Volmer equation:

$$j_c = ai_{o,T}^c \left[\exp\left(\frac{\alpha_a}{RT} \cdot F \cdot \eta \right) - \exp\left(-\frac{\alpha_c}{RT} \cdot F \cdot \eta \right) \right] \qquad \text{(Eq. 7-22)}$$

where $i_{o,T}^c$ is the exchange current density, which depends on the catalyst electrochemical kinetics, and a is the surface-to-volume ratio, which describes the electrode roughness. The following empirical formula is frequently used to account for ice coverage:

$$a = (1 - s_{ice})^{\tau a} a_0. \qquad \text{(Eq. 7-23)}$$

The coefficient τ_a is determined by the morphology of ice formation in the catalyst layer. Using the Tafel kinetics for the ORR yields

$$j_c = -a \, i_{0,T}^c \frac{C_{O2}}{C_{O2}^{ref}} \exp\left(-\frac{\alpha_c F}{RT} \cdot \eta\right). \qquad \text{(Eq. 7-24)}$$

The surface overpotential η is defined as

$$\eta = \Phi_s - \Phi_e - U_o, \qquad \text{(Eq. 7-25)}$$

where Φ_s and Φ_e are electronic and electrolyte phase potentials, respectively. The equilibrium potential U_o is a function of temperature:

$$U_o = 1.23 - 0.9 \times 10^{-3}(T - 298). \qquad \text{(Eq. 7-26)}$$

The exchange current density $i_{0,T}^c$ is a function of temperature, in the Arrhenius form as follows:

$$i_{0,T}^c = i_0^c \exp\left[-\frac{E_a}{R}\left(\frac{1}{T} - \frac{1}{353.15}\right)\right], \qquad \text{(Eq. 7-27)}$$

where E_a denotes the ORR activation energy at the Pt/Nafion electrode.

During transient operations, the output current density consists of the faradaic current of the electrochemical reactions and that of the double-layer charging or discharging:

$$I = j_c \delta_{cl} + I_{db}. \qquad \text{(Eq. 7-28)}$$

The double layer is a thin layer with thickness being on the order of nm at the reaction interface, acting as a capacitor during transient. The double layer in the cathode catalyst layer is analogous to the charge transfer reaction resistor. The double layer charging/discharging occurs quickly with a time constant less than 1 ms. The current contribution from the double-layer charging/discharging can be estimated using the following equation:

$$I_{db} \sim a C_{db} \frac{\Delta V_{cell}}{\Delta t} \delta_{CL}. \qquad \text{(Eq. 7-29)}$$

For the case with the capacity C_{db} being around 20 $\mu F/cm^2$ and specific area a about 10^3 cm^{-1}, $\Delta V_{cell} \sim 0.5$ V and Δt of ~ 100 s, Eq. 7-29 yields $I_{db} \sim 0.01$ A/cm^2. Thus, the effect of double layers is small, accounting for $\sim 10\%$ of the usual cold-start operating current. Consequently, double layer's effect is neglected in the analyses.

7.3.3 OXYGEN PROFILE IN THE CATALYST LAYER

Ice presence hinders oxygen transport to the triple-phase reaction sites. The general form of the oxygen transport equation in one dimension (x-direction) can be written as

$$\frac{\partial \varepsilon C_{O_2}}{\partial t} + \frac{\partial u\, C_{O_2}}{\partial x} = \frac{\partial}{\partial x}\left[D_{O_2}^{\text{eff}} \frac{\partial C_{O_2}}{\partial x} \right] + S_{O_2}. \qquad \text{(Eq. 7-30)}$$

Two types of diffusive processes are important in the catalyst layer: one is the molecular diffusion, occurring under the condition that the molecular mean free length is large as opposed to the pore size. In this case, the molecules collide with each other, and hence the molecular interactions become the limiting step in diffusion. It is well known that molecular diffusion depends on temperature and pressure:

$$D_{O_2,M} = D_{O_2,M,0}\left(\frac{T}{353}\right)^{3/2}\left(\frac{1}{P}\right). \qquad \text{(Eq. 7-31)}$$

The other type of diffusive process is Knudsen diffusion, which is important in situation in which gas molecules collide more frequently with pore walls than with each other. This mode of diffusion is encountered when the mean free path of gas molecules is on the order of the characteristic pore size. In the catalyst layer, the mean radius of micropores and the mean free path of the oxygen molecule $\left(\lambda_{\text{molecule}} = \dfrac{8RT}{\sqrt{2}\pi D^2\, NaP}\right)$ are both ca 0.1 µm; therefore, Knudsen diffusion plays an important role. The Knudsen diffusion coefficient for a long straight pore can be derived from the kinetic theory of gases, given by

$$D_{O_2,K} = \frac{2 r_{\text{pore}}}{3}\sqrt{\frac{8RT}{\pi M_{O_2}}}. \qquad \text{(Eq. 7-32)}$$

For PEM fuel cells, $D_{O_2,K}$ is around $2.64 \times 10^{-5}\,\text{m}^2/\text{s}$ for $r_{\text{pore}} = 0.1$ µm, comparable to the diffusivity for molecular diffusion. To combine the two mechanisms, the harmonic mean approach can be taken to estimate the average diffusivity. As the ice volume increases, the Knudsen diffusion becomes more important due to decreasing pore radius. To account for the effects of porosity ε and the tortuosity factor τ, the diffusivity is usually modified as follows:

$$D_{O_2}^{\text{eff}} = \frac{\varepsilon}{\tau} D_{O_2} = \varepsilon^{\tau_{d,0}} D_{O_2}, \qquad \text{(Eq. 7-33)}$$

where the Bruggeman factor $\tau_{d,0}$ is taken to be constant. The last term is referred to as the Bruggeman correlation. The Bruggeman factor $\tau_{d,0}$ is usually set to 1.5 when the structure information is absent.

Ice can attach to the pore wall, narrowing diffusive passage and changing the morphology of the solid matrix. The diffusion coefficient needs to be modified as follows to account for this situation:

$$D_{O_2}^{\text{eff}} = [\varepsilon(1-s_{\text{ice}})]^{\tau_d}\, D_{O_2}. \qquad \text{(Eq. 7-34)}$$

The tortuosity factor τ_d takes into account the effect of pore size change on the Knudsen diffusion. It can be a function of the ice volume fraction, depending on the morphology of ice crystals. The time constant of oxygen diffusion across a GDL can be estimated by

$$\tau_{\text{diff}} = \frac{\delta_{\text{GDL}}^2}{D_{O_2,\text{GDL}}^{\text{eff}}}. \qquad \text{(Eq. 7-35)}$$

This time constant is ~0.01–0.1 s for typical GDLs, which is fairly short. Thus, the transient term in Eq. 7-30 is neglected in cold-start analysis. Ignoring convection, the oxygen concentration drop across a GDL is obtained by

$$\Delta C_{O_2,c\text{GDL}} = C_{O_2,c\text{GDL}} - C_{O_2,c\text{CL}} = \frac{I}{4F} \frac{\delta_{\text{GDL}}}{D_{O_2,M}^{\text{eff}} \varepsilon_{\text{GDL}}^{\tau_{d,0}}}. \qquad \text{(Eq. 7-36)}$$

At 0.1 A/cm² during cold start, $\Delta C_{O_2,c\text{GDL}}$ is less than 0.1 mol/m³. Assuming that diffusion is dominant and the ORR rate is uniform in the cathode catalyst layer, the oxygen profile is obtained as [3]

$$C_{O_2} = C_{O_2,c\text{CL}} - \frac{I}{8F} \frac{\delta_{\text{CL}}^2 - (x - x_{c\text{CL}} + \delta_{\text{CL}})^2}{\delta_{\text{CL}} D_{O_2,K}[\varepsilon_{\text{CL}}(1 - s_{\text{ice}})]^{\tau_d}} = C_{c\text{CL}}^{O_2}\left[1 - Da\frac{1 - \left(\frac{x - x_{c\text{CL}}}{\delta_{\text{CL}}} + 1\right)^2}{\varepsilon_{\text{CL}}^{\tau_d - \tau_{d,0}}(1 - s_{\text{ice}})^{\tau_d}}\right],$$

$$\text{(Eq. 7-37)}$$

where the Damköhler number Da is defined as

$$Da = \frac{I}{8F} \frac{\delta_{\text{CL}}}{C_{O_2,c\text{CL}} D_{O_2,K} \varepsilon^{\tau_{d,0}}} = \frac{\text{Reaction rate}}{\text{Mass transport rate}}. \qquad \text{(Eq. 7-38)}$$

For $I = 0.1$ A/cm², $P = 1.0$ atm, and $T = -30°$C, Eq. 7–38 yields $Da\sim1.3 \times 10^{-4}$. Several key parameters, such as operating pressure, stoichiometric ratio, current density, and catalyst layer thickness, are lumped in Da. The oxygen concentration drop across the catalyst layer is then expressed as

$$\Delta C_{O_2} = C_{O_2,c\text{CL}} - C_{O_2,c\text{MEM}} = Da\frac{C_{O_2,c\text{CL}}}{\varepsilon_{\text{CL}}^{\tau_d - \tau_{d,0}}(1 - s_{\text{ice}})^{\tau_d}}, \qquad \text{(Eq. 7-39)}$$

which is small (< 0.5 mol/m³) at 0.1 A/cm², $\tau_{d,0} = \tau_d = 1.5$, and $s_{\text{ice}} < 0.98$. Under severe oxygen starvation (thus j_c is no longer uniform and the transient term cannot be neglected), the solution as presented in Eq. 7-39 is invalid.

Figure 7-13 shows the oxygen profiles in the catalyst layer at different ice volume fractions given a constant tortuosity ($\tau_{d,0} = \tau_d = 1.5$). The oxygen concentration undergoes a fairly small decrease (<0.5 mol/m^3 or $<5\%$) for the ice volume fraction up to 98% and $Da = 1.3 \times 10^{-4}$ at 0.1 A/cm^2. When the ice fraction reaches 99%, the oxygen concentration inside the catalyst layer starts to change considerably (~ 1 mol/m^3). $Da = 1.3 \times 10^{-3}$ (e.g., $I = 1.0$ A/cm^2) still yields a small drop at the ice volume fraction less than 90%. The oxygen concentration at the CL–GDL interface is lower at $Da = 1.3 \times 10^{-3}$ under which the operating current is larger.

Oxygen starvation in the catalyst layer starts at $\bar{x} = 0$. Setting $C_{O2} = 0$ yields

$$1 = Da \frac{1}{\varepsilon_{CL}^{\tau_d - \tau_{d,0}} \left(1 - s_{ice}^{starvation}\right)^{\tau_d}}. \qquad \text{(Eq. 7-40)}$$

Assuming $\tau_d = \tau_{d,0}$ the above equation is simplified as

$$s_{ice}^{starvation} = 1 - \sqrt[\tau_d]{Da}. \qquad \text{(Eq. 7-41)}$$

For $\tau_d = 2$ and $Da \sim 1.0 \times 10^{-4}$, $s_{ice}^{starvation}$ is $\sim 99\%$. Obviously, τ_d plays a critical role in determining $s_{ice}^{starvation}$.

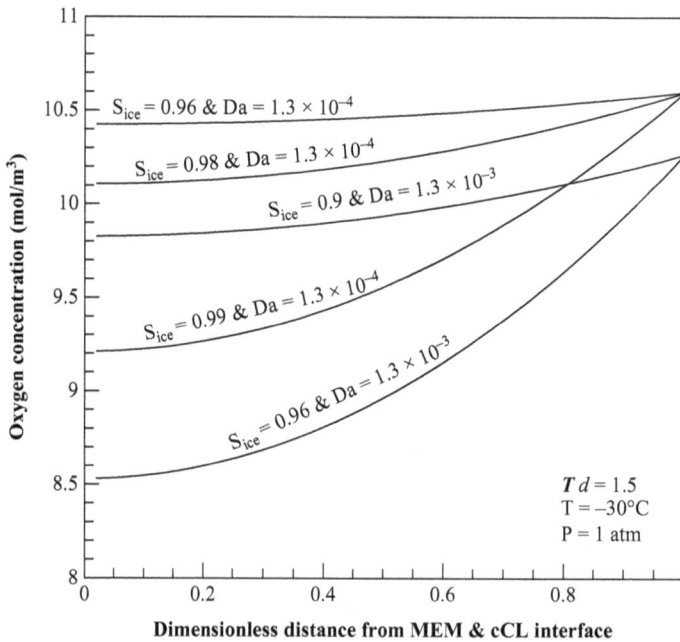

Figure 7-13. Oxygen profiles within the cathode catalyst layer [3] (Courtesy of the Electrochemical Society).

7.3.4 VOLTAGE LOSS DUE TO ICE FORMATION

Substituting the oxygen profile solution, Eq. 7-37, into the reaction rate using the Tafel equation yields [3]:

$$\eta(s_{\text{ice}},x) = -\frac{RT}{\alpha_c F} \ln \left\{ \frac{I\delta_{\text{CL}} C_{O2,\text{ref}}}{a_0 i_{0,c}^{\text{ref}} C_{O2,c\text{CL}}} \left[(1-s_{\text{ice}})^{\tau a} \left(1 - Da \frac{1 - \left(\frac{x-x_{c\text{CL}}}{\delta_{\text{CL}}} + 1 \right)^2}{\varepsilon_{\text{CL}}^{\tau d - \tau d,0} (1-s_{\text{ice}})^{\tau d}} \right) \right]^{-1} \right\} \quad \text{(Eq. 7-42)}$$

A dimensionless function can be defined as follows:

$$\Pi(s_{\text{ice}},x) = \ln \left[(1-s_{\text{ice}})^{\tau a} \left(1 - Da \frac{1 - \left(\frac{x-x_{c\text{CL}}}{\delta_{\text{CL}}} + 1 \right)^2}{\varepsilon_{\text{CL}}^{\tau d - \tau d,0} (1-s_{\text{ice}})^{\tau d}} \right) \right]. \quad \text{(Eq. 7-43)}$$

The function Π consists of two components: one is the decrease of the electrochemically active surface area due to ice formation and coverage; the other is the added oxygen concentration polarization due to the presence of ice, which narrows transport passage. The overpotential can then be written by

$$\eta(s_{\text{ice}},x) = -\frac{RT}{\alpha_c F} \ln \left(\frac{I\delta_{\text{CL}} C_{O2}^{\text{ref}}}{a_0 i_{0,c}^{\text{ref}} C_{O2,c\text{CL}}} \right) + \frac{RT}{\alpha_c F} \Pi(s_{\text{ice}},x) = \eta_{c,o} + \Delta\eta_c, \quad \text{(Eq. 7-44)}$$

where

$$\eta_{c,o} = -\frac{RT}{\alpha_c F} \ln \left[\frac{I\delta_{\text{CL}} C_{O2,c\text{CL}}^{\text{ref}}}{a_0 i_{0,c}^{\text{ref}} C_{O2,c\text{CL}}} \right] \text{ and } \Delta\eta_c = \frac{RT}{\alpha_c F} \Pi(s_{\text{ice}},x). \quad \text{(Eq. 7-45)}$$

Here, $\eta_{c,o}$ denotes the overpotential at the CL–GDL interface in the absence of ice. The overpotential change during cold start can be further expressed by

$$\Delta\eta_c(s_{\text{ice}},x) = \Delta\eta_{c,1} + \Delta\eta_{c,2}, \quad \text{(Eq. 7-46)}$$

where

$$\Delta\eta_{c,1} = \tau_a \frac{RT}{\alpha_c F} \ln(1 - s_{\text{ice}}) \ \text{and} \ \Delta\eta_{c,2} = \frac{RT}{\alpha_c F} \ln\left(1 - Da \frac{1 - \left(\frac{x - x_{c\text{CL}}}{\delta_{\text{CL}}} + 1\right)^2}{\varepsilon_{\text{CL}}^{\tau_d - \tau_{d,0}} (1 - s_{\text{ice}})^{\tau_d}}\right). \quad \text{(Eq. 7-47)}$$

Here, $\Delta\eta_{c,1}$ and $\Delta\eta_{c,2}$ represent the voltage losses arising from the reduction of reactive surface area and increase of oxygen transport resistance, respectively. A dimensionless parameter β_3 can be defined as the ratio of these two as follows:

$$\beta_3 = \frac{\Delta\eta_{c,1}}{\Delta\eta_{c,2}}. \quad \text{(Eq. 7-48)}$$

The voltage loss due to oxygen starvation at the CL–membrane interface is expressed by

$$\Delta\eta_{c,2}(s_{\text{ice}}, x_{c\text{CL}} - \delta_{\text{CL}}) = \frac{RT}{\alpha_c F} \ln\left(1 - \frac{Da}{\varepsilon_{\text{CL}}^{\tau_d - \tau_{d,0}} (1 - s_{\text{ice}})^{\tau_d}}\right). \quad \text{(Eq. 7-49)}$$

Another important factor is the sensitivities of $\Delta\eta_{c,1}$ and $\Delta\eta_{c,2}$ to s_{ice} defined as

$$k_{\eta c,1} = \left|\frac{\partial\eta_{c,1}}{\partial s_{\text{ice}}}\right| \ \text{and} \ k_{\eta c,2} = \left|\frac{\partial\eta_{c,2}}{\partial s_{\text{ice}}}\right|. \quad \text{(Eq. 7-50)}$$

Figure 7-14 presents the profiles of $\Delta\eta_{c,1}$ $\Delta\eta_{c,1}$ and $\Delta\eta_{c,2}$ as well as their ratio β_3 at $-30°C$, $\tau_d = 1$, and $\tau_d = 1.5$. The magnitude of $\Delta\eta_{c,1}$ increases steadily with the ice volume fraction, whereas $\Delta\eta_{c,2}$ is fairly small except approaching $s_{\text{ice}} = 1$. $\Delta\eta_{c,1}$ is several-order-of-magnitude larger than $\Delta\eta_{c,2}$ in most of the space in s_{ice}. Only at high ice fraction (> 95%), the magnitude of $\Delta\eta_{c,2}$ starts to increase dramatically. As a result, β_3 exhibits an initial increase to about thousands, followed by a decrease at the latter stage. The reactive surface area reduction due to ice coverage is a major cause for the voltage loss at the initial stage.

The above conclusion is drawn under a tortuosity factor $\tau_d = 1.5$. Figure 7-15 shows $\Delta\eta_{c,2}$ at different tortuosities and Da. It is seen that Da has a small effect on $\Delta\eta_{c,2}$: a larger Da yields a slightly earlier starting point of oxygen starvation, whereas the effect of tortuosity is significant. For a tortuosity of 4.0, $\Delta\eta_{c,2}$ starts to dramatically drop at s_{ice} of around 0.8.

Figure 7-16 displays the sensitivities of $\Delta\eta_{c,1}$ and $\Delta\eta_{c,2}$, that is, $k_{\eta c,1}$ and $k_{\eta c,2}$. τ_d and τ_d are assumed to be independent of s_{ice}. Both $k_{\eta c,1}$ and $k_{\eta c,2}$ increase dramatically as s_{ice} approaches unity and $k_{\eta c,2}$ is much smaller than $k_{\eta c,1}$ in most range except when s_{ice} approaches unity or $s_{\text{ice}} = 1$.

7.3.5 A MODEL OF COLD-START CELL VOLTAGE

The cell output voltage can be modeled by accounting for the major voltage losses:

$$V_{\text{cell}} = E - |\eta_c| - |\eta_a| - R_\Omega I. \quad \text{(Eq. 7-51)}$$

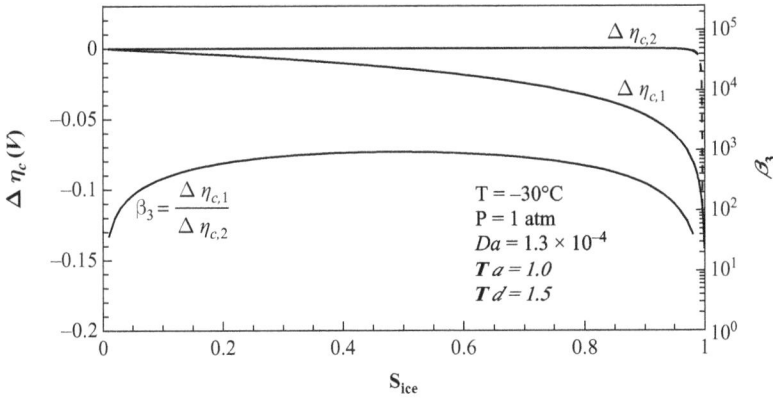

Figure 7-14. $\Delta\eta_{c,1}$, $\Delta\eta_{c,2}$ (at the membrane–catalyst layer interface), and their ratio β_3 as a function of S_{ice} [3] (Courtesy of the Electrochemical Society).

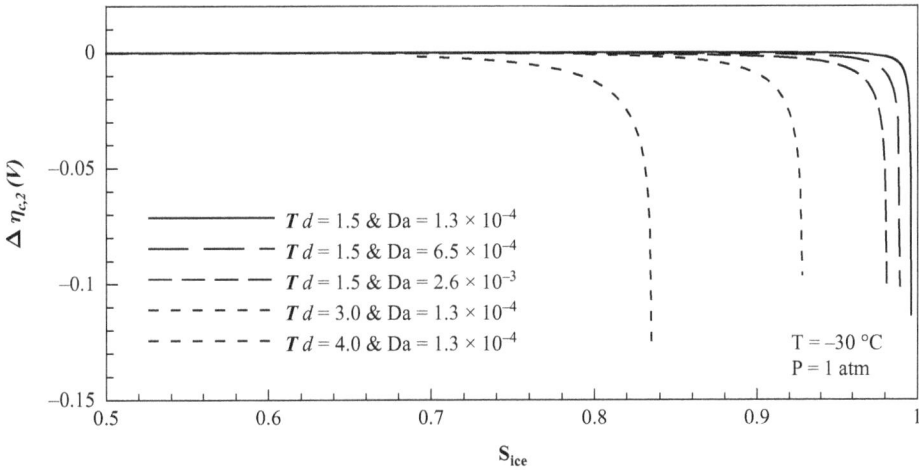

Figure 7-15. $\Delta\eta_{c,2}$ at the interface of membrane and cathode catalyst layer [3] (Courtesy of the Electrochemical Society).

Assuming ice being absent in the anode, η_a can be neglected at low current. The last term on the right-hand side represents the ohmic voltage loss, consisting of those by the ionic, and electronic as well as contact resistances. R_Ω can be estimated as follows:

$$R_\Omega = \frac{\delta_m}{\kappa_m} + \frac{\delta_{aCL}}{2\kappa_{aCL}\varepsilon_m^{\tau k}} + \frac{\delta_{cCL}}{2\kappa_{cCL}\varepsilon_m^{\tau k}} + R_{e-} + R_{contact}. \qquad \text{(Eq. 7-52)}$$

Usually, R_{e-} is small and negligible. Several ionic conductivity correlations were given in Chapter 4.

Figure 7-17 shows evolutions of temperature, ice volume fraction, and cell voltage for cold start from $-30°C$. In this case, $\beta_2 = \dfrac{\tau_{T,1}}{\tau_{S_{ice}}} > 1$, that is, the ice volume fraction reaches unity

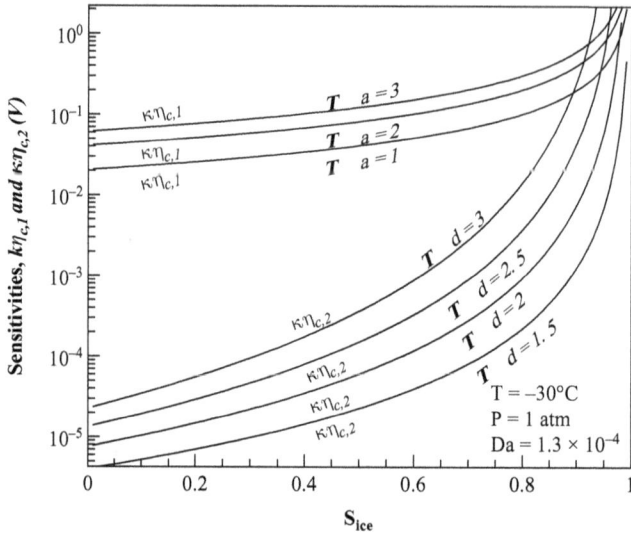

Figure 7-16. Sensitivities $k_{\eta c,1}$ and $k_{\eta c,2}$ as a function of s_{ice} [3] (Courtesy of the Electrochemical Society).

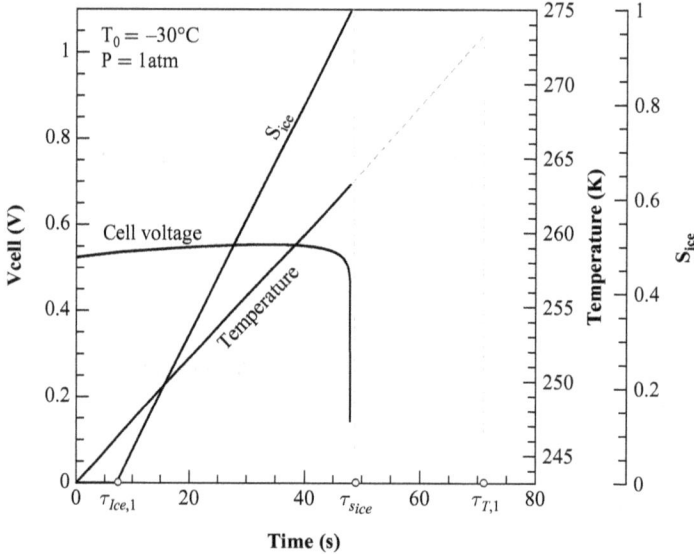

Figure 7-17. Temperature, ice volume fraction, and cell voltage during a cold start from $-30°C$ [3] (Courtesy of the Electrochemical Society).

before the cell temperature reaches $0°C$, resulting in startup failure. A constant current density of 0.1 A/cm^2 is applied; thus the ice volume fraction increases almost linearly after the membrane is hydrated at $t = \tau_{ice,1}$. As for the cell voltage, it first increases due to temperature increase as a result of the waste heat production, benefiting both the reaction kinetics and transport property. As the ice volume fraction further increases, the effect of ice formation becomes important, leading to a rapid drop in cell voltage.

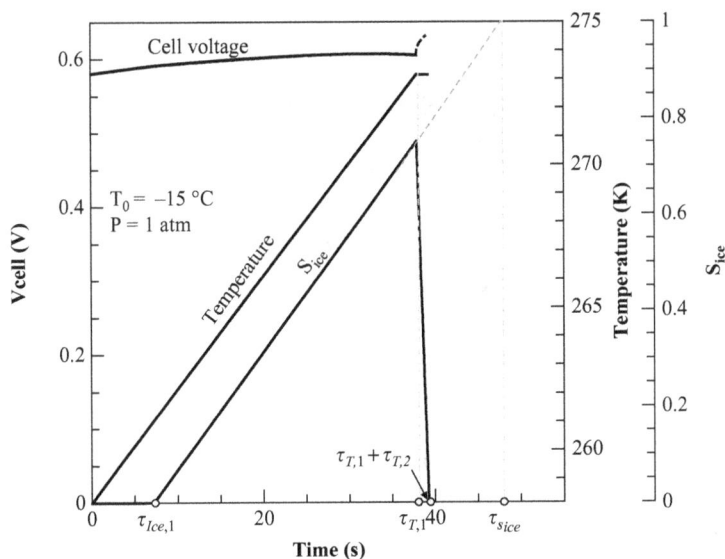

Figure 7-18. Temperature, ice volume fraction, and cell voltage during a cold start from $-15°C$ [3] (Courtesy of the Electrochemical Society).

Figure 7-18 shows the key variables of a cold start from $-15°C$, in which $\beta_2 < 1$, that is, the cell temperature reaches $0°C$ before the ice volume faction is unity. The temperature stays constant at $0°C$ in a short duration as a result of fusion of the produced ice. The ice volume fraction decreases with time, resulting in voltage increase. In this case, liquid water is assumed to be drained immediately after it emerges. During operation, residual liquid water can stay in the catalyst layer, and a portion of liquid water may enter the ionomer phase. Voltage change is complex during this period.

7.4 STATE OF SUBFREEZING WATER

The state of subfreezing water inside the pores of catalyst layers significantly affects freezing operation, specifically, the reduction in reaction surface coverage and increase in oxygen transport resistance caused by product water. For example, the ice crystal shape determines the contacting area between the reaction surface and ice. Supercooled water may be present in PEM fuel cells and able to move out of the cathode electrode and into the MPL, GDL, and even gas flow channels. Figure 7-19 shows an experimental visualization of product water on the catalyst layer surface, along with the evolution of cell voltage. The MEA was cooled to $-10°C$ using a Peltier element and the voltage was fixed at 0.5 V. In the first 380 s, liquid water was generated, and the current increased with time; after water froze, the fuel cell performance underwent a quick drop. Thus, product water did not freeze immediately inside the catalyst layer in this case, and the supercooled liquid water was able to flow to the surface of the catalyst layer.

Oberholzer et al. [9] carried out high-resolution neutron imaging to study ice formation under $-10°C$ and 0.1 A/cm^2, as shown in Figure 7-20. At $t = 0$ min, the cell was at open circuit voltage (OCV) with no product water; thus the corresponding image is used as a reference. With applied current load, the voltage rapidly decreased down to around 0.75 V, afterward the

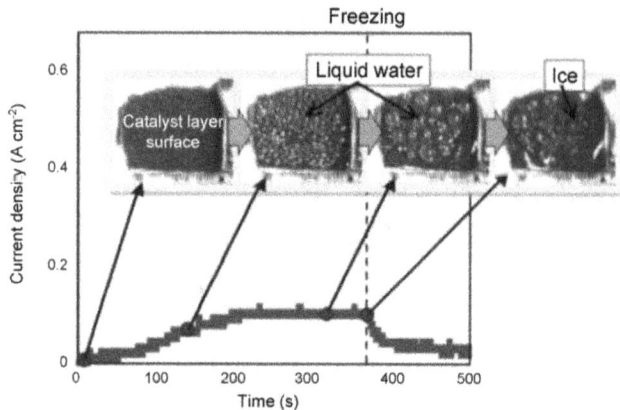

Figure 7-19. Supercooled liquid water formation and freezing on the cathode catalyst layer surface under −10°C [8] (Courtesy of Elsevier).

decrease slowed down until 100 min. After 20 min, the radiogram shows that water accumulated not only in the MEA but also in the GDL. It is likely that a capillary flow of supercooled liquid water emerged and flew to the GDL. At $t = 40$ min, water was observed over the channel walls, which connects to the water under the adjacent rib. At 60 and 80 min, two of the four cathode channels were getting filled with water, which was likely in the ice phase as liquid water usually would spread over the channel wall and is removable by the channel gas stream. At $t = 100$ min, the voltage dropped quickly and the current and the reactant gases were automatically shut down.

In addition, it has been observed that ice crystals exhibit different morphology under different conditions. Figure 7-21 displays the diagram of snow crystal morphology, showing different types of snow crystals that grow in the air, as a function of temperature and water vapor supersaturation. Note that the morphology diagram in Figure 7-21 is not necessarily indicative of all natural snow crystals, which are often dominated by polycrystalline forms, but rather it reflects the growth of ice single-crystals [12]. For natural crystals, it is also important to note that ice growth is enhanced when air flows over the surface, a phenomenon called the ventilation effect [13, 14]. In PEM fuel cells, it is likely that the ice morphology changes with operating conditions.

Figure 7-22 shows the voltage evolution during subfreezing operation at −10°C. Both cases indicate a moderate impact of ice formation on cell voltage at the initial stage, and a rapid voltage drop near the final stage of the cold start. The ice coverage factor τ_a may not be constant, instead it can depend on the ice content and have a large value determined by ice morphology. To best match the experimental data, a constant τ_a of 1.0 for $s_{ice} < 0.7$, then a linear increase to 5 from $s_{ice} = 0.7$ to 0.1, is selected; see Figure 7-22 (d). A possible explanation for adopting the varying τ_a is that at low ice content ice exists as small isolated islands over the ionomer film surface, allowing oxygen access to the catalyst in the lateral direction; see Figure 7-22 (b). As ice accumulates, isolated islands merge and eventually a thin layer of ice is formed. The layer can effectively resist oxygen transport toward the catalyst, shutting down local catalyst activity. In this form, a thin layer of ice can effectively reduce local catalyst activity. Figure 7-23 displays $\Delta\eta_{C,1}$, along with β_3 for the adopted profile of τ_a, indicating that τ_a around 5 results in a considerable effect on voltage loss.

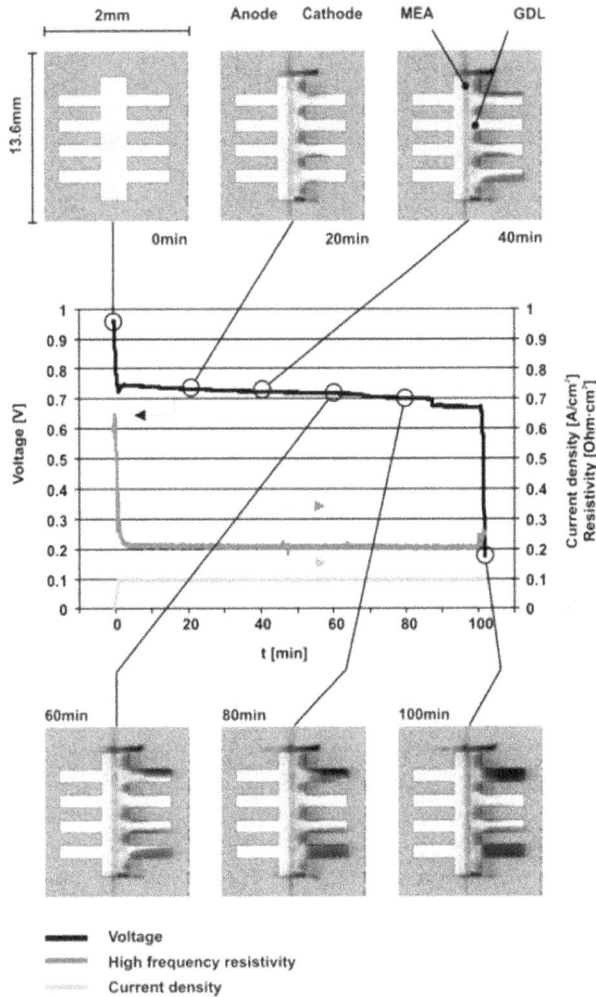

Figure 7-20. PEM fuel cell operation at $-10°C$ and 0.1 A/cm^2 and water visualized by neutron imaging [9] (Courtesy of the Electrochemical Society).

7.5   CHAPTER SUMMARY

In this chapter, several important issues are discussed regarding PEM fuel cell operation under subfreezing conditions, in which ice is produced by the oxygen reduction reaction (ORR) and adversely affects fuel cell performance. Highlights of this chapter are summarized as follows:

- Under subfreezing conditions, PEM fuel cells are able to produce water through the ORR, and the freezing water or ice may lead to the operation failure of PEM fuel cells. At subzero temperatures, the channel gas flow removes a small portion of water produced by the ORR; ice forms in the cathode catalyst layer, as evidenced from high-resolution neutron radiography; and a dimensionless parameter $\hbar$ is introduced to measure the nonuniformity of the ORR and found to be small (which means that ORR is uniform) under low current density operation (<0.01 A/cm^2).

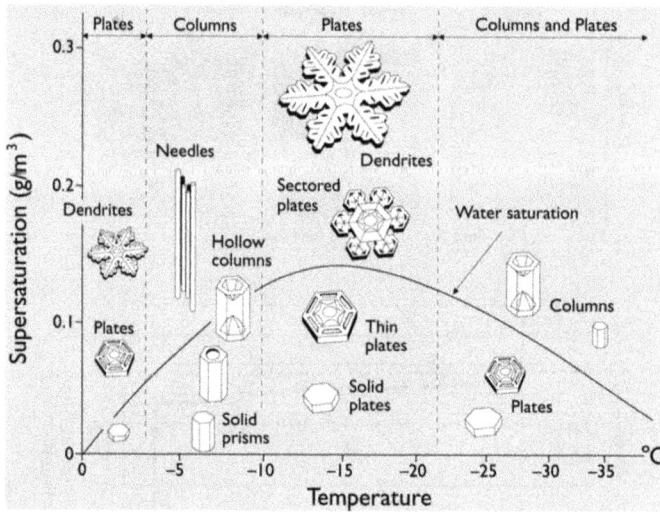

Figure 7-21. Snow crystal morphology diagram. Note the morphology switches from plates ($T \approx -2°C$) to columns ($T \approx -5°C$) to plates ($T \approx -15°C$) to predominantly columns ($T < -30°C$) as temperature is decreased. Temperature mainly determines whether snow crystals will grow into plates or columns. High supersaturations produce more complex structures [10, 11] (Courtesy of Institute of Physics).

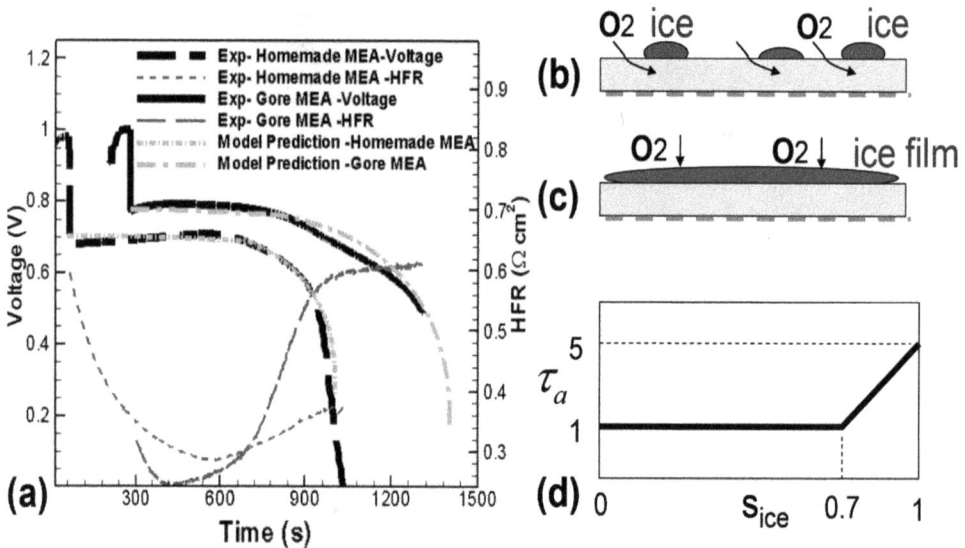

Figure 7-22. (a) Dynamic response of cell voltages and HFR (high-frequency resistance) during subfreezing operation at 0.02 A/cm^2 and −10°C; (b) ice islands scatter over the thin ionomer film, allowing oxygen access to the catalyst (dashed line); (c) ice islands connect, forming thin ice layer, and blocking oxygen access to local catalyst; and (d) the factor of the ice coverage τ_a used in the model validation [15].

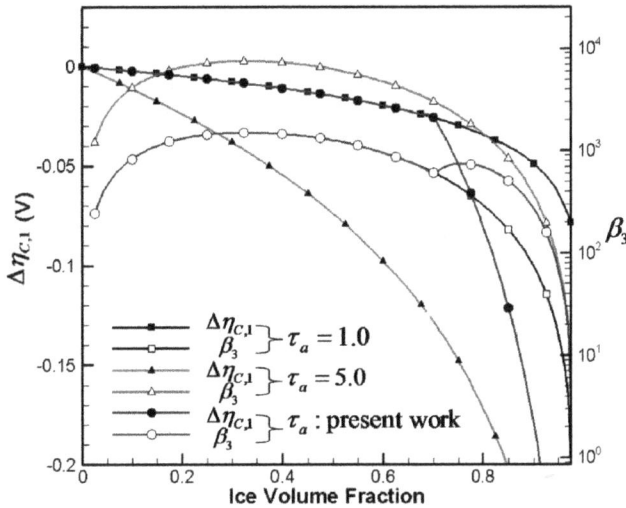

Figure 7-23. The values of $\Delta\eta_{C,1}$ and β_3 as a function of τ_a and ice volume fraction at = 2.5. "Present work" refers to Ref. [15] or Figure 7-22 (d).

- Three stages in cold start are identified and their respective time scales are analyzed and estimated. The first stage is the hydration of the electrolyte phase with time constant ranging from a fraction of a second to tens of seconds, depending on fuel cell's geometric and material parameters as well as initial membrane and cold-start conditions; the second stage is ice formation in the void space of cathode catalyst layer and a model for predicting the ice volume fraction is presented; and the third stage is ice melting with a time constant about 2 s at 0.1 A/cm^2.
- A model is proposed to account for the voltage losses due to ice formation, and two mechanisms are evaluated and compared. The oxygen-concentration profile across the cathode catalyst layer is estimated and related to a dimensionless number Da; the surface coverage effect is found to be significant at the initial stage of cold start, whereas the oxygen starvation caused by ice formation becomes important at the very end of operation failure; and ice formation morphology plays an important role in the voltage loss caused by ice formation.

REFERENCES

1. Mishler, J., Wang, Y., Mukherjee, P.P., Mukundan, R., and Borup, R.L. An experimental study of polymer electrolyte fuel cell operation at sub-freezing temperatures, *J. Electrochem Soc.*, *160* (6) F1–F8 (2013). doi: 10.1149/2.051306jes
2. Wang, Y., Mukherjee, P.P., Mishler, J., Mukundan, R., and Borup, R.L. Cold start of polymer electrolyte fuel cells: three-stage startup characterization. *Electrochim. Acta 55* (2010): 2636–44. doi: http://dx.doi.org/10.1016/j.electacta.2009.12.029

3. Wang, Y. Analysis of the key parameters in the cold start of polymer electrolyte fuel cells. *J. Electrochem. Soc.* 154 (2007): B1041–48. doi: http://dx.doi.org/10.1149/1.2767849

4. Thompson, E.L., Jorne, J., Gu, W., and Gasteiger, H.A. PEM fuel cell operation at −20°C: I. Electrode and membrane water (charge) storage. *J. Electrochem. Soc.* 155 (2008): B625–34. doi: http://dx.doi.org/10.1149/1.2905857

5. Ge, S., and Wang, C.Y. *In situ* imaging of liquid water and ice formation in an operating PEFC during cold start. Electrochem. *Solid-State Lett.* 9 (2006): A499–503. doi: http://dx.doi.org/10.1149/1.2337860.

6. Wang, Y., and Feng, X. Analysis of reaction rates in the cathode electrode of polymer electrolyte fuel cells: Part I. Single-layer electrodes. *J. Electrochem. Soc.* 155 (2008): B1289–95. doi: http://dx.doi.org/10.1149/1.2988763

7. Springer, T.E., Zawodinski, T.A., and Gottesfeld, S. Polymer electrolyte fuel cell model. *J. Electrochem. Soc.* 138 (1991): 2334. doi: http://dx.doi.org/10.1149/1.2085971

8. Ishikawa, Y., Morita, T., Nakata, K., Yoshida, K., and Shiozawa, M. Behavior of water below the freezing point in PEFCs. *J. Power Sources* 163 (2007): 708–12. doi: http://dx.doi.org/10.1016/j.jpowsour.2006.08.026

9. Oberholzer, P., Boillat, P., Siegrist, R., Perego, R., Kästner, A., Lehmann, E., Scherer, G.G., and Wokaun, A. Cold-start of a PEFC visualized with high resolution dynamic in-plane neutron imaging. *J. Electrochem. Soc.* 159 (2012): B235–45. doi: http://dx.doi.org/10.1149/2.085202jes

10. Libbrecht, K.G. The physics of snow crystals. *Rep. Prog. Phys.* 68 (2005): 855–95. doi: http://dx.doi.org/10.1088/0034-4885/68/4/R03

11. Furukawa, Y., and Wettlaufer, S. Snow and ice crystals, physics today. *Am. Inst. Phys.* (2007): S-0031-9228-0712-350-4.

12. Bailey, M., and Hallett, J. Growth rates and habits of ice crystals between −20°C and −70°C. *J. Atmos. Sci. 61* (2004): 514–44. doi: http://dx.doi.org/10.1175/1520-0469(2004)061<0514:GRAHOI>2.0.CO;2

13. Keller, V.W., and Hallett, J. Influence of air velocity on the habit of ice crystal growth from the vapor. *J. Cryst. Growth* 60 (1982): 91–106. doi: http://dx.doi.org/10.1016/0022-0248(82)90176-2

14. Hallett, J. How snow crystals grow. *Am. Sci. 72* (1984): 582–89.

15. Mishler, J., Wang, Y., Mukherjee, P.P., Mukundan, R., and Borup, R.L. Subfreezing operation of polymer electrolyte fuel cells: ice formation and cell performance loss. *Electrochim. Acta 65* (2012): 127–33.

CHAPTER 8

THERMAL TRANSPORT AND MANAGEMENT

Waste heat is released during the energy conversion by PEM fuel cells due to process inefficiency. Heat must be removed efficiently to maintain optimal fuel cell temperature and thus performance. To start-up, fuel cell temperature needs to be increased to its operating temperature through external sources or its own waste heat. In this chapter, heat generation, removal, and management in PEM fuel cells are discussed. The chapter begins with an overview of heat transfer fundamentals, followed by discussions on the major mechanisms of waste heat generation, heat transfer analysis, transient phenomenon, and experimental measurement. Phase-change phenomena involving the interactions between heat transfer and multiphase transport will be discussed in Chapter 9.

In addition, Appendix VIII.A is provided for this chapter, which lists thermal properties of selected materials, including metallic solids, nonmetallic solids, and common solids.

8.1 HEAT TRANSFER OVERVIEW

Heat is an important and common type of energy, generally called thermal energy. Heat is defined as the energy given up by the elemental particles, such as atoms, molecules, or free electrons from the hotter regions of a body to those in cooler regions. In heat transfer, the flow of heat is studied. There are three modes of heat transfer, namely, conduction, convection, and radiation. Readers who are interested in detailed discussions on general heat transfer are referred to consult with the reference texts by Özisik [1], Carslaw and Jaeger [2], Kays and Crawford [3], and Incropera et al. [4].

8.1.1 HEAT TRANSFER AND ITS IMPORTANCE

Conventional power generation, for example, the heat-engine-based technologies, utilizes heat as intermediate in energy conversion, releasing unused waste heat to the environment through cooling units such as cooling towers. Heat removal is critical to effective operation of heat engines. Differing from conventional technologies, PEM fuel cells operate in the absence of combustion; its energy conversion is realized through the electrochemical

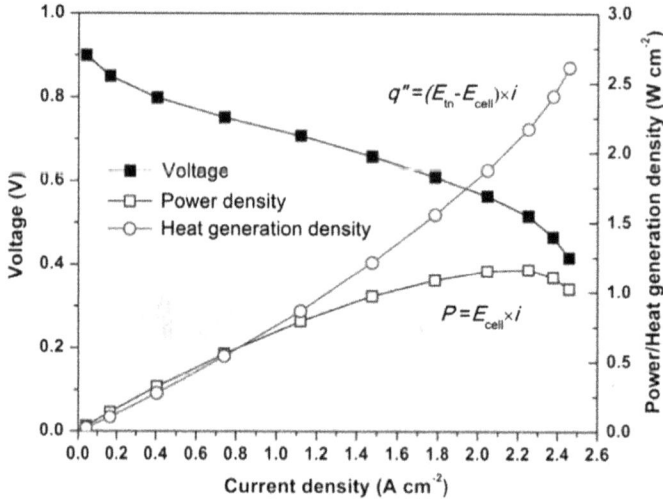

Figure 8-1. Power and waste heat generation of a PEM fuel cell. E_{cell} and i are the cell operating voltage and the current density, respectively. E_{tn} ($=E'$) represents the thermodynamic potential assuming that all the enthalpy change of reactions is converted to electric energy [5] (Courtesy of the Electrochemical Society).

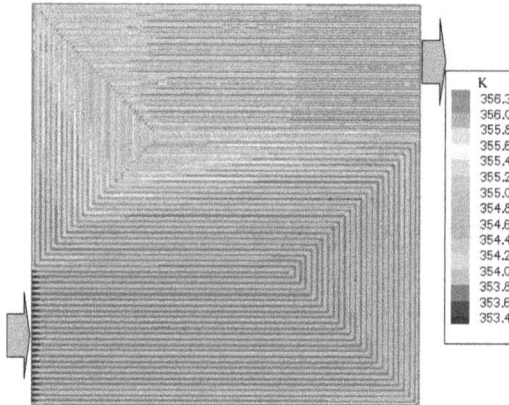

Figure 8-2. Temperature contours in the membrane of a 200 cm² commercial-size fuel cell [6].

reactions. Though the efficiency of PEM fuel cells is not limited by the Carnot cycle and can be theoretically as high as 80%, the waste heat of PEM fuel cells is not negligible; instead it is still a considerable amount: given a 50% efficiency of the electric energy conversion, the remaining 50% is released in form of waste heat. In other words, for a power density of 1 W/cm², heating power is 1 W/cm² as well. A PEM fuel cell stack of 100 kW (e.g., in vehicular applications) releases heat at a rate of 100 kW. In operation, the waste heat generation varies with operating conditions. Figure 8-1 sketches the waste heat and power generation rates along with the I–V curve, showing that the heat generation rate exceeds the electric power output at high current. This imposes a great challenge in the cooling of automobile PEM fuel cells, which frequently operate under high current and power.

PEM fuel cells operate in relatively low temperature as opposed to SOFCs, ranging from −40°C to 120°C: most studies investigate operation around 80°C. This range of operating temperature is determined by the material properties of polymer electrolytes, usually Nafion-based materials. Nafion exhibits high ionic conductivity when the material is adequately hydrated. Under dry condition such as low relative humidity (RH), the electrolyte membrane suffers from low ionic conductivity, yielding considerable ohmic voltage loss. Dryness can arise from high temperature, in which the vapor saturation temperature is high. However, low temperature around room temperature or lowers may cause "flooding" (which refers to the excessive presence of liquid water), which raises transport resistance for hydrogen and oxygen and subsequently voltage loss. Also under low temperature, the ionic conductivity is low, and transport properties and electrochemical kinetics yield poor performance. In addition, thermal stress and fatigue accelerates material degradation, leading to membrane cracking or pin-hole formation. Figure 8-2 presents temperature contours in the membrane of a large-scale PEM fuel cell, showing that the temperature varies spatially, in particular, the effect of the land–channel structure is evident: that is, the contours help distinguish the locations of land from that of channel. It shows a temperature variation about 3 K between the regions under land and under channel. Figure 8-3 presents one cooling strategy to create a temperature gradient from the inlet to the outlet. About 10 K temperature increase is shown from the inlet to the outlet. This cooling flow significantly affects the temperature distribution inside the membrane. Near the outlet, the membrane temperature is about 10 K higher than the inlet one. Similarly, the land–channel structure's effect is still evident.

Most of the waste heat is generated in the membrane electrode assembly (MEA), that is, the catalyst layers plus the membrane. In addition to the reversible heating, the irreversible process of the reaction charges overcoming the overpotential contributes a major source of waste heat at high current densities. The waste heat needs to be efficiently removed from MEAs in order to avoid hot spot formation and sustain the fuel cell energy conversion, and this is accomplished through conduction in GDLs/MPLs and bipolar plates toward the cooling units. The cooling units remove heat out of fuel cells by either liquid coolant flow or air convection. In addition, electric current flows in PEM fuel cell components encounter ohmic resistance, giving rise to

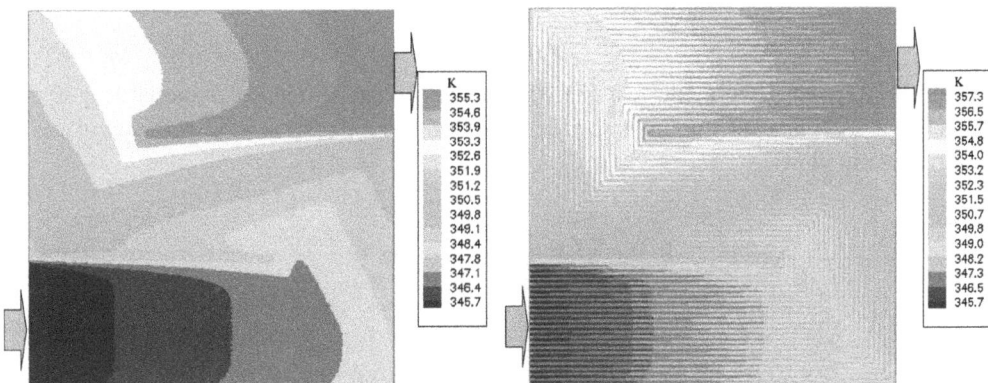

Figure 8-3. One cooling strategy: temperature contours in the cathode cooling plate (left) and the membrane (right) of a large-scale PEM fuel cell, respectively [6].

the Joules heating, which is another major source of waste heat. This portion of waste heat generation corresponds to the ohmic voltage loss, as discussed in Chapter 2.

8.1.2 HEAT TRANSFER MODES

8.1.2.1 Heat Conduction

Heat conduction is a major mode of heat transfer, which is realized through the random movement of molecules. At the atomic or molecular level, the kinetic energy of particles is transferred from more-energetic particles to less-energetic ones through their collisions. In gases, more-energetic particles are able to travel to cooler sites without requiring bulk flow, leading to heat transfer. Fourier's law states that heat flux is proportional to temperature gradient as follows:

$$\vec{q}''_{\text{cond}} = -\vec{k} \cdot \nabla T. \tag{Eq. 8-1}$$

For anisotropic materials, such as carbon paper-based GDLs, $\vec{k}$ is a tensor, describing conductivities in different directions. In many situations, isotropic media are encountered, and the flux is usually expressed by $\vec{q}''_{\text{cond}} = -\vec{k} \cdot \nabla T$, where k is a scalar. In general, k varies with temperature but is usually assumed to be constant, when the temperature range of interest is small. Appendix VIII.A lists thermal properties of selected materials, including metallic solids, nonmetallic solids, and common solids. Appendix V.A lists thermal properties of typical gases in PEM fuel cells.

In heat transfer analysis, the concept of thermal resistance is frequently used to account for both the physical dimension and thermal conductivities, defined as

$$R_{\text{cond}} = \frac{L}{kA}. \tag{Eq. 8-2}$$

With the thermal resistance, the heat flow rate is directly related to the temperature difference:

$$Q = \frac{\Delta T}{R_{\text{cond}}} = k \frac{\Delta T}{L} A, \tag{Eq. 8-3}$$

where ΔT, L, and A represent, respectively, the temperature difference, material length, and heat flow area of interest. For multiple layers, the total thermal resistance is additive and can be expressed as follows:

$$R_{\text{tot}} = \sum_{i=1}^{N} R_i = \sum_{i=1}^{N} \frac{L_i}{k_i A_i} \quad \text{and} \quad Q = \frac{\Delta T}{R_{\text{tot}}}. \tag{Eq. 8-4}$$

In multi-layer configuration, contact resistance among layers arises, as a result of interfacial roughness or lack of compression (the void space at the interface is usually filled with reactant gas or liquid water). Likewise, the electric contact resistance among layers is present as well. Current flow through the interface yields the Joules heating. Applying the energy balance at the interface yields

$$k_1 \frac{\partial T_1}{\partial n} - k_2 \frac{\partial T_2}{\partial n} = \dot{q}'', \qquad \text{(Eq. 8-5)}$$

where n is the unit vector normal to the interface; $\dot{q}''$ is the interfacial heating flux; and k_1 and k_2 are the thermal conductivities of the two materials, respectively. Contact resistance, in general, can be mitigated by increasing compression.

In PEM fuel cells, conductive heat transfer occurs in all fuel cell components such as the electrolyte membrane and bipolar plates. The energy conservation equation, in the absence of bulk flow, is given by

$$\rho C_p \frac{\partial T}{\partial t} = k \nabla^2 T + \dot{q}''. \qquad \text{(Eq. 8-6)}$$

The terms in the above represent the rates of change in stored thermal energy, conductive transfer, and heat generation, respectively. Heat transfer in the bipolar plates and electrolyte membrane follows the above governing equation, provided that no bulk flows are present in these components.

8.1.2.2 Convective Heat Transfer

Convection is another common mode of heat transfer arising from fluid motion: thermal energy contained in a fluid matter can be delivered by bulk fluid motion. The convective flux in a fluid is the product of thermal energy and flow velocity: $\rho C_p T \vec{u}$. The convective flux between a solid surface to a flowing fluid is proportional to the temperature difference between the solid surface T_s and bulk flow T_∞, given by Newton's law of cooling:

$$q''_{\text{conv}} = h \Delta T = h(T_s - T_\infty). \qquad \text{(Eq. 8-7)}$$

The heat transfer coefficient h is determined by fluid properties (e.g., specific heat and surface properties) and flow velocity. The local heat transfer coefficient $h(x, z)$ is defined in the same manner where x and z are the coordinates on the solid surface A. It varies spatially depending on local flow conditions. The overall coefficient is then evaluated by averaging the local heat transfer coefficient:

$$\bar{h} = \frac{1}{A} \int_A h(x, z) \, dx \, dz, \qquad \text{(Eq. 8-8)}$$

where A is the surface area of solid surface A. The convective transfer resistance is defined by

$$R_{\text{conv}} = \frac{1}{hA}. \qquad \text{(Eq. 8-9)}$$

Due to the similarity between heat and momentum transfers, the boundary-layer theory, originally developed for fluid flow, is conventionally extended to heat transfer. When fluid flows over a solid surface, the no-slip condition at the surface satisfies macroscopically; a thin fluid layer (called the boundary layer) develops near the wall. In this layer, a sharp change in velocity is present, with its value rapidly increasing from zero to that of the free stream. The boundary-layer thickness δ is conventionally defined as the distance from the surface to a location where the surface influence is negligible (e.g., local velocity reaches 99% of the free-stream velocity):

$$u = 0.99u_\infty \text{ at } y = \delta. \tag{Eq. 8–10}$$

Likewise, in convective heat transfer near a surface, a thermal boundary layer develops, in which a sharp temperature change is encountered, with fluid temperature rapidly changing from the wall temperature to the free stream one. Its thickness is defined as the distance where local temperature reaches 99% of the free stream temperature, that is,

$$\frac{T_s - T}{T_s - T_\infty} = 0.99 \text{ at } y = \delta_T. \tag{Eq. 8-11}$$

Figure 8-4 presents schematically the velocity and temperature profiles in the momentum and thermal boundary layers near a plate, respectively.

At the solid surface, the energy balance yields

$$q'' = -k\frac{\partial T}{\partial y}\Big|_{y=0} = h(T_s - T_\infty) \text{ and } h = \frac{-k\dfrac{\partial T}{\partial y}\Big|_{y=0}}{T_s - T_\infty}. \tag{Eq. 8-12}$$

The x and y dimensions are shown in Figure 8-4. Two dimensionless parameters, the Nusselt and Prandtl numbers, are fundamentally important in understanding convective heat transfer and thermal boundary layer (see Chapter 3 for details). The Nusselt number ($Nu = \dfrac{hL}{k_f}$) evaluates the relative importance of heat convection to conduction in a fluid. The Prandtl number ($Pr = \dfrac{\nu}{\alpha}$) is defined as the ratio of the momentum diffusivity ν to the thermal diffusivity α; thus, it quantitatively compares the thermal and momentum boundary-layer thicknesses. For fluid flow and convective heat transfer over a plate, the two boundary layers develop simultaneously, and the ratio of their thicknesses, r, can be analytically derived, when $r < 1$:

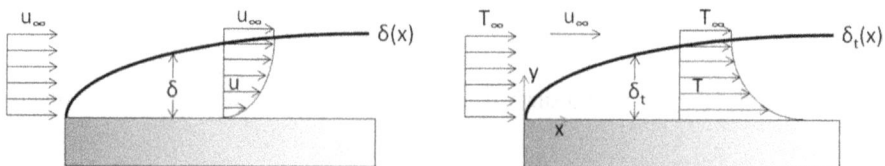

Figure 8-4. Schematic of the momentum (left) and thermal (right) boundary layers.

$$r = \frac{\delta_T}{\delta} = \frac{1}{1.026 Pr^{1/3}}. \tag{Eq. 8-13}$$

For convective transfer over a flat plate, the Nusselt numbers can be determined through the Reynolds and Prandtl number for laminar flows:

$$Nu = 0.664\, Re^{1/2} Pr^{1/3}. \tag{Eq. 8-14}$$

The Reynolds number $(Re = \frac{\rho u_\infty L}{\mu})$ determines flow regimes: the laminar flow versus turbulent flow. For the external flow over a flat plate, the critical Re is known as about 5×10^5 when fluid flow changes from the laminar regime to the turbulent regime. For the internal flows in pipes, the critical Re is about 2300. In turbulent flows, local fluctuations of flow and temperature play an important role in heat transfer. In PEM fuel cells, both reactant gas and cooling flow are kept low in their velocities to minimize pumping power consumption. Thus, Re is usually low, and turbulent flow is rarely encountered.

For convective heat transfer in channels or tubes, similar to fluid flow, the fully developed condition (i.e., there exists a generalized temperature profile that is invariant with tube length) can be defined for heat transfer:

$$\frac{\partial}{\partial x}\left(\frac{T_s - T}{T_s - T_m}\right) = 0, \tag{Eq. 8-15}$$

where x is in the along-tube direction, and T_s and T_m are the surface and mass-averaged fluid temperatures, respectively. Figure 8-5 shows the development of fluid flow inside a 2D channel.

Different from fluid flow, two standard boundary conditions can be encountered in heat transfer: fixed temperature and heat flux at the surface. Figure 8-6 shows the fully developed temperature profiles for these two boundary conditions. The Nusselt numbers $(Nu = hD/k$ where D is the tube diameter) are constant in the fully developed regimes [3].

8.1.2.3 Heat Radiation

Heat radiation is the emission of energy by the surface of a matter or body that is at a finite temperature in form of electromagnetic waves. All bodies radiate energy in the form of photons in all the directions at random phase and frequency. When arriving at a surface, the photons are either absorbed or reflected or transmitted. This mode of heat transfer can occur in the absence of matter or through a vacuum. The rate of radiation energy release can be expressed by the Stefan–Boltzmann law:

$$q'' = \varepsilon \sigma T_s^4, \tag{Eq. 8-16}$$

where ε is the radiation property of a surface, called the emissivity. The value of emissivity ranges between 0 and 1, quantifying the efficiency that the surface emits energy relative to

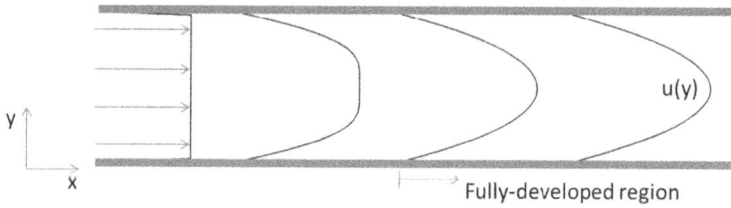

Figure 8-5. The velocity profile of a 2D channel flow.

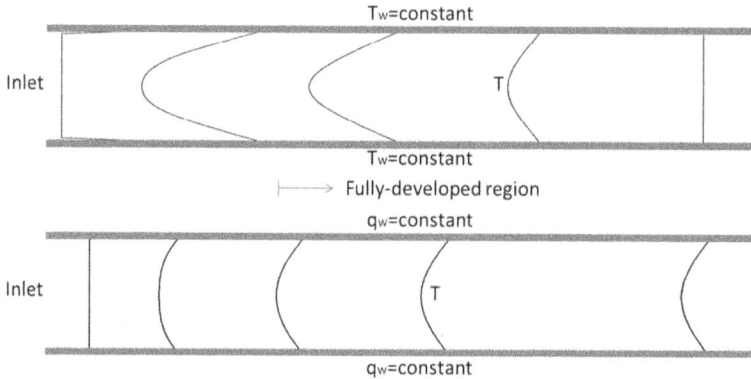

Figure 8-6. Temperature profiles in a 2D channel with two types of thermal boundary conditions.

blackbody. Blackbody is an ideal radiator with maximum radiation emission, that is, $\varepsilon = 1$. T_s is the surface temperature and σ is the Stefan–Boltzmann constant (= 5.67×10^{-8} W/m^2 K^4). Note that absolute temperature needs to be used when radiation is involved. At a surface, radiation energy is emitted and absorbed simultaneously. The rate at which radiation is incident on unit area of a surface from its surroundings is defined as the irradiation rate. Some portion of irradiation is reflected to surroundings again and some is transmitted into the surface, for example, a semi-transparent surface. The energy gained at the surface is determined by the surface radiation property called the absorptivity α, which ranges from 0 to 1. Kirchhoff's law states that good radiators are good absorbers, and the absorptivity and emissivity are equal to each other, that is, $\alpha = \varepsilon$. Then, the net rate of radiation heat transfer from the surface is expressed by

$$q''_{rad} = \varepsilon \sigma (T_s^4 - T_\infty^4). \tag{Eq. 8-17}$$

This equation shows the energy balance as a result of emission and absorption at a surface. In the cases that $(|T_s - T_\infty|)/T_\infty \ll 1$, the radiation flux can be linearized by defining the heat transfer coefficient for radiation h_r:

$$q''_{rad} = h_r(T_s - T_\infty) \text{ where } h_r = \varepsilon \sigma T_\infty^3 (T_s^2/T_\infty^2 + 1)(T_s/T_\infty + 1) \approx 4\varepsilon \sigma T_\infty^3. \tag{Eq. 8-18}$$

Under this condition, radiation mathematically is similar to a convection term. Radiation is only significant at high temperature, for example, in SOFCs radiation plays an important role in heat transfer. Under PEM fuel cell operating conditions, radiation is unimportant and negligible.

8.1.3 HEAT TRANSFER IN POROUS MEDIA

In a PEM fuel cell, the GDL and CL components are porous media, enabling transport of both electrons and gaseous reactants. The waste heat produced in the MEA is removed via these porous components. Most waste heat is generated within the porous components, including the Joules heat by electric current in the catalyst layer and GDLs, and the reaction heat in the catalyst layers. Thus, heat transfer in porous media is critical to the thermal management of PEM fuel cells. Figure 8-7 displays the structures of several porous media.

A porous medium consists of void space (in which fluid occupies) and solid matrix. At the microscopic level, the fluid in void space can contribute both convection and conduction, and local heat transfer occurs between the fluid in voids and solid matrix. Thus, the structural detail determines the exact heat flow routes. In many engineering problems, it suffices to use just one macroscopic variable for temperature without distinguishing temperatures of solid matrix and void fluid at a pore level, that is, revoking local thermal equilibrium. As a result, only one

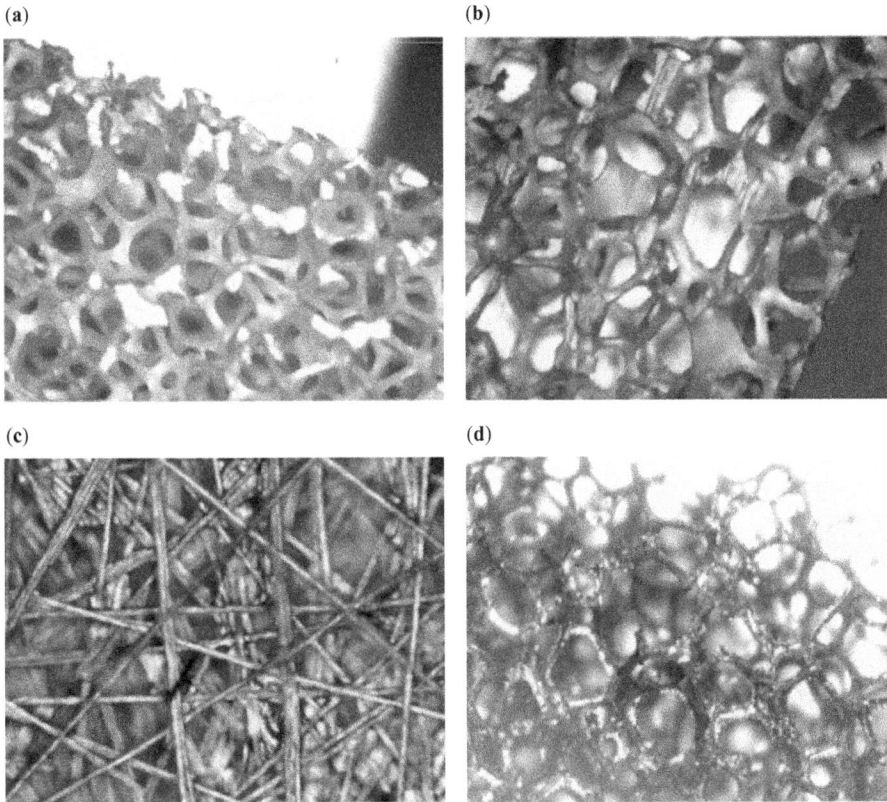

Figure 8-7. Structures of several porous media: (a) Al foam, (b) copper foam, (c) carbon paper, and (d) carbon foam.

energy equation is required. The effective heat conductivity k^{eff} is defined to account for the thermal conductivities of both phases. For a gas–solid matrix system, the general energy equation can be written as

$$\frac{\partial\left[\varepsilon\rho_g c_{p,g}+(1-\varepsilon)\rho_s c_{p,s}\right]T}{\partial t}+\nabla\cdot\left(\rho_g c_{p,g}uT\right)=\nabla\cdot\left(k^{\text{eff}}\nabla T\right)+S_T. \quad \text{(Eq. 8-19)}$$

To evaluate the convective heat flux in the second term on the left, local bulk flow u needs to be known. Darcy's law directly relates the flow velocity to the pressure gradients and medium's permeability K as

$$\rho u = -\frac{K}{\nu}\nabla P \quad \text{where} \quad K = \frac{\varepsilon^3}{180(1-\varepsilon)^2}d^2, \qquad \text{(Eq. 8-20)}$$

where the Carman–Kozeny model is adopted as the correlation of permeability K. Conductive heat transfer is via both solid matrix and fluid in void space. The effective thermal conductivity is evaluated using the volume fraction of each phase as follows:

$$k^{\text{eff}} = (1-\varepsilon)k^s + \varepsilon k^f. \qquad \text{(Eq. 8-21)}$$

The above equation neglects the tortuosities of constituent phase networks. In PEM fuel cells, the solid phase is highly conductive as opposed to the fluids in the void, that is, $k^s \gg k^f$; thus the overall conductivity can be evaluated based on the solid matrix only:

$$k^{\text{eff}} = k^s(1-\varepsilon)^\tau, \qquad \text{(Eq. 8-22)}$$

where τ is the tortuosity of the solid matrix.

8.2 HEATING MECHANISMS

In PEM fuel cells, the chemical energy stored in hydrogen gas is converted to electricity by the electrochemical HOR and ORR reactions at efficiency possibly over 60%. The rest is released as waste heat through four major mechanisms: the reversible entropic heat, irreversible reaction heat, ohmic or Joules heat, and latent heat release/absorption during phase change. The details on latent heat release or absorption during ice–liquid, liquid–vapor, vapor–ice phase changes, or membrane hydration/dehydration will be discussed in Chapter 9. The amount of the entropic heat is evaluated from the entropy change of the electrochemical reaction in a reversible way, generally referred to as the reversible heating. Any irreversibility in the reaction causes waste heat generation; in PEM fuel cells, it is primarily referred to the part caused by charges overcoming the interfacial overpotential. The ohmic heat is generated when either the protonic or electronic current flows through resistors, such as the catalyst layers, membranes, or GDLs. In addition, internal current or H_2 cross-over

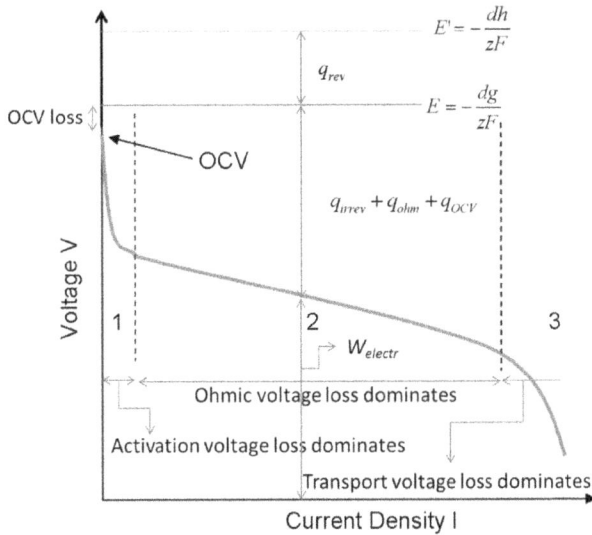

Figure 8-8. Breakdown of energy conversion in a PEM fuel cell.

reduces PEM fuel cell efficiency, thus leading to the waste heat generation. It is, however, neglected in most heat transfer study because the cross-over is dependent on the specific design of a fuel cell, instead of general fundamentals of PEM fuel cells. In this book, we skip a detailed discussion on this portion of heating. The waste heat is not uniformly distributed throughout a PEM fuel cell; instead it differs among the components and varies spatially within each component. A major portion of the waste heat is generated in the MEA, which may account for over 90% of total heat generation. A small portion of ohmic heat can be produced in GDLs and bipolar plates as well as the contact area among components. As a result, the temperature within a PEM fuel cell is nonuniform. In the thermal management of PEM fuel cells, local temperature needs to be well managed to avoid issues such as hot spot formation, membrane dehydration, and local material failure due to thermal stress.

The polarization (I–V) curve provides information that directly evaluates waste heat generation. Providing that the chemical energy stored in hydrogen gas is constant under given conditions and the I–V curve shows the useful output energy, the difference between these two is emitted in form of waste thermal energy. The breakdown of energy conversion in PEM fuel cells is shown schematically in Figure 8-8.

8.2.1 THE ENTROPIC HEAT

In the thermodynamic process involving the PEM fuel cell reactions, the second law states that a 100% energy conversion is impossible even without any irreversibility. The inefficiency in a reversible process releases waste heat, termed as the reversible heat. The part of heat can be evaluated through the entropy change as discussed in Chapter 2 in a reversible process of the energy conversion:

$$Q_{\text{rev}} = TdS. \qquad \text{(Eq. 8-23)}$$

The thermodynamic relation between G and S gives

$$dG = VdP - SdT \text{ thus } -dS = \left(\frac{dG}{dT}\right)_P. \qquad \text{(Eq. 8-24)}$$

The change in the Gibbs free energy will yield the reversible potential change as

$$dE = -\frac{dG}{znF} \text{ thus } -znF\left(\frac{dE}{dT}\right)_P = \left(\frac{dG}{dT}\right)_P. \qquad \text{(Eq. 8-25)}$$

For the heating rate per volume q_{rev}, zn is replaced by the transfer current density, yielding

$$q_{rev} = j \cdot T\frac{dU_o}{dT}, \qquad \text{(Eq. 8-26)}$$

where j is the transfer current density, obtained from the Butler–Volmer equation, and U_o is the reversible voltage as a function of temperature.

8.2.2 IRREVERSIBILITY OF THE ELECTROCHEMICAL REACTIONS

In energy conversion, that is, the process of breaking the molecular bonds of the reactant species and forming new lowerenergy bond of product species, irreversibility occurs. Irreversibility in the electrochemical reactions arises from the interfacial overpotential, which gives rise to voltage loss and hence the release of waste heat. The reaction rate depends exponentially on the interfacial overpotential, as seen in the Butler–Volmer equation. The amount of irreversible heat q_{irrev} equals to the product of interfacial overpotential and reaction current:

$$q_{irrev} = j \cdot \eta. \qquad \text{(Eq. 8-27)}$$

This heating mechanism contributes a major portion of the waste heat under high current, in which a large overpotential is encountered.

8.2.3 THE JOULES HEAT

The Joules heat is generated when the protonic or electric current flows through a resistor. This amount of heat is determined by the electric and ionic resistances and currents. In electrolyte membranes, the ionic conductivity is a function of temperature and water content, as discussed in Chapter 4. Thus, water management is essential to reducing the Joules heat. For a membrane thickness of L_m, the ionic resistance is given by

$$R_{ohm} = \int_0^{L_m} \frac{dz}{\kappa}, \qquad \text{(Eq. 8-28)}$$

where z is in the thickness direction. In general, the ionic conductivity is around 0.1 S/cm at 80°C, yielding a resistance of about 0.05 Ω cm^2 for the Nafion membrane of N112. This value is usually much larger than the electric resistances; thus protonic current generates a major portion of the Joules heating. The Joules heating rate is evaluated by

$$Q_{\text{ohm},H+} = R_{\text{ohm}} I^2, \qquad \text{(Eq. 8-29)}$$

where I represents the protonic current flow. In a membrane, the ionic conductivity varies spatially, yielding the Ohmic heat that changes from place to place. The local Ohmic heating (per volume) arising from the ionic resistance is calculated by

$$q_{\text{ohm},H+} = \frac{\vec{i}_e^{\,2}}{\kappa}. \qquad \text{(Eq. 8-30)}$$

The Joules heat is also produced in the catalyst layer, GDLs, and bipolar plates arising from the electric current flow. Similar to the heating by the protonic current, the heat generation rate is written as follows:

$$q_{\text{ohm},e-} = \frac{\vec{i}_s^{\,2}}{\sigma}. \qquad \text{(Eq. 8-31)}$$

The currents $\vec{i}_e$ and $\vec{i}_s$ vary locally, and they are determined by the phase potential gradient and conductivity, respectively, as follows:

$$\vec{i}_e = -\kappa \nabla \phi^e \quad \text{and} \quad \vec{i}_s = -\sigma \nabla \phi^s. \qquad \text{(Eq. 8-32)}$$

In addition to bulk resistance, contact resistances are present at the interfaces between components, giving rise to local Joules heating.

8.3   STEADY-STATE HEAT TRANSFER

Figure 8-9 sketches a representative temperature profile in a PEM fuel cell and lists the thermal characteristics in each component. Heat conduction occurs in all the fuel cell components and plays an important role in the heat transfer of these components except the gas and cooling flow channels. Heat conductivity varies among the components as shown in the figure. Mass flows in the porous components are usually very slow; thus, heat convection is unimportant in GDLs and CLs, and most waste heat is removed via heat conduction through solid matrices.

In addition, many studies are focused on macroscopic phenomena for the thermal management of PEM fuel cells. To understand the microstructure of a material and its effect on thermal properties, knowledge of heat transfer at the microscopic level needs to be established. In the following sections, both macroscopic and microscopic heat transfer are discussed.

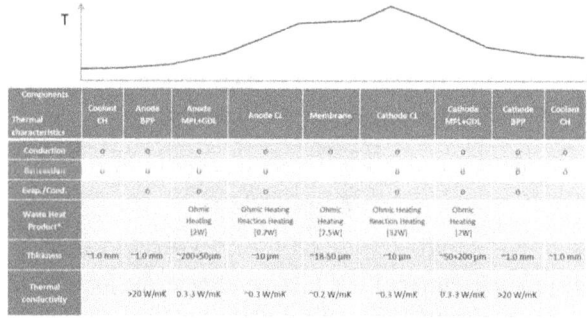

Figure 8-9. A representative temperature profile in a PEM fuel cell, along with thermal properties in individual component. *Data are based on a case study in Ref. [7].

8.3.1 ONE-DIMENSIONAL (1D) HEAT TRANSFER ANALYSIS

Most of the waste heat is generated inside the MEA and removed by the cooling flow in the bipolar plates or surface cooling. Thus, the through-plane heat transfer plays an important role in waste heat removal and hot spot formation. For simplicity, all of the waste heat is assumed to be generated in the catalyst layer. Temperature gradients develop in the catalyst layer, GDL, and bipolar plate, driving conductive heat transfer. Temperature variation can be evaluated through the conductive resistance R:

$$\Delta T \sim \frac{Q}{R}. \tag{Eq. 8-33}$$

In the catalyst layer, the temperature variation can be estimated as follows:

$$\Delta T_{\mathrm{CL}} \sim \frac{\delta_{\mathrm{CL}}^2 S_T}{k_{\mathrm{CL}}^{\mathrm{eff}}} = \frac{\delta_{\mathrm{CL}} I (E' - V_{\mathrm{cell}})}{k_{\mathrm{CL}}^{\mathrm{eff}}}, \tag{Eq. 8-34}$$

where E' is defined as $-\dfrac{\Delta h}{2F}$, representing the EMF (electromotive force) that all the energy from hydrogen/oxygen, the "calorific value," heating value, or enthalpy of formation, were transformed into electrical energy with ice as the final reaction product. The EMF accounts for both the latent heat of phase change and sensible heat. It is the same as E' in Figure 8-1. For the effective conductivity k^{eff} of 1.0–3.0 W/m·K, $\Delta T_{\mathrm{CL}} \sim 0.01$ K at 0.1 A/cm^2 and 0.1 K at 1.0 A/cm^2, respectively [8]. A more accurate estimate can be formulated by assuming uniform heating inside the catalyst layer, leading to about half of the above estimated value. Appendix VIII presents theoretical solutions to several important heat conduction problems.

Similarly, ΔT_{GDL} and ΔT_{BP} are evaluated using the same expression based on their respective thermal properties. Given a GDL conductivity of ~3.0 W/m·K, the variation is ~1 K at 1 A/cm^2. As for bipolar plates (BPs), the graphite plates are common BP materials for PEM fuel cells. Metallic BPs are a viable alternative due to their excellent mechanical strength and the ease of machining. Both graphite and metallic materials are good thermal conductors, leading to small temperature variations within them.

8.3.2 TWO-DIMENSIONAL (2D) HEAT TRANSFER ANALYSIS

The above 1D analysis presents a rough estimate: effects of the land–channel structure and lateral heat transfer are neglected. The channel gas flows usually carry a small portion of the waste heat generated. A parameter β_1 is defined as the ratio of the heat removed by the reactant channel flow to the total heat generation by a PEM fuel cell:

$$\beta_1 = \frac{\left(\rho_g c_{p,g} \Delta T A_{ch} u\right)_a + \left(\rho_g c_{p,g} \Delta T A_{ch} u\right)_c}{I(E' - V_{cell}) A_m} \approx \frac{\rho_{g,a} c_{p,g,a} \dfrac{2\xi_a}{C_{H_2}} + \rho_{g,c} c_{p,g,c} \dfrac{\xi_c}{C_{O_2}}}{4F(E' - V_{cell})} \Delta T, \quad \text{(Eq. 8-35)}$$

where ΔT_a and ΔT_c are the temperature increments from the inlet to outlet on the anode and cathode sides, respectively. The last term on the right-hand side is approximated by setting $\Delta T_a = \Delta T_c = \Delta T$. The common values for PEM fuel cells, for example, $\xi = 1.5$ and full inlet humidification, lead to $\beta_1 < 5\%$ even for $\Delta T = 10\ K$, which is small and therefore can be safely neglected in analysis.

Neglecting the heat removal by reactant flows, heat generation under GFCs is conducted laterally toward the land, followed by the through-plane removal in both the GDL and the land. Thus, the heat transfer pathway consists of two routes or resistors, an in-plane thermal resistor plus a through-plane one; see Figure 8-10. The through- and in-plane temperature variations can be evaluated by [9]

$$\Delta T_{max}^{\text{thr-plane}} = \frac{\frac{1}{2}I(E' - V_{cell})}{\dfrac{k_{GDL,H}^{eff}}{H_{GDL}}} \quad \text{and} \quad \Delta T_{max}^{\text{in-plane}} \approx \frac{\frac{1}{2}I(E' - V_{cell})W_{ch}^2}{2H_{GDL}\,k_{GDL,W}^{eff}}. \quad \text{(Eq. 8-36)}$$

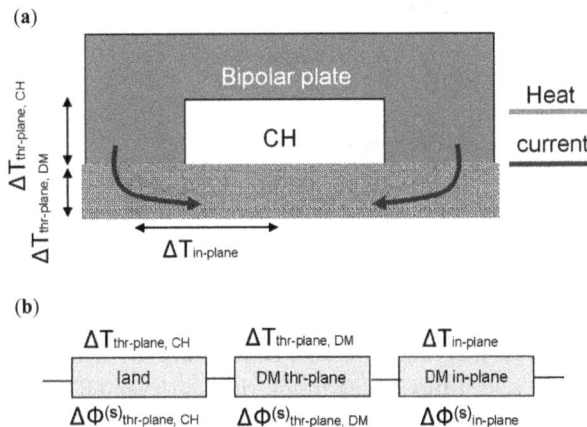

Figure 8-10. Schematics of (a) heat and current flows on the cathode side, and (b) the connections of thermal and electrical resistors. Note: DM (diffusion media) refers to GDL plus MPL.

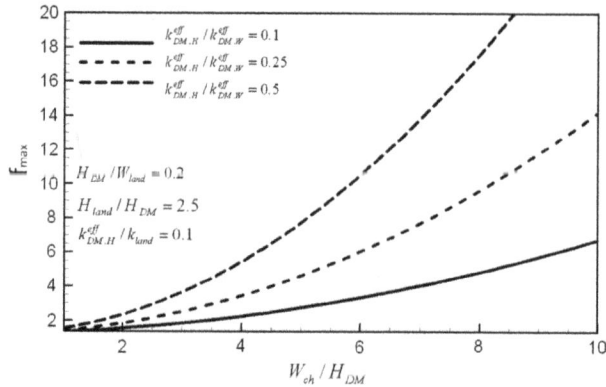

Figure 8-11. f_{max} versus the ratio of the characteristic channel in-plane length to the GDL (or DM) thickness $\dfrac{W_{ch}}{H_{DM}}$ [9] (Courtesy of the Electrochemical Society).

Assuming the two resistors are arrayed in series, the total temperature variation is the sum of the two:

$$\Delta T_{max} = \Delta T_{max}^{\text{in-plane}} + \Delta T_{max}^{\text{thr-plane}} = \Delta T_{max}^{\text{thr-plane}} f_{max}\left(\frac{W_{ch}}{H_{GDL}}, \frac{k_{GDL,H}^{eff}}{k_{GDL,W}^{eff}}\right), \qquad \text{(Eq. 8-37)}$$

where $f_{max} = 1 + \dfrac{W_{ch}^2 k_{GDL,H}^{eff}}{2H_{GDL}^2 k_{GDL,W}^{eff}}$, which lumps both the geometrical parameters and thermal properties; see Figure 8-11. This parameter measures the effect of in-plane heat transfer relative to the through-plane one. It is seen that the geometric parameters are important in determining the in-plane temperature variation. Figure 8-12 compares 2D numerical prediction computed by solving the partial differential equation of the energy conservation with the above analytical solution and shows acceptable agreement.

8.3.3 NUMERICAL ANALYSIS

8.3.3.1 Macroscopic Model Prediction

The energy conservation equation, a partial differential equation, can be numerically solved to obtain temperature distribution and heat flow in PEM fuel cells. To account for the coupling of the electrochemical reaction, species transport, and heat transfer, the partial differential equations of species and charge conservation need to be solved as well. One way to simplify the problem is to decouple heat transfer from the electrochemical reaction kinetics and treat the waste heat generation as a heat flux added to the catalyst layer–GDL interface. Figure 8-13 displays the numerically computed temperature contours inside a cathode GDL, bipolar plate, and GFC. Peak temperatures appear at the central line under the channel, as a result of the waste heat generation and small heat removal capability by channel gas flow.

A large-scale PEM fuel cell exhibits complex geometry, in which heat transfer takes place, imposing a challenge to the thermal management. Figure 8-14 displays the structure of a commercial-size PEM fuel cell with cooling flow channels embedded in the bipolar plates.

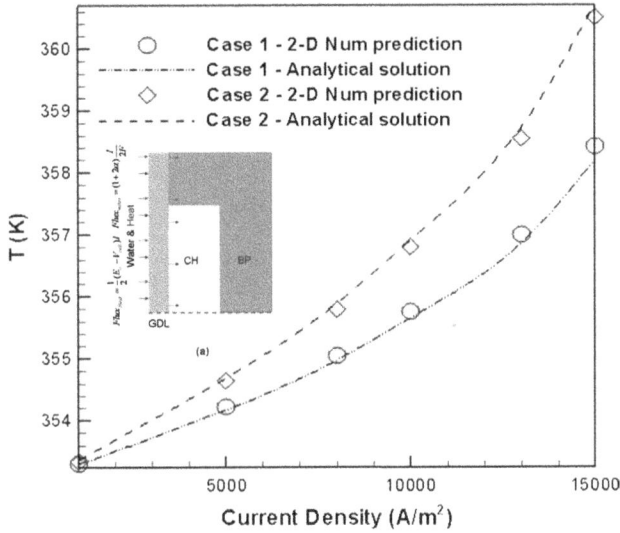

Figure 8-12. Comparison of the maximum temperatures between a 2D numerical prediction and analytical solution for: a GDL conductivity of 3 W/m·K (Case 1) and 2 W/m·K (Case 2) [9] (Courtesy of the Electrochemical Society).

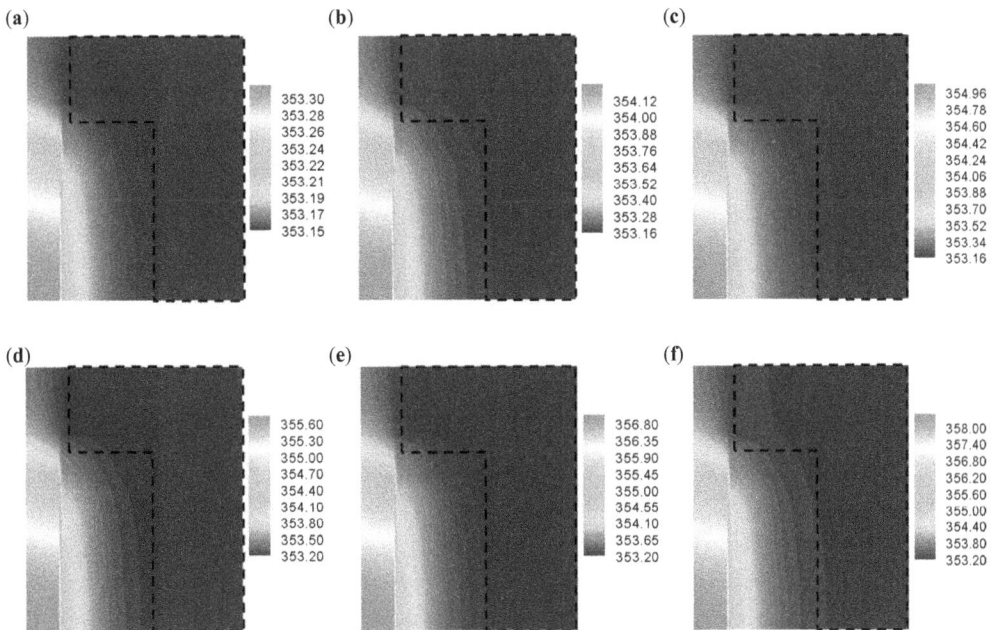

Figure 8-13. Temperature contours in the GDL, GFC, and bipolar plate, under: (a) 0.1 A/cm²; (b) 0.5 A/cm²; (c) 0.8 A/cm²; (d) 1.0 A/cm²; (e) 1.3 A/cm²; and (f) 1.5 A/cm². The BP outer surface is set at 353.15 K and thermal conductivity is 3 W/m·K.

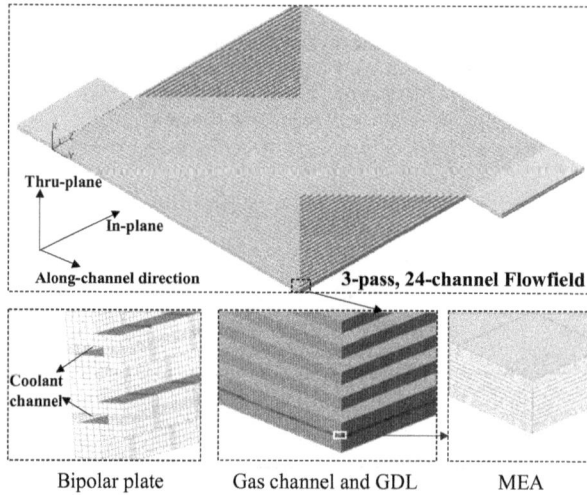

Figure 8-14. Configuration of a 200 cm² commercial-size PEM fuel cell with embedded cooling channels.

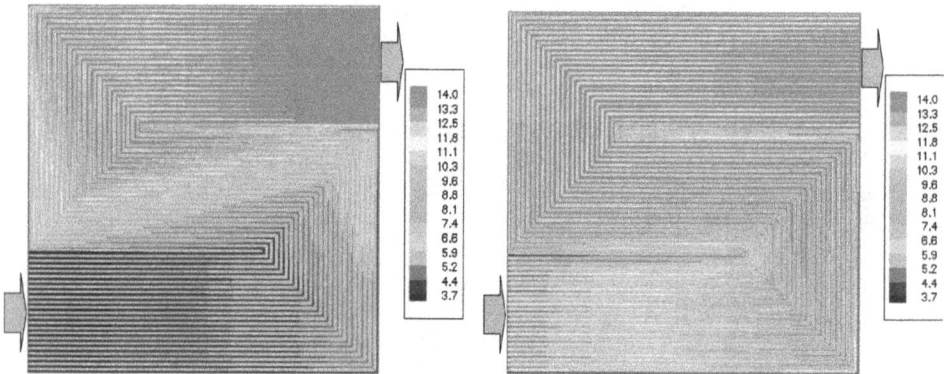

Figure 8-15. Water content at the membrane-cathode interface in a PEM fuel cell without cooling strategy (left) and with a cooling strategy (right), respectively. The temperature contours of the PEM fuel cell are given by Figure 8-2 and Figure 8-3, respectively [6].

The gas flow channel size is 1 mm × 1 mm in the cross-section; each flow plate consists of 24 parallel channels in the serpentine configuration (two U-turns).

Two conflicting issues are encountered in automobile fuel cells for water management in the along-GFC direction. First, dry reactant feed, preferred in the automotive application, causes membrane dryout near the inlet. Secondly, the outlet gas stream is over-saturated, causing liquid water formation and "flooding." These two issues can be resolved by designing proper control strategy of cooling flow. Figure 8-3 presents the temperature contours in a large-scale fuel cell with the following cooling channel design: the cooling inlet temperature is set 343.15 K with its inlet velocity of 2.9×10^{-6} m s^{-1}. As a result, a temperature gradient from inlet toward outlet develops, as opposed to the case with a constant temperature at the BP surface of 353.15 K (see Figure 8-2). Thus, the water vapor removal capability near the outlet is higher than that near the inlet. In an extreme case, the outlet temperature can be increased to a level that all the local liquid water evaporates, which eliminates any "flooding" concern. Figure 8-15 compares

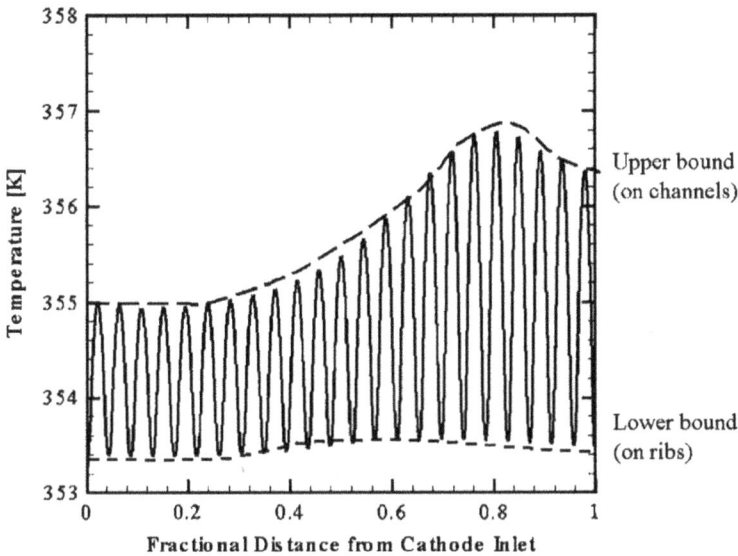

Figure 8-16. Temperature profile in the plane cutting across the middle of a PEM fuel cell under RHa/c = 75%/0% and 0.7 V [10] (Courtesy of the Electrochemical Society).

the membrane water contents for the two cases, clearly showing that the proposed cooling flow strategy alleviates the dryness near the inlet and flooding near the outlet. Discussions on phase change and coupling of thermal and water management are to be presented in Chapter 9.

From Figures 8-2 and 8-3, it is clear that the land–channel structure affects the temperature distribution. Figure 8-16 shows the cross-sectional temperature profile along the center line of the membrane in a cross section of a PEM fuel cell. The upper and lower bounds correspond to the channel and land regions, respectively. The spatial temperature fluctuation between the channel and land results from the relative ineffective cooling under the channel, as opposed to that under the land. As all the heat generation sources increase with the current density, the temperature bounds increase along the cathode flow path and then decreases near the outlets, in correspondence to local current variation (not shown).

8.3.3.2 Pore-Level Heat Transfer

GDLs are typical porous material, consisting of both the solid matrix and pore space. Inside the medium, heat is primarily transported via the solid matrix through conduction because the fluids in the void space are poor thermal conductors. The solid-matrix structure of GDLs is random, featured by randomly distributed fibers and other constituent materials such as binders. As a result, locally high thermal resistance may be present in places where solid phase is small in its content or high in its tortuosity, leading to the formation of hot spots. In addition, the effective conductivity of GDLs is determined by microscale heat transfer: the microscopic phenomena provide information to evaluate the macroscopic properties and structure–property correlations. Figure 8-17 shows pore-level temperature distributions at three cross-sections. Because the fiber's thermal conductivity is several-order-of-magnitude higher than that of the reactant gas (air), most heat is conducted via the randomly distributed fibers. The temperature

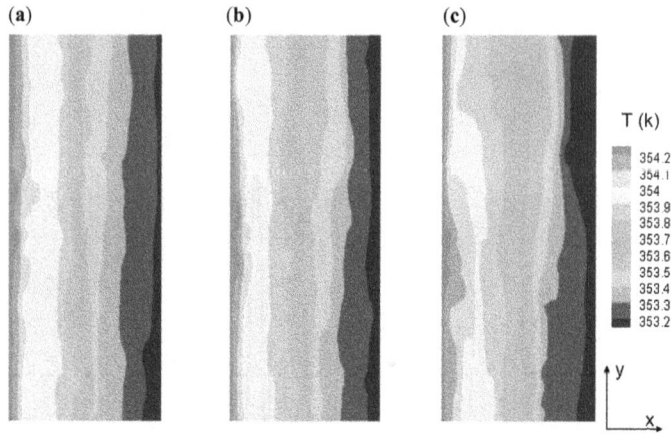

Figure 8-17. Pore-level temperature contours at three GDL cross-sections: (a) $\bar{z}$=0.25; (b) $\bar{z}$=0.5; (c) $\bar{z}$=0.75. $\bar{z}$ is the dimensionless distance in the z direction [11].

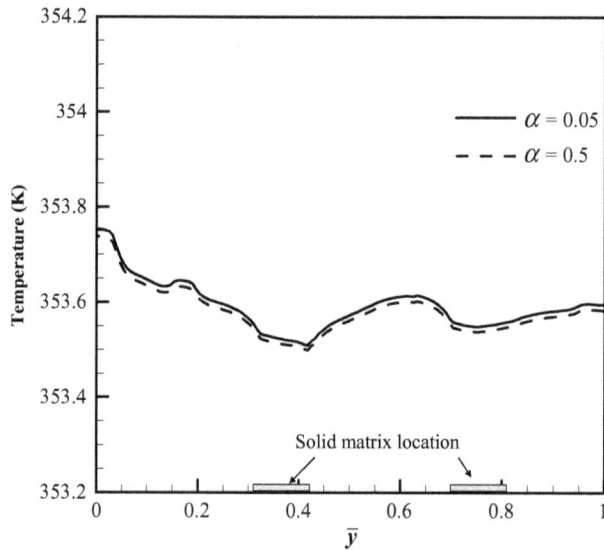

Figure 8-18. Pore-level temperature profiles at the cross line of the planes $\bar{z}$ 0.5 and $\bar{x} = 0.5$. $\bar{z}$ and $\bar{x}$ are the dimensionless distances in the z and x directions, respectively. The geometry is the GDL is shown in Figure 5-15.

change within a pore (ΔT_d) is ~0.2 K. The solid matrix is indistinguishable from the surrounding fluid in terms of temperature contours because the pore dimension is small and the heat flux at the fiber surface rapidly raises local air temperature. The temperature variation across the GDL sample, that is, the system level ΔT_L, is only ~1 K, which is comparable to the pore level ΔT_d, raising a concern of local thermal non-equilibrium.

Figure 8-18 presents the temperature profiles between two cases with different net water transfer coefficients and hence various mass flows in the pore space. The two profiles are very close with the case of $\alpha = 0.5$ having slightly lower temperatures. The mass flow induced by the mass exchange between anode and cathode delivers thermal energy through convection. The amount of convective heat transfer is, however, small, as seen from the figure, which shows that temperature variation near the solid–gas interface is moderate and thus a small heat flux is transferred between the solid and fluid.

8.4 TRANSIENT PHENOMENA

In the transportation or portable applications of PEM fuel cells, rapid variation in load is frequently encountered. In transient, temperature varies upon any changes in the waste heat generation or cooling conditions. Common transient operation, frequently encountered in PEM fuel cells, includes startup, shut-down, and regular load variation.

8.4.1 GENERAL TRANSIENT OPERATION

Figure 8-19 presents the temperature evolution when scanning current density from 0 to 750 mA/cm^2 with 20 mA increments, showing temperature changes (with a fluctuation around 5°C) simultaneously with the current scanning. Heat transfer is usually rapid inside a PEM fuel cell, due to the large thermal diffusivity of fuel cell components. Thus, the thermal energy in the reaction area or any hot spots can usually be quickly dissipated.

Figure 8-20 shows the temperature and current density when varying the voltage in steps: from 0.63 to 0.54 V, then to open circuit potential (≈1.0 V), and finally back to 0.54 V. The current increases from approximately 0.5 to 0.68 A/cm^2, then to 0 A/cm^2, and back to 0.67 A/cm^2. All the temperatures change rapidly with current density. Under open-circuit condition, temperatures are stable around 65°C. The temperature at the anode catalyst surface (T_2)

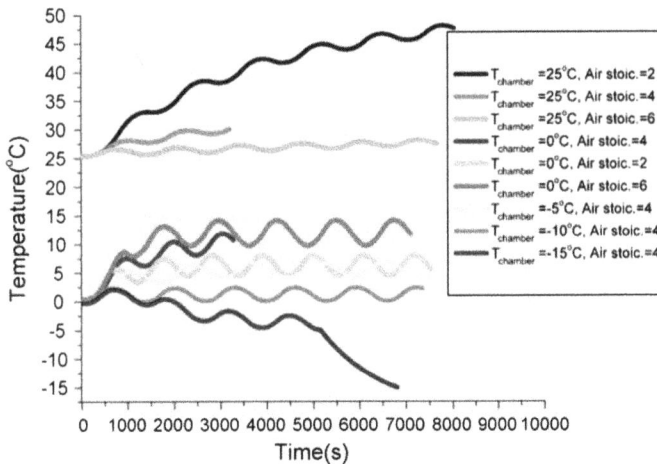

Figure 8-19. Temporal variations of the cathode temperature in a PEM fuel cell when scanning current density [12] (Courtesy of Elsevier).

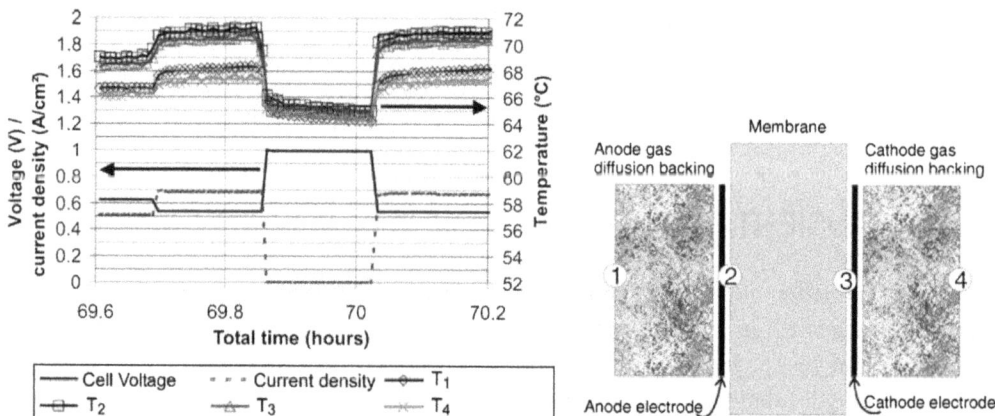

Figure 8-20. Temperatures in a PEM fuel cell (left) and the positions of the four thermocouples (right) [13] (Courtesy of Elsevier).

increases slightly more than that at the cathode catalyst surface (T_3), while the temperatures in the channels stay lower.

In Chapter 2, we discussed the output voltage, which is determined by several major voltage losses:

$$V_{cell} = E - \frac{RT}{2\alpha_c F}\ln(\frac{i}{i_{0,c}}) - \Delta V_\Omega - \Delta V_{trans},$$ (Eq. 8-38)

where V_{cell}, E, i, $i_{0,c}$, and α_c are, respectively, the electrode potential, thermodynamic potential, current density, cathode exchange current density, and the cathodic transfer coefficient, $\dfrac{RT}{2\alpha_c F}$ is the Tafel slope for the cathode ORR. In the anode, the HOR's reaction kinetics is much faster than the ORR's, thus the corresponding voltage loss is neglected.

In the above equation, it is obvious that temperature affects both ΔV_Ω and ΔV_{trans}. ΔV_Ω ($= R_\Omega I$) is proportional to the ohmic resistance (R_Ω), a major portion of which is the membrane ionic resistance. The membrane conductivity, the reciprocal of the resistance, is a strong function of temperature and membrane water content, as given in the correlations developed in Chapter 4. The membrane hydration level (or water content) is determined by the surrounding water activity, which is defined as the ratio of water vapor concentration to the saturated value for partially saturated condition. The saturated vapor pressure is determined by temperature; thus the membrane water content and hence the ionic resistance is temperature-dependent. In a latter section, a case evaluation will be presented to discuss ΔV_Ω under subfreezing operation.

ΔV_{trans} is determined by the species transport resistance and physical dimension of the transport passage. In the diffusion-dominated regime, diffusivity plays a critical role in ΔV_{trans}. Both molecular and Knudsen diffusions are dependent on temperature; thus, any temperature variation affects ΔV_{trans}. However, the effect is usually small in PEM fuel cells because of the small range of temperature variation in common operation. Instead, liquid water or ice presence can significantly reduce the effective diffusivity of reactant species, raising ΔV_{trans}. In Chapter 7, analysis is presented to show the effect of ice formation on cell voltage loss. Both ice formation and liquid water presence are significantly affected by the operating temperature because the saturated vapor pressure is a strong function of temperature.

In what follows, a brief discussion is presented on the variation of E, the Tafel slope, and i_0 with temperature:

a. Temperature Dependence of the Thermodynamic Potential E

The thermodynamic potential E is determined by the Gibbs free energy:

$$E = -\frac{dG}{znF} = -\frac{dg}{zF} \qquad \text{(Eq. 8-39)}$$

$$E = \frac{1}{zF}\left(-dg_o - \frac{\partial(dg)}{\partial T}\Big|_o (T - T_o)\right)_{P=\text{const}} = E_o + \frac{ds_o}{zF}(T - T_o). \qquad \text{(Eq. 8-40)}$$

Choosing $T_o = 298$ K as reference temperature gives

$$E = E_o(298\text{K}) + \frac{ds_o}{zF}(T - 298\text{K}). \qquad \text{(Eq. 8-41)}$$

Below 100°C the change in entropy (ds_o) for the H_2/O_2 reaction is ca. -163.5 J K^{-1} mol^{-1}, while above 100°C ds_o is approximately -44.5 JK^{-1} mol^{-1}. The corresponding $\partial E/\partial T$ are, respectively, -0.85 mV/K (<100°C) and -0.23 mV/K (>100°C). Both $E = 1.23 - 0.9 \times 10^{-3}(T - 298)$ and $E = 1.482 - 0.000845T$ have been proposed [14] to account for temperature dependence.

In addition, the Nernst equation states, for the reaction of $2H_2 + O_2 \rightarrow 2H_2O$, E can be written as

$$E = E_o + \frac{RT}{2F}\ln\left(\frac{a_{H_2} a_{O_2}^{1/2}}{a_{H_2O}}\right), \qquad \text{(Eq. 8-42)}$$

where the species activity a_i is defined as p_i/p_0 (for ideal gases), C_i/C_0 (for ideal solutions), or unity (for pure substance). Clearly, changing temperature affects a_{H_2O} and hence E.

b. Temperature Dependence of the Tafel Slope b $(= \dfrac{RT}{2\alpha_c F})$

For the ORR, the dependence of b on temperature is derived as follows:

$$\frac{\partial b}{\partial T} = -\frac{R}{\alpha_c F}, \qquad \text{(Eq. 8-43)}$$

b varies linearly with temperature with $db/dT \approx 0.2$ mV/K for $\alpha = 1$. Two sets of Tafel slopes have been reported for the low and high current density regimes, respectively: one Tafel parameter corresponds to the ORR at a Pt oxide-covered surface (Temkin adsorption conditions, low current density, $b = 60$ mV/decade) and the other, at a Pt oxide-free surface (Langmuirian conditions, high current density, and $b = 120$ mV/decade) [14]. Experimentally, the Tafel slope was found to increase with temperature in the low current density regime, whereas independent of temperature in the high current density regime [15].

c. Temperature Dependence of the Exchange Current Density i_0

The exchange current density i_0 for the ORR ($\sim 10^{-8}$ to 10^{-9} A/cm^2) is much smaller than that of the HOR ($\sim 10^{-3}$ to 10^{-4} A/cm^2). As a result, the overpotential for HOR is negligibly small. The slow ORR kinetics is a major limiting factor for achieving high fuel cell performance. It was found that i_0 increases with temperature: Beattie et al. [16] reported i_0 values for Nafion® 117 and BAM® 407 (an experimental sulfonated polytrifluorostyrene-based membrane), which range from 2.08×10^{-10} to 3.71×10^{-9} A/cm^2 and 8.80×10^{-11} to 4.38×10^{-10} A/cm^2, respectively, over the temperature range of 303–343 K. Parthasarathy et al. [15] reported that i_0 at Pt/Nafion® 117 interfaces in the low current density region increases from 1.69×10^{-10} to 5.54×10^{-9} A/cm^2 as the temperature is raised from 303 to 343 K. The following correlation in the Arrhenius form is frequently adopted for describing the temperature dependence of i_0:

$$i_0(T) = i_0(353.15)\exp\left[-\frac{E_a}{R}\left(\frac{1}{T} - \frac{1}{353.15}\right)\right], \qquad \text{(Eq. 8-44)}$$

where E_a is the activation energy.

8.4.2   TRANSIENT SUBFREEZING OPERATION

8.4.2.1   Temperature Evolution and Voltage Loss

To start-up, a PEM fuel cell increases its temperature from the ambient to operating temperature, thus experiences complex conditions in combination of vapor condensation at low temperatures and liquid evaporation at high temperatures. In a startup from normal ambient temperature (~ 25°C), there appear few technical difficulties. One reason is that the startup is usually quick (in order of 10–100 s), resulting in small amount of liquid water production. The liquid water is removable through the intrinsic capabilities designed for water management in PEM fuel cells. A challenging subject in automobile applications is cold-start, that is, startup under subfreezing conditions. In this situation, product water freezes, resulting in ice accumulation in the catalyst layer. Because ice cannot be removed properly, it accumulates to a level that causes startup failure. In order to start successfully from subfreezing conditions, fuel cell temperature must increase to the freezing point before product ice causes considerable voltage loss and subsequent operation shutdown.

Time constant characterizes the duration for a system responding to an input, for example, a step change in conditions. For PEM fuel cells, the time constant of startup, that is, the duration taken to raise a PEM fuel cell to its operating temperature, is determined by its thermal properties and heating rate. For un-assisted cold start, the waste heat generated by a PEM fuel cell is utilized for warm up. The time constant of cold start is critical to the automobile applications. By adding external heating sources, the duration can be significantly reduced. The technical target set by the Department of Energy (DOE) for 2010 is the ability of a PEM fuel cell stack to reach 50% rated power in 30 s starting from $-20°C$, and the un-assisted start-up temperature may be as low as $-40°C$ [17].

Assuming uniform temperature throughout a PEM fuel cell during cold start, a governing equation can be derived from energy balance to obtain the dynamic temperature:

$$T = \frac{\int_0^t \left[I(E' - V_{cell})A_m + \Delta Q_{gasflow} + \Delta Q_{coolant} - Q_{loss} \right] dt}{m_m Cp_m + m_{CL} Cp_{CL} + m_{GDL} Cp_{GDL} + m_{BP} Cp_{BP}} + T_o, \qquad \text{(Eq. 8-45)}$$

where T_o is the initial temperature. In the above, $\Delta Q_{coolant}$ also accounts for any external heating. For un-assisted start-up without heat loss to the surrounding, the time constant for warming the fuel cell up to $0°C$ can be estimated by

$$\int_0^{\tau_{T,1}} (E' - V_{cell}) dt = \frac{(m_m Cp_m + m_{CL} Cp_{CL} + m_{GDL} Cp_{GDL} + m_{BP} Cp_{BP})(273.15 - T_o)}{A_m I}. \qquad \text{(Eq. 8-46)}$$

Among the thermal capabilities of fuel cell components, the thermal mass of the bipolar plates (BP), $m_{BP} Cp_{BP}$, is a major contributor, and contribution from other components are usually small and negligible. For constant heat generation, the time constant can be estimated by

$$\tau_{T,1} = \frac{\rho_{BP} Cp_{BP} \delta_{BP}}{I(E_{tn} - V_{cell})}(273.15 - T_o), \qquad \text{(Eq. 8-47)}$$

where the BP effective length δ_{BP} is defined as $\dfrac{m_{BP}}{A_m \rho_{BP}}$. For $\rho_{BP} Cp_{BP}$ of $\sim 1600 \text{ kJ/Km}^3$, δ_{BP} of ~ 0.02 m, I of 0.1 A/cm^2, and cold start from $-30°C, \tau_{T,1} \sim 100$ s. As a comparison, $m_{GDL} Cp_{GDL}/m_{BP} Cp_{BP} \sim 0.03$ and thus it takes only a few seconds (~ 3 s) to heat up the GDLs. Using this time constant, the temperature evolution is derived as

$$T = \frac{t}{\tau_{T,1}}(273.15 - T_o) + T_o. \qquad \text{(Eq. 8-48)}$$

Temperature influences cell voltage through several ways, such as the activation and the ohmic voltage losses, as discussed previously. For the former, the exchange current density is dependent on temperature, and the dependence can be substituted into the Butler–Volmer equation to evaluate the overpotential. The overpotential during cold start is discussed and derived in Chapter 7. The solution can be rewritten by accounting for temperature change as follows [18]:

$$\eta(s_{\text{ice}},T,\bar{x}) = -\frac{RT}{\alpha_c F}\ln\left(\frac{IC^{O_2,\text{ref}}}{a_0 i_{0,c}^{\text{ref}}\exp\left[-\frac{E_a}{R}\left(\frac{1}{T_o}-\frac{1}{353.15}\right)\right]C_{c\text{CL}}^{O_2}\delta_{\text{CL}}}\right) - \frac{TE_a}{\alpha_c F}\left(\frac{1}{T}-\frac{1}{T_o}\right)$$

$$+\frac{RT\tau_a}{\alpha_c F}\ln(1-s_{\text{ice}})+\frac{RT}{\alpha_c F}\ln\left(1-Da\frac{1-\bar{x}^2}{\varepsilon_{\text{CL}}^{\tau_d-\tau_{d,0}}(1-s_{\text{ice}})^{\tau_d}}\right)$$

$$=\eta_o+\Delta\eta_T+\Delta\eta_{c,1}+\Delta\eta_{c,2},$$

(Eq. 8-49)

where

$$\eta_o = -\frac{RT}{\alpha_c F}\ln\left(\frac{IC^{O_2,\text{ref}}}{a_0 i_{0,c}^{\text{ref}}\exp\left[-\frac{E_a}{R}\left(\frac{1}{T_o}-\frac{1}{353.15}\right)\right]C_{c\text{CL}}^{O_2}\delta_{\text{CL}}}\right),\quad \Delta\eta_T = -\frac{TE_a}{\alpha_c F}\left(\frac{1}{T}-\frac{1}{T_o}\right).$$

$$\Delta\eta_{c,1}=\frac{RT\tau_a}{\alpha_c F}\ln(1-s_{\text{ice}})\ \text{and}\ \Delta\eta_{c,2}=\frac{RT}{\alpha_c F}\ln\left(1-Da\frac{1-\bar{x}^2}{\varepsilon_{\text{CL}}^{\tau_d-\tau_{d,0}}(1-s_{\text{ice}})^{\tau_d}}\right).$$

(Eq. 8-50)

8.4.2.2 Activation Voltage Loss

The physical meaning of η_o is the overpotential at the cathode catalyst layer–GDL interface at T_o in the absence of ice whereas $\Delta\eta_T$ represents the overpotential change due to the temperature dependence of the exchange current density. $\Delta\eta_{c,1}$ and $\Delta\eta_{c,2}$ arise from the presence of ice in the catalyst layer. The Tafel slope $-\frac{RT}{\alpha_c F}$ is a function of temperature; however, its tempera ture dependence is usually not dominant. Rather, the exchange current density plays a much more important role. The temperature dependence of the Tafel slope has been discussed extensively in the literature. Some studies claimed that the Tafel slope is independent of temperature in certain cases, provided that the cathodic transfer coefficient α_c is a function of temperature. In the expression of $\Delta\eta_T$, $\frac{T}{\alpha_c}$ is assumed to be constant. The maximum value $\Delta\eta_T^{\text{max}}$ during cold-start is estimated by

$$-\frac{T_o E_a}{\alpha_c F}\left(\frac{1}{273.15}-\frac{1}{T_o}\right).$$

(Eq. 8-51)

Figure 8-21 displays the sum of the two overpotentials caused by ice formation, $\Delta\eta_{c,1}+\Delta\eta_{c,2}$, in the cathode catalyst layer. Between these two components, only the second directly depends

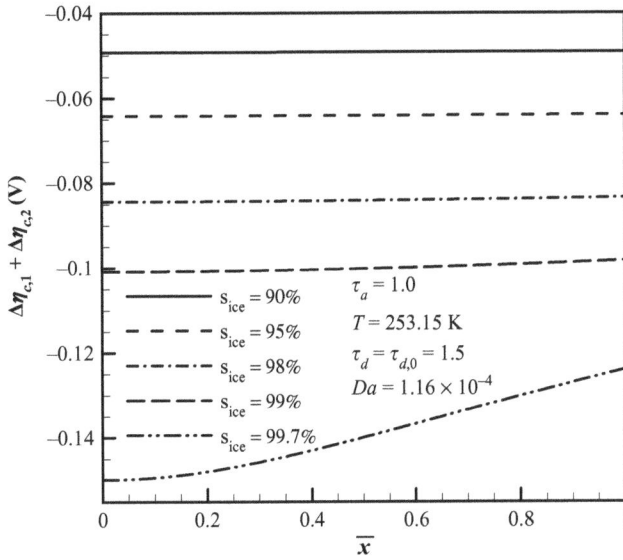

Figure 8-21. $\Delta\eta_{c,1} + \Delta\eta_{c,2}$ in the cathode catalyst layer under different ice volume fractions [18].

on location, whereas the first is a function of local ice volume fraction only. The predicted transport polarization starts to become important at $s_{ice} = 99.7\%$. Even at $s_{ice} = 99\%$ the variation is small.

Figure 8-22 compares $\Delta\eta_T^{max}$ with $\Delta\eta_{c,1}.\Delta\eta_{c,1}$ characterizes the effect of ice coverage on the electrochemical reaction surface and increases quickly with s_{ice}. Ice coverage deactivates the active reaction surface; on the other hand, the ice-covered interface is able to recover after the ice is removed. Figure 8-22 shows that temperature has little impact on $\Delta\eta_{c,1}$, instead the effect of the coefficient τ_α is evident. Furthermore, the magnitudes of $\Delta\eta_T^{max}$ are indicated in Figure 8-22 using two values of the activation energy. These values were obtained from different Tafel slopes with the higher value of 73.3 kJ/mol corresponding to -60 mV/decade under the low-current regime, whereas the other value leading to -120 mV/decade under the high-current regime. $\Delta\eta_T^{max}$ in Figure 8-22 excludes the temperature difference of the Tafel slope. The value shows the magnitude of the overpotential change due to the temperature dependence of the exchange current density. $\Delta\eta_{c,1}$ is comparable with $\Delta\eta_T$ at low and intermediate ice volume fraction regimes, whereas $\Delta\eta_{c,1}$ becomes dominant under large values of s_{ice}.

8.4.2.3 Ohmic Voltage Loss

As for the ohmic voltage loss, temperature is a major factor determining the ionic conductivity and hence the ohmic voltage loss ($\Delta V_{ohm} = R_\Omega I$). The ionic resistance R_Ω^{H+} can be estimated by

$$R_\Omega^{H+} = \frac{\delta_m}{\kappa} + \frac{\delta_{aCL}}{2\kappa^{aCL,eff}} + \frac{\delta_{cCL}}{2\kappa^{cCL,eff}}. \qquad \text{(Eq. 8-52)}$$

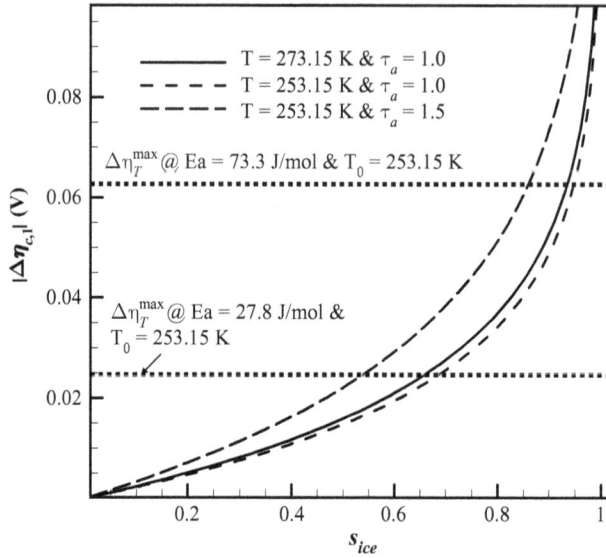

Figure 8-22. Magnitudes of $\Delta\eta_T^{max}$ and $\Delta\eta_{c,1}$ [18].

The above ionic conductivities κ, $\kappa^{aCL,eff}$, and $\kappa^{cCL,eff}$ are the values averaged over the membrane, anode and cathode catalyst layers, respectively. The latter two can be calculated using the Bruggeman correlation:

$$\kappa^{aCL,eff} = \varepsilon_m^{1.5}\kappa^{aCL} \text{ and } \kappa^{cCL,eff} = \varepsilon_m^{1.5}\kappa^{cCL}. \tag{Eq. 8-53}$$

The membrane ionic conductivity κ is a function of water content and temperature. Under subfreezing temperatures, a portion of the membrane water freezes, reducing the ionic conductivity. One correlation of the ionic conductivity for Nafion® 117 from −30 to 0°C is as follows [18]:

$$\kappa = (0.01862\lambda - 0.02854)\exp\left[4029\left(\frac{1}{303} - \frac{1}{T}\right)\right] \text{ or}$$

$$= (0.004320\lambda - 0.006620)\exp\left[4029\left(\frac{1}{273} - \frac{1}{T}\right)\right] = \kappa_0(\lambda)\exp\left[4029\left(\frac{1}{273} - \frac{1}{T}\right)\right] \text{ for } \lambda \leq 7.22$$

$$\kappa = \kappa(\lambda = 7.22) \text{ for } \lambda > 7.22 \tag{Eq. 8-54}$$

For the ohmic voltage loss:

$$\Delta V_{ohm}^{H+} = R_\Omega^{H+}I = \left(\frac{\delta_m}{\kappa_0(\lambda^m)} + \frac{\delta_{aCL}}{2\kappa_0(\lambda^{aCL})\varepsilon_m^{1.5}} + \frac{\delta_{cCL}}{2\kappa_0(\lambda^{cCL})\varepsilon_m^{1.5}}\right)\exp\left[-4029\left(\frac{1}{273} - \frac{1}{T}\right)\right]I$$

$$= R_{\Omega,273K}^{H+}(\overline{\lambda})\exp\left[-4029\left(\frac{1}{273} - \frac{1}{T}\right)\right]I, \tag{Eq. 8-55}$$

where $\bar{\lambda}$ is averaged over the entire MEA. To assess the temperature effect, one can differentiate the above equation with respect to temperature:

$$d(\Delta V_{\text{ohm}}^{\text{H+}}) = \left[\begin{array}{c} \dfrac{dR_{\Omega,273K}^{\text{H+}}}{d\lambda} \exp\left[-4029\left(\dfrac{1}{273} - \dfrac{1}{T}\right)\right] d\lambda \\[2mm] -\dfrac{4029}{T^2} R_{\Omega,273K}^{\text{H+}} \exp\left[-4029\left(\dfrac{1}{273} - \dfrac{1}{T}\right)\right] dT \end{array} \right] I. \qquad \text{(Eq. 8-56)}$$

For the case that the ionomer hydration is always above 7.22, the first term in the bracket is zero. The second term provides an estimate of the ohmic voltage loss variation when integrating from T to $T + dT$. For $\lambda = 14$, $T = -30°C$ and $dT = 30$ °C, $R_{\Omega,273K}^{\text{H+}}$ is ~700 mΩ cm^2, and $d(\Delta V_{\text{ohm}}^{\text{H+}})\Big|_{-30°C}^{0°C}$ is ~0.18 V at 0.1 A/cm^2. The computed $d(\Delta V_{\text{ohm}}^{\text{H+}})\Big|_{-30°C}^{0°C}$ is much higher than $\Delta\eta_T^{\max}$. Using Springer et $al.$ [19]; however, $d(\Delta V_{\text{ohm}}^{\text{H+}})\Big|_{-30°C}^{0°C}$ is ~ 0.03V, significantly underestimating the temperature dependence of the Ohmic polarization.

8.5 EXPERIMENTAL MEASUREMENT OF THERMAL CONDUCTIVITY

Thermal conductivity is an important physical property of PEM fuel cells with respect to thermal management because conduction is the major mode of waste heat removal from the MEA. The conductivity is a scalar for isotropic materials but a tensor for anisotropic materials such as fibrous GDLs. Experimental thermal conductivity and contact resistance measurements have been reported in the literature. Figure 8-23 schematically shows two sets of experimental setups for measuring thermal conductivity of thin-layer materials.

Khandelwal and Mench [20] measured the thermal conductivities of several PEM fuel cell components, including carbon paper materials and Nafion membrane. Their measurement shows a decreasing trend of thermal conductivity with temperature for carbon papers, and a thermal conductivity of 0.22 ± 0.04 W/(m·K) for SIGRACET® 20 wt% PTFE carbon paper and 1.80 ± 0.27 W/(m°C) at 26°C for Toray papers. Zamel et $al.$ [21] experimentally measured the through-

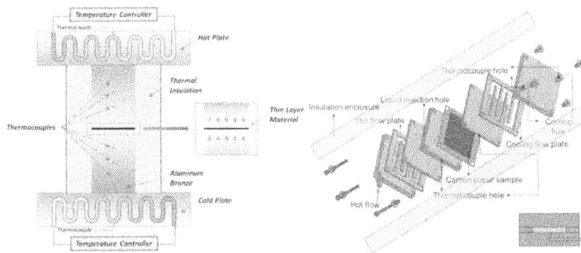

Figure 8-23. Schematic of two setups for measuring thermal conductivity of thin-layer material (redrawn from [20] and [24]).

Table 8-1. Experimentally measured thermal conductivity of PEM fuel cell components [20, 24]

Material	Thickness	PTFE load-ing	Measured thermal conductivity, k (W/m·K)
DuPont Nafion membrane	–	–	0.16 ± 0.03
Toray carbon-fiber paper GDL (TGP-H-60)	190 μm	–	1.80 ± 0.27
Toray carbon-fiber paper GDL (TGP-H-60) with 15 ± 3% compression	190 μm	0%	~0.6
Toray carbon-fiber paper GDL (TGP-H-60) with 18 ± 3% compression	190 μm	30%	~0.55
SIGRACET carbon-fiber paper GDL	190 μm	0 wt%	0.48 ± 0.09
SIGRACET carbon-fiber paper GDL	280 μm	5 wt%	0.31 ± 0.06
SIGRACET carbon-fiber paper GDL	280 μm	20 wt%	0.22 ± 0.04
E-Tek ELAT GDL (LT1200-W)	–	–	0.22 ± 0.04
Catalyst layer (0.5 mg/cm^2 platinum on carbon)	25 μm	–	0.27 ± 0.05

plane thermal conductivity of TORAY carbon paper from 50°C to 120°C. They reported that the media exhibit a thermal conductivity of about 0.8–1.8 W/(m.K) at high deformation and 0.2–0.4 W/(m·K) at low deformation for a wide range of temperature. Burheim *et al.* [22] measured the thermal conductivity of SolviCore porous media under various compaction pressures. For the dry media at 4.6, 9.3, and 13.9 bar, the thermal conductivity was found to be 0.27, 0.36, and 0.40 W/m·K, respectively. The thermal conductivity of the media was found to increase by about 0.17 W/m·K when adding about 25% liquid saturation into the media. Table 8-1 lists the measurements of several PEM fuel cell components.

The compact pressure over GDLs affects the effective thermal conductivity. Figure 8-24 shows the changes in the thicknesses of Toray carbon papers TGP-H-060 and TGP-H-120 under various compression loads, showing that GDL deformation increases with the pressure as expected. Figure 8-25 presents the effective thermal conductivity under different compact pressures in vacuum and atmospheric air conditions, respectively. The effective conductivity increases with the compressive load due to better contact among fibers. A small difference (less than 3%) can be observed between thermal conductivity values obtained under atmospheric and vacuum conditions, indicating that the air in the void space contributes to heat conduction but its effect is small.

The presence of liquid water improves GDL's effective thermal conductivity. Direct numerical simulation results show that the through-plane heat conduction occurs via a route that combines the lateral path along fibers and through-plane path at the contact points [11] (see Fig. 8-26(a)), resulting in a tortuosity value about 13 for the solid matrix. In reality, liquid water attaches to the hydrophilic matrix; liquid may not follow the matrix morphology completely. Instead, it exists preferentially at fiber joints, considerably reducing local thermal resistance (see Fig. 8-26(b)). In addition, the morphology of liquid water presence is certainly important to the effective thermal properties. Using the hydrophobic and hydrophilic surfaces as an example, liquid tends to form isolated bulk droplets on a hydrophobic surface, whereas it tends to form a film on a hydrophilic surface (see Figs. 8-26(e) and (f)). Isolated droplets contribute

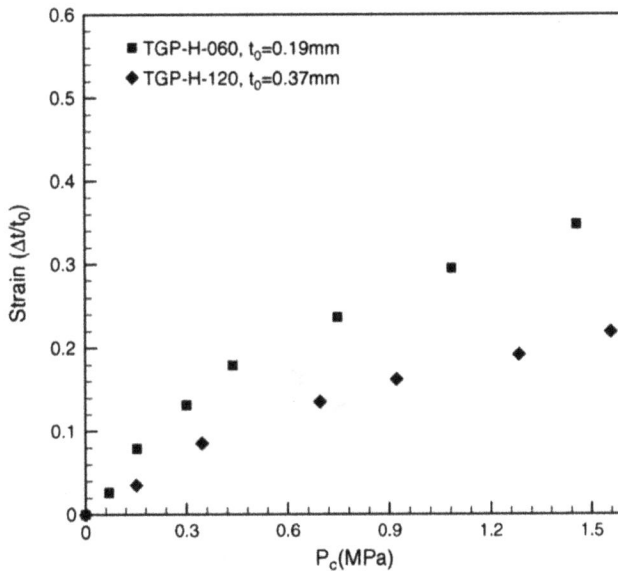

Figure 8-24. Thickness change of Toray carbon paper TGP-H-060 and TGP-H-120 upon compact pressure [23] (Courtesy of Elsevier).

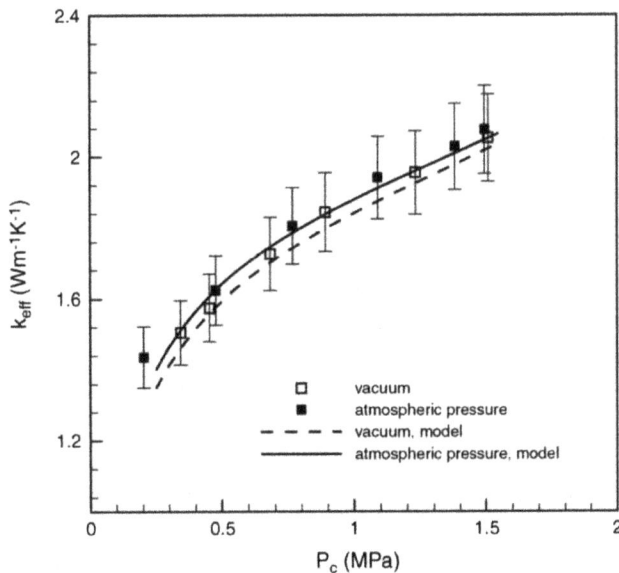

Figure 8-25. Effective thermal conductivities of Toray carbon papers in vacuum and atmospheric air environments [23] (Courtesy of Elsevier).

smaller conductance; however, the water droplets serve as bridges and thus make it easier to link neighboring fibers, consequently reducing tortuosity (see Fig. 8-26(c)). The net effect is determined by the balance of the two factors. Furthermore, Figure 8-26(d) shows the fibers and their contacts with a plate: the contact resistance arises not only between the carbon paper and

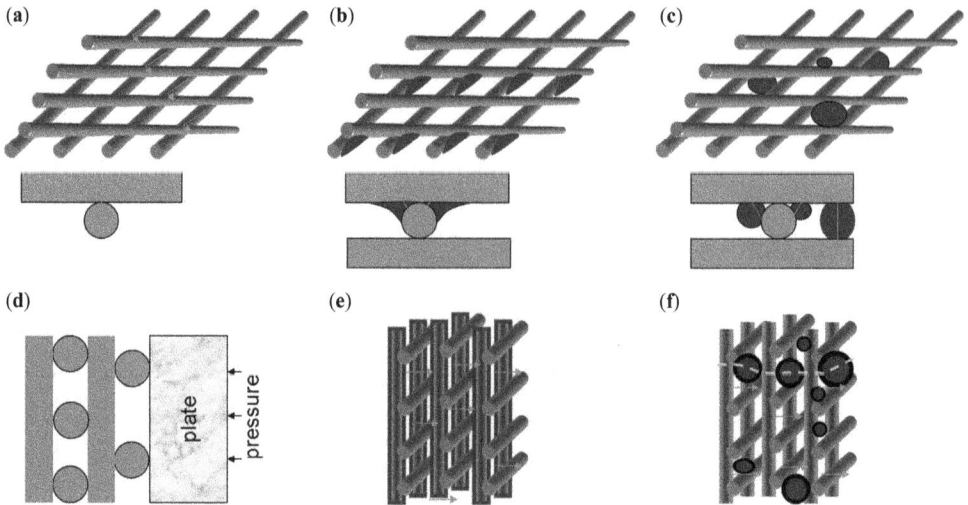

Figure 8-26. Schematics of heat transfer in different GDL substrates and configurations: (a) dry carbon paper, (b) hydrophilic carbon paper, and (c) hydrophobic carbon paper; (d) contacts between the sample and plate and among fibers, (e) heat flow by heat pipe effects in hydrophilic carbon paper, and (f) heat flow in hydrophobic carbon paper. Note: the carbon fibers are intentionally drawn in a structural fashion for illustration purposes [24].

plate (which is the conventional meaning of contact resistance), but also among the constituent fibers. High compression yields a better contact (that is, larger contact areas both between the sample and plate, and among fibers). As explained in the pore-level investigation discussed [11], the fiber-to-fiber contact area can be a limiting factor for the through-plane conductivity. Thus, the overall conductivity is dependent on the pressure imposed over its surface (see Fig. 8-25). This is similar to unsolidified packed beds having thermal conductivity increasing with compression due to the enhanced contacts among particles.

The contact thermal resistance is present at the interface between the GDL and bipolar plate, due to the imperfect contact (surface roughness and the lack of compression). This interfacial resistance raises the temperature at the GDL and hence the MEA side; consequently, its effect needs to be kept sufficiently low. In addition to interfacial roughness and compression, erosion or corrosion at the interface also attributes to thermal contact resistance. Nonetheless, compression is a dominating factor determining the contact resistance and can be controlled during fuel cell assembly. The experimental setup of measuring contact resistance can follow that shown in Figure 8-23 by placing two materials with the interface of interest between the pillars. Figure 8-27 shows the testing result of thermal contact resistance as a function of compression, showing a rapid decrease of the resistance with compression pressure.

8.6   COOLING METHODS

Effective and efficient cooling is critical to water and thermal management of PEM fuel cells, particularly for high-power applications. Cooling PEM fuel cells is more challenging than the internal combustion engines (ICEs) for two major reasons: (1) the exhaust gas of PEM fuel cells, unlike that of ICEs, does not carry along significant amount of waste heat.

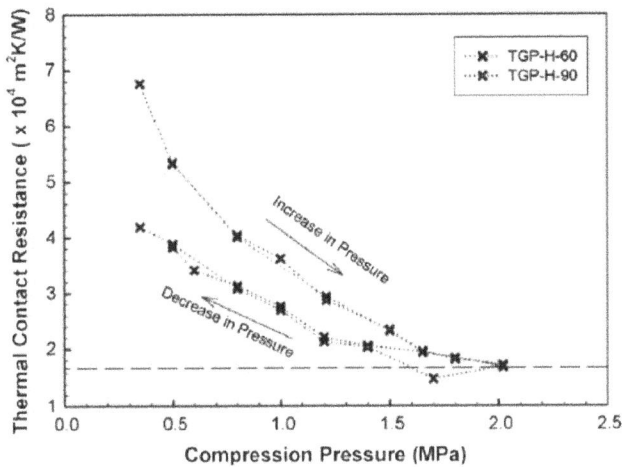

Figure 8-27. Thermal contact resistance between a Toray carbon paper and smooth aluminum bronze under various compression pressure [20] (Courtesy of Elsevier).

As a result, the rate of waste heat needed to be dissipated by the cooling system is much more (approximately doubled) than that of ICEs. (2) The difference in temperature available for heat transfer in PEM fuel cells is significantly lower (by approximately 50%) than that in ICEs. This is because the working temperature of coolants in a PEM fuel cell (and thus the temperature difference between the coolant and ambient) is smaller than that of an ICE. The consequence is that the cooling pumping power required is nearly four times higher for an automobile fuel cell stack as compared with the corresponding ICE. Figure 8-28 shows the breakdown of energy for a PEM fuel cell in comparison with an ICE. Figure 8-29 presents two examples of the thermal-management systems for PEM fuel cells: conventional and passive thermal managements. The passive cooling has no coolant circulation, therefore eliminating the need for coolant pumps that are used in the conventional systems. There are various cooling techniques developed for PEM fuel cells, including (i) heat spreaders based on high thermal conductivity materials or heat pipes, (ii) cooling flow of air or liquid (water or antifreeze coolants), and (iii) phase-change based cooling. Their advantages and disadvantages are listed in Table 8-2.

8.6.1 HEAT SPREADERS COOLING

Heat spreaders cooling removes heat primarily through the in-plane direction, that is, from the central region to the edges; see Figure 8-30 as an example. Thus, heat is more easily removed from the edge area, resulting in spatial temperature variation. High in-plane conductivity of the cooling plates is usually chosen to diminish the temperature variation from the central region to the edge, such as carbon-based materials, metals, or heat pipes. Examples are graphite-based materials, for example, expanded graphite and pyrolytic graphite, and metals such as aluminum or copper foils. Pyrolytic graphite sheets (PGS) exhibit a high in-plane conductivity, ranging from 600 to 800 W m^{-1} K^{-1} [29]. Highly oriented pyrolytic graphite (HOPG) can reach a

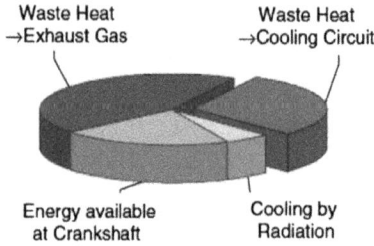

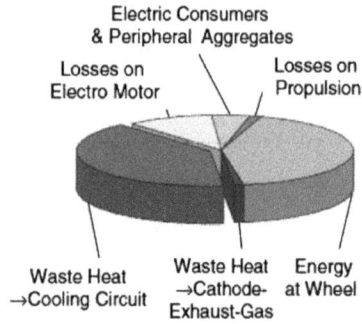

Figure 8-28. Energy breakdowns of an internal combustion engine and a PEM fuel cell engine [26] (Courtesy of John Wiley and Sons).

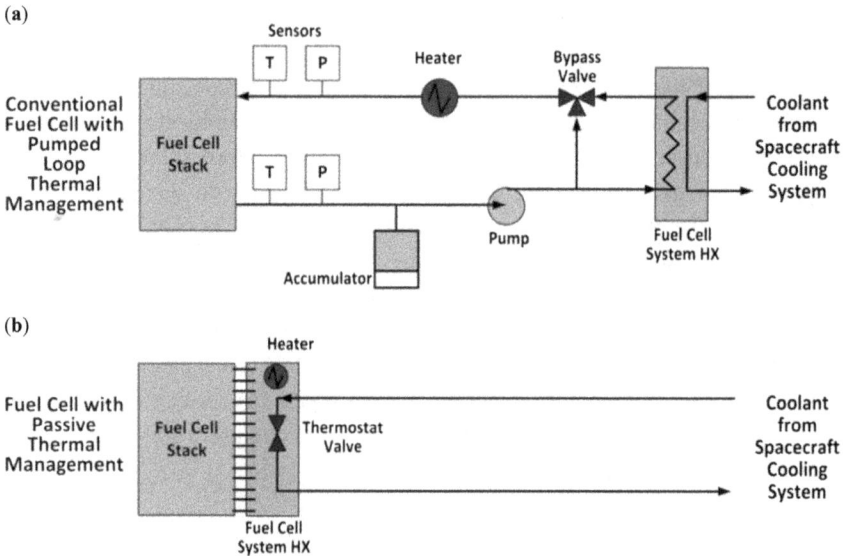

Figure 8-29. Two thermal-management systems: (a) a conventional system and (b) a passive (cooling with heat spreaders) system for space applications [27] (Courtesy of Elsevier).

conductivity up to 1500–1700 W m^{-1} K^{-1}. Adding metal foils (stainless steel or copper foils with a thickness about 0.05 mm) improves the TPG's (thermal pyrolytic graphite, one type of HOPG) mechanical strength. Their effective thermal conductivity were measured to be 962 W m^{-1} K^{-1} (316 SS clad TPG) and 1105 W m^{-1} K^{-1} (Copper clad TPG) [27].

Table 8-2. Summary of important cooling strategies for PEM fuel cells

Cooling strategy	Method	Advantages	Disadvantages
Heat spreaders /edge cooling	Using highly conductive material as spreaders	– Simple system – No internal coolant – No parasitic power	– Limited lateral dimension – Non-uniform temperature – High cost
	Using heat pipes as spreaders	– Small parasitic power – High thermal conductance	– Integration of heat pipes
Cooling by fluid flow	Air cooling	– Large parasitic power	– Small cooling capability
	Cooling by liquid flow (DI water/ antifreeze coolant)	– Large cooling capability – Small parasitic power	– Coolant maintenance (leakage or degradation)
Phase-change cooling	Evaporative cooling	– Enable combined water and heat management (cooling and internal humidification)	– Complicated system – Need to precisely control water injection
	Cooling through boiling	– Simplified system – Large cooling capability	– Extra coolant addition or recirculation
Thermoelectric cooling	Use the Peltier effect between the junction of two materials	– Simplified system: no fluids and moving parts – Flexible shape – Easy control through input voltage/current	– Small cooling capability
Heat sink (surface cooling)	Passive heat exchanger that dissipates heat into surrounding air	– Simplified system: no fluids and moving parts – Flexible shape	– Small cooling capability

For large-scale PEM fuel cells, the heat conduction distance becomes large, for example, ≥ 10 cm, and heat pipes are frequently adopted. Heat pipes are effective heat-transfer devices based on the phase changes (evaporation and condensation) and transport of a working fluid. Heat pipes can transport a large amount of heat over a considerable distance without additional power input, exhibiting high effective thermal conductivity. Cost-effective integration of heat pipes into PEM fuel cells is critical but challenging. One approach is to embed additional small channels into bipolar plates for heat pipes, as shown schematically in Figure 8-31(a). This requires extra manufacturing efforts in fabricating channels and sealing the heat pipes within the channels. Another approach is to integrate flat heat pipes with the bipolar plates, as shown in Figure 8-31(b). There are different types of heat pipes, including micro and miniature heat pipes (1–10 W), loop heat pipes (10–100 W), pulsating, and sorption heat pipes (100–1000 W) [30]. Burke *et al.* [27] measured the effective thermal conductivity of planar heat pipes based on titanium and copper, respectively. The titanium planar heat pipe was thinner and lighter with a thickness of 1.19 mm and

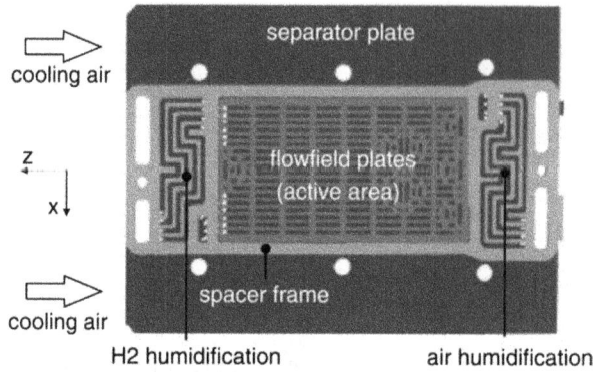

Figure 8-30. Cooling and bipolar plates of a PEM fuel cell with edge cooling [28] (Courtesy of Elsevier).

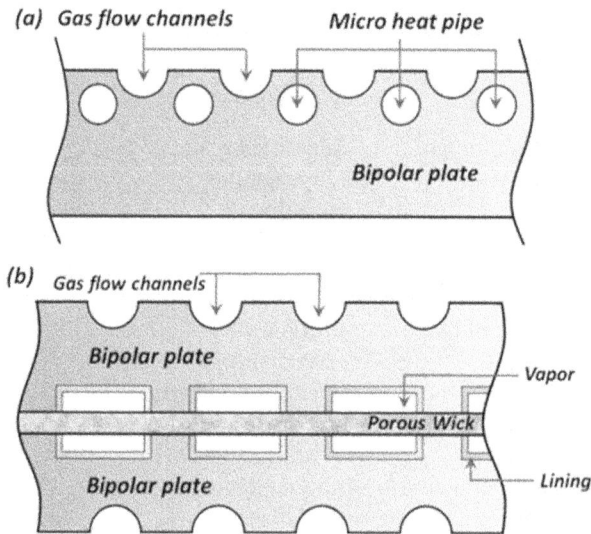

Figure 8-31. Microheat pipe: (a) embedded in a bipolar plate [31] and (b) integrated in a bipolar plate [32].

thermal conductivity up to 20,447 m^{-1} K^{-1}, an order of magnitude higher than that of HOPG composite cooling plates.

8.6.2 COOLING BY AIR OR LIQUID FLOW

Separate flow channels can be machined for cooling air or liquid (water or antifreeze coolants) to flow through. Similar to heat pipes, the cooling units can be fabricated in the bipolar plates or in separate cooling plates placed between the bipolar plates. As for cooling air, its low heat capacity requires a relatively high flow rate in the cooling channel, comparing with liquid coolants. Thus, the auxiliary pumping power is a concern when the need for heat removal is high.

Air flow is usually suitable for PEM fuel cells with power lower than 5 kW. The portion of the heat removal by air flow can be evaluated using the following equation:

$$\beta_1 = \frac{\rho_{air} c_{p,air} \Delta T \forall_{coolant}}{I(E' - V_{cell}) A_m} \approx \frac{\rho_{air} c_{p,air} \dfrac{\xi}{C_{O_2}}}{4F(E' - V_{cell})} \Delta T, \qquad \text{(Eq. 8-57)}$$

where ΔT is the temperature increment from the inlet to outlet of the cooling channel and $\forall_{coolant}$ the coolant flow rate. The last term on the right uses the reactant air flow rate as reference: for cooling air flow at $\xi = 1.5$, $\beta_1 < 30\%$ for $\Delta T = (80 - 20)°C = 60°C$.

For a PEM fuel cell stack with output power over 5 kW, liquid cooling is preferred; consequently, it is widely used in high-power fuel cells. Coolant fluid can be deionized water which has the advantage of high heat capacity, or antifreeze coolant for subfreezing operation, for example, mixture of ethylene glycol and water. The coolant flow rate for PEM fuel cells can be evaluated by

$$\beta_1 = \frac{\rho_{coolant} c_{p,coolant} \Delta T \forall_{coolant}}{I(E' - V_{cell}) A_m} \xrightarrow{\beta_1 = 1} \forall_{coolant} = \frac{I(E' - V_{cell}) A_m}{\rho_{coolant} c_{p,coolant} \Delta T}. \qquad \text{(Eq. 8-58)}$$

In addition, cooling flow can be sophisticatedly designed to improve the internal water management in a PEM fuel cell: this can be done by creating a temperature gradient from the inlet to the outlet of reactant gas flows to take advantage of the waste heat generation. Higher temperature near the outlet increases local vapor removal capability, thereby minimzing any "flooding" risk. Figures 8-2, 8-3, and 8-15 show the temperature contours in PEM fuel cells and corresponding membrane water contents, respectively.

8.6.3 PHASE-CHANGE-BASED COOLING

Unlike air/liquid cooling which utilizes the sensible heat of the fluid, phase-change based cooling takes advantage of the coolant's latent heat. The amount of latent heat is usually very large, for example, the latent heat of water is 2250 kJ/kg at 1 atm, more than 500 times higher than the sensible heat absorbed by liquid water with a 1°C increase. As a result, a much lower rate of coolant flow is required, which may be sustained by capillary action (e.g., wicking) so as to eliminate coolant pumps. Thus, this cooling strategy has several attractive advantages, for example, reducing coolant flow rate, simplifying system layout, and eliminating pumping units.

There are basically two approaches of phase-change cooling: through evaporation or boiling. With the former, the coolant's boiling temperature is higher than that of fuel cells whereas with the latter the boiling temperature is lower. When using water as the coolant, cooling strategy can be integrated with fuel cell water management: a wicking material can be employed as the lands (as shown in Fig. 8-32(a)) or placed around the channels (as shown in Fig. 8-32(b)) on the cathode side. The bipolar plate is impervious to both water and reactant gas, so as to allow only the in-plane transport of liquid water for evaporative cooling and internal humidification. Figure 8-33 shows the conceptual layout of a PEM fuel cell with a cooling system for both internal humidification and cooling. However, using only water product of fuel cells has limited

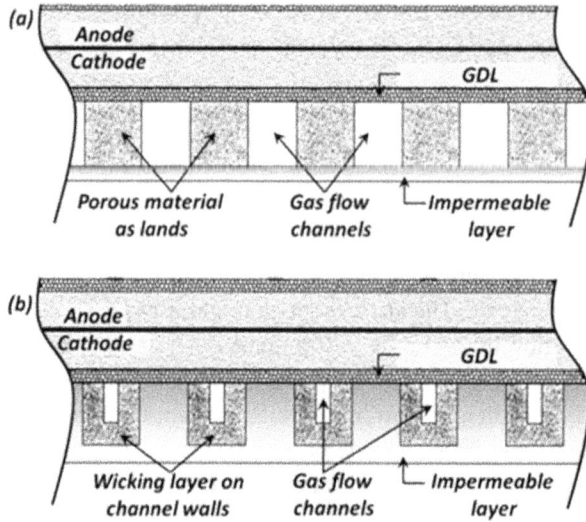

Figure 8-32. Evaporative cooling using wicking material (a) as lands or (b) placed around the channels on the cathode side [33].

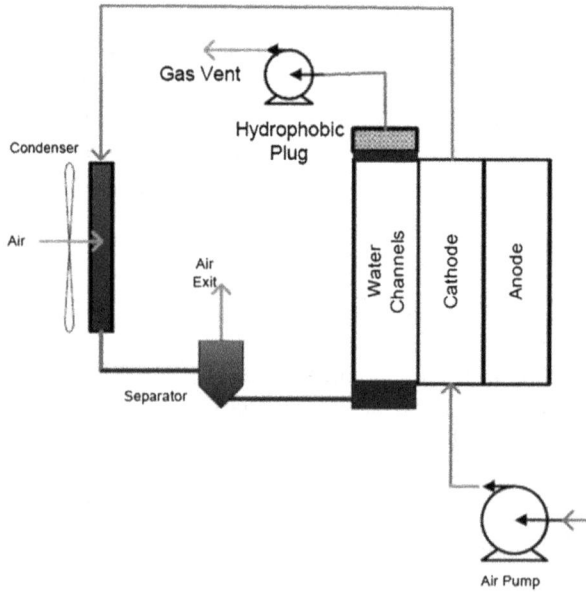

Figure 8-33. Conceptual layout of a PEM fuel cell with the evaporative cooling system for water management [34] (Courtesy of Elsevier).

impact on heat removal. This can be directly seen from the following two reactions with heat releases, relevant to the LHV and HHV:

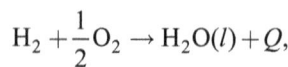

$$H_2 + \frac{1}{2}O_2 \rightarrow H_2O(l) + Q,$$

$$H_2 + \frac{1}{2}O_2 \rightarrow H_2O(v) + Q'. \qquad \text{(Eq. 8-59)}$$

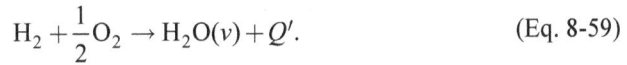

The difference between Q and Q' is the latent heat of water evaporation/condensation. The second equation is equivalent to that for combustion of hydrogen gas and oxygen with water vapor as the product. Unless the exhaust wet air from the fuel cell is further cooled, additional cooling is needed to remove Q' out of the system.

8.7 EXAMPLE: A THERMAL SYSTEM OF AUTOMOTIVE FUEL CELLS

This section is based on the work of Nolan and Kolodziej [35], and the example of a PEM fuel cell stack consists of 440 cells with an active area of 360 cm². The net power output by the stack is approximately 120 kW. This system is based on that for General Motor's (GM) Equinox FC vehicle. The PEM fuel cell stack is schematically shown in Figure 8-34, along with three main subsystems [35].

THE ANODE SUBSYSTEM

This subsystem supplies hydrogen gas from an onboard storage tank to the fuel cell. A pressure regulator maintains the pressure inside the stack. The anode heat exchanger warms up the inlet hydrogen gas. Note that the hydrogen temperature is below the tank temperature due to the adiabatic expansion of the gas through the pressure regulator. A recycle unit is used to both

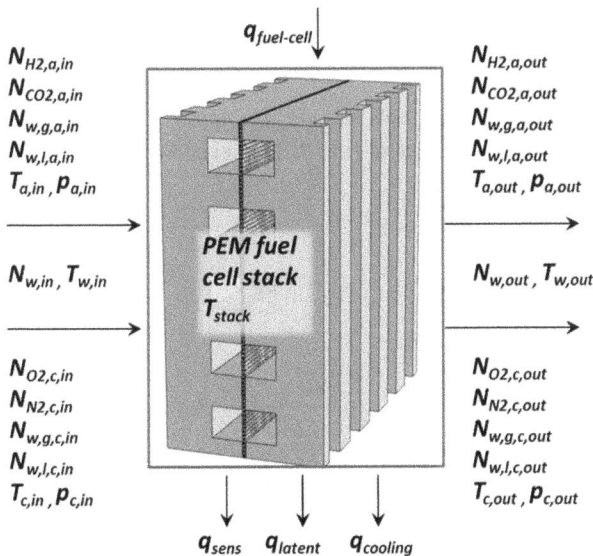

Figure 8-34. Schematic of an automotive PEM fuel cell system with thermal management [35] (Courtesy of Elsevier).

humidify the incoming hydrogen gas and recirculate unused hydrogen. A vent valve is used to purge water and waste gas (N_2) from the anode.

THE CATHODE SUBSYSTEM

In this subsystem, a compressor feeds pressurized air to the fuel cell. Reactant air flows through an intercooler to cool or warm up and through a humidifier to humidify the inlet air stream using the water in the cathode exhaust stream. The cathode intercooler is needed due to that the temperature of the compressed air exiting the compressor can be above the stack operating temperature. A bypass leg around the humidifier is used both to purge the stack and to control the inlet humidity.

THE THERMAL SUBSYSTEM

This subsystem regulates the stack's temperature. In this subsystem, a bypass valve is used to direct flow either to the heat exchanger or through a bypass leg. A pump circulates the coolant throughout the system. The flow resistance may be significant in the path through the bypass valve to radiator and then pump, leading to a significant lag in the temperature response. Three heat exchangers are employed in the thermal subsystem: the main radiator for dissipating heat to the environment; the cathode intercooler; and the anode heat exchanger. Three external inputs are provided to the thermal subsystem: the system power load, environmental temperature, and vehicle speed. Power load determines the fuel cell current, air flow and hydrogen flow rates, waste heat generation, and cooling loads for the anode and cathode heat exchangers. The vehicle speed determines the airflow velocity across the radiator and thus the heat transfer coefficient.

8.7.1 A LUMPED-SYSTEM MODEL OF A PEM FUEL CELL

The fuel cell stack generates waste heat during operation. Following the lumped-system approach, the energy balance of the system (the PEM fuel cell plus coolant flow) yields

$$\dot{q}_{store} = \dot{q}_{in} - \dot{q}_{out} + \dot{q}_{gen},$$ (Eq. 8-60)

where

$$\dot{q}_{store} = \frac{d}{dt}\left(\rho V_{eff} C_p T_{s,o}\right)$$

$$\dot{q}_{in} = \dot{m}_{s,i} C_p T_{s,i}$$

$$\dot{q}_{out} = \dot{m}_{s,o} C_p T_{s,o},$$ (Eq. 8-61)

where ρ and C_p are, respectively, the density and specific heat of the coolant; $T_{s,o}$ and $T_{s,i}$ are, respectively, the fuel cell coolant outlet and inlet temperatures; $\dot{q}_{in}$ and $\dot{q}_{out}$ are, respectively, the energy rates carried into and out of the system by the coolant; and $\dot{q}_{store}$ is the accumulation rate of thermal energy. With the lumped-system approach, the temperature is assumed to be well-mixed and uniform within the control volume; thus, any thermal gradient within the fuel cell is neglected. Thus, $T_{s,o}$ is used to represent the fuel cell temperature. V_{eff} is the effective volume of the fuel cell, which accounts for all of the fuel cell components and the coolant flow: that is, $\rho V_{eff} C_p$ accounts for all the thermal mass of the PEM fuel cell and coolant flow. For incompressible flow, density is constant. In addition, mass balance of the open system states that the mass flow is equal to the mass flow out. Additionally, the specific heat of the coolant is assumed to be constant given the small range of temperature variation, yielding

$$\frac{d}{dt}\left(\rho V_{eff} C_p T_{s,o}\right) = \dot{m}_s C_p \left(T_{s,i} - T_{s,o}\right) + \dot{q}_{gen}. \qquad \text{(Eq. 8-62)}$$

The generation term, $\dot{q}_{gen}$, accounts for inefficiency of the PEM fuel cell in generating power and is the summation of all the heating mechanisms and viscous dissipation heating of flows. The latter is usually small and negligible in PEM fuel cells. The total heating is then written by, see Figure 8-1,

$$\dot{q}_{gen} = \left(E' - E_{cell}\right) \cdot n \cdot I, \qquad \text{(Eq. 8-63)}$$

where E_{cell} is the cell voltage, n is the number of cells in the stack, and I is the stack current. This amount of heat needs to be rejected to keep the fuel cell at its operating temperature. Combining Eqs. 8-62 and 8-63 yields

$$\frac{d}{dt}\left(\rho V_{eff} C_p T_{s,o}\right) = \dot{m}_s C_p \left(T_{s,i} - T_{s,o}\right) + \left(E' - E_{cell}\right) \cdot n \cdot I. \qquad \text{(Eq. 8-64)}$$

For $\rho V_{eff} C_p$ =constant, the equation above reduces to

$$\dot{T}_{s,o} = \frac{\dot{m}_s}{\rho V_{eff}}\left(T_{s,i} - T_{s,o}\right) + \frac{1}{\rho V_{eff} C_p}\left(E' - E_{cell}\right) \cdot n \cdot I. \qquad \text{(Eq. 8-65)}$$

8.7.2 BYPASS VALVE

The bypass valve is used to control the coolant flow between the heat exchanger and the bypass leg. The valve is assumed to be ideal in the model without any delay in response. The flow through the valve is treated as a linear function of the valve command k; a k value of 0% indicates that all the coolant flow enters the radiator and a value of 100% signals that all the flow is through the bypass:

$$\dot{m}_{bypass} = \dot{m}_s k,$$
$$\dot{m}_r = \dot{m}_s \left(1 - k\right), \qquad \text{(Eq. 8-66)}$$

where $\dot{m}_{bypass}$ is the flow rate through the bypass leg and $\dot{m}_r$ is the mass flow rate through the radiator.

0.7.3 RADIATOR

The radiator is a standard automotive heat exchanger. The main function of the radiator is dissipating waste heat to the environment in the final stage of heat removal. The radiator model consists of two parts: the first part describes the heat transfer between the radiator and the environment, which is a function of ambient temperature, air mass flow, coolant flow rate, and coolant inlet temperature. The second part describes the heat transfer between the coolant and the radiator. Energy balance around the radiator, along with an accumulation term, yields

$$\dot{q}_{store} = \dot{q}_{in} - \dot{q}_{out} + \dot{q}_{thermal},$$ (Eq. 8-67)

where

$$\dot{q}_{store} = \frac{d}{dt}\left(\rho V c_p T_{r,o}\right)$$

$$\dot{q}_{in} = \dot{m}_{r,i} c_p T_{r,i}$$

$$\dot{q}_{out} = \dot{m}_{r,o} c_p T_{r,o}$$ (Eq. 8-68)

where $\left(\dot{q}_{in} - \dot{q}_{out}\right)$ is the net energy carried into the radiator by the coolant and $\dot{q}_{store}$ is the heat storage rate. Mass is balanced; thus, $\dot{m}_{r,i}\dot{m}_{r,o} = \dot{m}_r$. $\dot{q}_{thermal}$ represents the heat transfer rate between the coolant and radiator, given by

$$\dot{q}_{thermal} = G\left(T_{r,w} - T_{r,o}\right),$$ (Eq. 8-69)

where G represents the thermal transfer coefficient of the radiator. Combining Eq. 8-67 through Eq. 8-69, along with the assumption of constant specific heat, yields

$$\frac{d}{dt}\left(\rho V c_p T_{r,o}\right) = \dot{m}_r c_p \left(T_{r,i} - T_{r,o}\right) + G\left(T_{r,w} - T_{r,o}\right).$$ (Eq. 8-70)

Setting $C = \rho V c_p$ and denoting $\dfrac{dT_{r,o}}{dt}$ as $\dot{T}_{r,o}$ the above changes to

$$\dot{T}_{r,o} = \frac{\dot{m}_r c_p}{C}\left(T_{r,i} - T_{r,o}\right) + \frac{G}{C}\left(T_{r,w} - T_{r,o}\right).$$ (Eq. 8-71)

Eq. 8-71 is the final dynamic equation governing the coolant outlet temperature of the radiator. As for the radiator, energy balance yields

$$\dot{q}_{store} = \dot{q}_{in} - \dot{q}_{out},$$ (Eq. 8-72)

where

$$\dot{q}_{store} = \frac{d}{dt}\left(\rho V c_p T_{r,w}\right),$$
$$\dot{q}_{in} = G\left(T_{r,o} - T_{r,w}\right),$$ (Eq. 8-73)
$$\dot{q}_{out} = f_{exp}\left(T_{amb} - T_{r,i}\right).$$

Here $\dot{q}_{in}$ is the heat transfer rate from the coolant to the radiator and $\dot{q}_{out}$ is that to the environment. Combining Eq. 8-72 and Eq. 8-73 yields

$$\dot{T}_{r,w} = \frac{G}{C}\left(T_{r,o} - T_{r,w}\right) + \frac{1}{C}f_{exp}\left(T_{r,i} - T_{amb}\right),$$ (Eq. 8-74)

where f_{exp} is a coefficient in the experimentally derived look-up table, relating the heat transfer to the inlet temperature difference between the coolant and ambient air.

8.7.4 TRANSPORT DELAY

The flow path, connecting the stack, the bypass valve, the pump, and the heat exchanger, introduces transport delay in the system model. Assuming no heat loss in the path, the lag in temperature is a function of the passage length and coolant flow rate:

$$\theta = \frac{V}{v},$$ (Eq. 8-75)

where θ is the delay, V is the volume of the plumbing, and v is the volumetric flow rate of the coolant. Using a (1/1) Pade approximation allows the delay to be modeled in a transfer function form as [36]:

$$G(s) = \frac{1 - (\theta/2)s}{1 + (\theta/2)s}.$$ (Eq. 8-76)

There are two transport delays to be considered: the first is for the plumbing from the bypass valve to the radiator, and the second is from the radiator back to the fluid mixer. It is important to include these delays in the system model because they are not constant and may impose a significant impact on the controller performance. For a radiator with a volume of 0.5 l and a piping with an approximate volume of 0.6 l, a coolant flow rate of 20–196 l m^{-1} yields a transport delay ranging from 0.2 to 2 s.

8.7.5 FLUID MIXER

The fluid mixer is the connector where the bypass and the heat exchanger legs join together before entering the pump. Physically, the mixer is no more than a T junction, where two fluid streams mix. Assuming its capacitance is zero, energy balance gives

$$
\begin{aligned}
\dot{q}_{\text{out}} &= \dot{q}_{\text{bypass},s} + \dot{q}_{\text{bypass},r}, \\
\dot{q}_{\text{out}} &= \dot{m}_s c_p T_{m,o}, \\
\dot{q}_{\text{bypass},s} &= \dot{m}_{\text{bypass}} c_p T_{s,o}, \\
\dot{q}_{\text{bypass},r} &= \dot{m}_r c_p T_{r,o}.
\end{aligned}
\tag{Eq. 8-77}
$$

For constant specific heat, the mixed stream outlet temperature can be written by

$$
T_{m,o} = \frac{\dot{m}_{\text{bypass}} T_{s,o} + \dot{m}_r T_{r,o}}{\dot{m}_{\text{bypass}} + \dot{m}_r}.
\tag{Eq. 8-78}
$$

8.7.6 CATHODE INTERCOOLER

In the cathode subsystem, air is compressed and fed into the fuel cell using a compressor. The added mechanical work raises the air temperature. An air-to-water intercooler can be used to reduce the temperature rise. The oxygen molar consumption rate is determined by the fuel cell current as follows:

$$
\dot{m}_{O_2} = \frac{I\,n}{4F} \xi_c,
\tag{Eq. 8-79}
$$

where I is the current, n is the number of cells, F is the Faraday's constant, and ξ_c is the stoichiometric ratio. This equation gives the mass flow requirement in moles per second. Since ambient air is used to provide oxygen, assuming the oxygen content in the air is β_{O_2} ($\sim$21% in the stand condition), the air mass flow rate becomes

$$
\dot{m}_{\text{air}} = \frac{I n \xi_c M_{\text{air}}}{4F \beta_{O_2}},
\tag{Eq. 8-80}
$$

where M_{air} is the molecular weight of air. The temperature rise can be computed by assuming adiabatic compression [36]:

$$
T_{\text{air},o} = T_{\text{amb}} + \frac{T_{\text{amb}}}{\eta_{\text{cp}}} \left[P_r^{(\gamma-1)/\gamma} - 1 \right],
\tag{Eq. 8-81}
$$

where γ is the specific heat ratio (1.4 for air), η_{cp} is the compressor efficiency, and p_r is the pressure ratio for a given mass flow. Using the compressor map provided by the manufacturer, look-up tables for efficiency and pressure ratio can be constructed versus mass flow.

The energy balance over the heat exchanger results in

$$\frac{d}{dt}\left(\rho V c_p T_{c,o}\right) = \dot{m}_s c_p \left(T_{m,o} - T_{c,o}\right) + \dot{q}_c. \qquad \text{(Eq. 8-82)}$$

The generation term $\dot{q}_c$ represents the heat transfer rate from the incoming air to the coolant. The amount of energy transfer to the coolant is evaluated using [4]

$$\dot{q}_c = \varepsilon \dot{m}_{air} c_{p,air} \left(T_{air,o} - T_{m,o}\right), \qquad \text{(Eq. 8-83)}$$

where ε is the effectiveness of the heat exchanger, which can be reasonably approximated as 0.88 for all operating conditions. Assuming a constant coolant volume and density, the above two equations yields

$$\dot{T}_{c,o} = \frac{\dot{m}_s}{\rho V}\left(T_{m,o} - T_{c,o}\right) + \frac{\varepsilon \dot{m}_{air} c_{p,air}}{\rho V c_p}\left(T_{amb} + \frac{T_{amb}}{\eta_{cp}}\left[P_r^{(\gamma-1)/\gamma} - 1\right] - T_{m,o}\right). \qquad \text{(Eq. 8-84)}$$

8.7.7 ANODE HEAT EXCHANGER

For a pressurized gaseous hydrogen tank, a regulating control valve is frequently used to control the flow of hydrogen gas into the fuel cell. Due to the high pressure in the tank, the temperature of the incoming gas after the valve becomes lower as a result of adiabatic expansion. Injecting cold hydrogen gas affects fuel cell performance, for example, low anode relative humidity. To raise the incoming hydrogen temperature, an air-to-water heat exchanger is used to warm up hydrogen gas. The heat transfer in the heat exchanger is given by [4]

$$\dot{q} = \varepsilon C_{min}\left(T_{c,o} - T_{H_2,in}\right), \qquad \text{(Eq. 8-85)}$$

where

$$C_{min} = c_{p,H_2} \dot{m}_{H_2}. \qquad \text{(Eq. 8-86)}$$

ξ is the efficiency of the heat exchanger. The heat transfer rate from the coolant flow is given by

$$\dot{q} = \dot{m}_s c_p \left(T_{c,o} - T_{a,o}\right). \qquad \text{(Eq. 8-87)}$$

Equating Eqs. 8-85 and 8-87 yields the coolant outlet temperature of the anode heat exchanger as

$$T_{a,o} = T_{c,o} - \frac{\varepsilon c_{p,H_2} \dot{m}_{H_2}}{c_p \dot{m}_s}\left(T_{c,o} - T_{H_2,in}\right). \qquad \text{(Eq. 8-88)}$$

The hydrogen consumption rate of the PEM fuel cell is given by

$$\dot{m}_{H_2} = \frac{In}{2F} \xi_a M_{H_2},$$ (Eq. 8-89)

where ξa is the anode stoichiometric ratio, usually ranging from 1.2 to 1.5. Thus,

$$T_{a,o} = T_{c,o} - \frac{\varepsilon c_{p,H_2} In \xi_a M_{H_2}}{2Fc_p \dot{m}_s} \left(T_{c,o} - T_{H_2,in} \right).$$ (Eq. 8-90)

8.8 CHAPTER SUMMARY

In this chapter, heat transfer fundamentals in PEM fuel cells are presented and discussed when phase change is absent. The key points in this chapter are summarized as follows:

- Heat conduction and convection are the two major modes of heat transfer in PEM fuel cells and radiation is not important or negligible. Conduction dominates heat transfer in membrane, catalyst layers, gas diffusion layers (GDLs), and bipolar plates (BPs), whereas convection is the major mechanism of thermal transport in reactant and cooling flow channels; the reactant gas flow removes less than 5% of total heat generated; the relative importance of convection to conduction can be evaluated by the dimensionless Nusselt number; and thermal conductivity and contact resistance are important thermal properties relevant to PEM fuel cells and they have been measured for various PEM fuel cell components and interfaces, respectively.
- Three major mechanisms of waste heat generation, the reversible heat of the reactions, irreversible heat of the reactions, and Joules heat, are discussed. Majority of the waste heat comes from the reaction irreversibility and the Joules heating occurs when electric currents (either via protons or electrons) pass a resistor; both the reversible and irreversible heats are released in the catalyst layers in which the reactions occur whereas the Joules heat is produced in the membrane electrode assembly (MEA), GDLs, and BPs; and the MEA produces the majority of the Joules heat among the three components (MEA, GDLs, BPs).
- Heat transfer is analyzed for the major components of PEM fuel cells. The through-plane temperature variation in the catalyst layer and GDLs are around 0.01–0.1 K and 1 K, respectively, when current density is under 1 A/cm^2; the maximum temperature variation due to the channel-land configuration can be estimated by assuming the through-plane and in-plane resistors being connected in series; coolant flow can be designed to alter the along-channel temperature distribution, and hence improve the local vapor removal capability; and at the pore level, most heat is conducted via the solid matrix due to the highly conductive fibers and the mass flow in the pore network contributes only a small portion of heat transfer.
- Transient analysis of heat transfer in PEM fuel cells is presented, particularly on cold start. The thermal mass of bipolar plates accounts for a major portion of a PEM fuel cell's thermal mass, and it strongly affects the time constant for cold start; the voltage change due to temperature variation in cold start stems from two factors: the temperature

dependence of the exchange current density (which affects the activation polarization) and the temperature dependence of ionic conductivity (which affects the ohmic voltage loss); the thermal system of an automotive PEM fuel cell along with its cooling unit is analyzed as an example; and the fuel cell, radiator, bypass valve, fluid mixer, and cathode/anode intercooler are all modeled in the analysis.

REFERENCES

1. Özisik, M.N. *Heat Conduction*, 2nd Ed. John Wiley & Sons Inc., (1993).

2. Carslaw, H.S., and Jaeger, J.C. *Conduction of Heat in Solids*, 2nd Ed. Oxford University Press, (1986).

3. Kays, W.M., Crawford, M.E., and Weigand, B. *Convective Heat and Mass Transfer*, 4th Ed., McGraw-Hill Inc., (2004).

4. Incropera, F.P., DeWitt, D.P., Bergman, T.L., and Lavine, A.S. *Fundamentals of Heat and Mass Transfer*, 6th Ed., John Wiley & Sons, (2006).

5. Owejan, J.P., Owejan, J.E., Gu, W., Trabold, T.A., Tighe, T.W., and Mathias, M.F. Water transport mechanisms in PEMFC gas diffusion layers. *J. Electrochem. Soc.* 157 (2010): B1456–64. doi: http://dx.doi.org/10.1149/1.3468615

6. Wang, Y., and Wang, C.Y. Ultra large-scale simulation of polymer electrolyte fuel cells. *J. Power Source* 153 (2006) 130–35. doi: http://dx.doi.org/10.1016/j.jpowsour.2005.03.207

7. Kandlikar, S.G., and Lu, Z. Fundamental research needs in combined water and thermal management within a proton exchange membrane fuel cell stack under normal and cold start conditions. *J. Fuel Cell Sci. Technol.*, 6 (2009): 044001. doi: http://dx.doi.org/10.1115/1.3008043

8. Wang, Y. Analysis of the key parameters in the cold start of polymer electrolyte fuel cells. *J. Electrochem. Soc.*, 154 (2007): B1041–48. doi: http://dx.doi.org/10.1149/1.2767849

9. Wang, Y. Porous-media flow fields for polymer electrolyte fuel cells: I. Low humidity operation. *J. Electrochem. Soc.*, 156 (2009): B1124–33. doi: http://dx.doi.org/10.1149/1.3183781

10. Ju, H., Wang, C.Y., Cleghorn, S., and Beuscher, U. Nonisothermal modeling of polymer electrolyte fuel cells: I. Experimental validation. *J. Electrochem. Soc.* 152 (2005): A1645–53. doi: http://dx.doi.org/10.1149/1.1943591

11. Wang, Y., Cho, S.C., Thiedmann, R., Schmidt, V., Lehnert, W., and Feng, X. Stochastic modeling and direct simulation of the diffusion media for polymer electrolyte fuel cells. *Int. J. Heat Mass Transf.* 53 (2010): 1128–38. doi: http://dx.doi.org/10.1016/j.ijheatmasstransfer.2009.10.044

12. Yan, Q., Toghiani, H., Lee, Y.W., Liang, K., and Causey, H. Effect of sub-freezing temperatures on a PEM fuel cell performance, startup and fuel cell components. *J. Power Sources* 160 (2006): 1242–50. doi: http://dx.doi.org/10.1016/j.jpowsour.2006.02.075

13. Vie, P.J.S., and Kjelstrup, S. Thermal conductivities from temperature profiles in the polymer electrolyte fuel cell. *Electrochimica Acta*, 49 (2004): 1069–77. doi: http://dx.doi.org/10.1016/j.electacta.2003.10.018

14. Zhang, J. *et al.* High temperature PEM fuel cells. *J. Power Sources* 160 (2006): 872–91. doi: http://dx.doi.org/10.1016/j.jpowsour.2006.05.034

15. Parthasarathy, A., Srinivasan, S., Appleby, A.J., and Martin, C.R. Temperature dependence of the electrode kinetics of oxygen reduction at the platinum/nafion® interface—a microelectrode investigation. *J. Electrochem. Soc.* 139 (1992): 2530–37. doi: http://dx.doi.org/10.1149/1.2221258

16. Beattie, P.D., Basura, V.I., and Holdcroft, S. Temperature and pressure dependence of O_2 reduction at Pt | Nafion® 117 and Pt | BAM® 407 interfaces. *J. Electroanal. Chem.* 468 (1999): 180–92. doi: http://dx.doi.org/10.1016/S0022-0728(99)00164-3

17. Energy Efficiency and Renewable Energy (EERE), Hydrogen, Fuel Cells & Infrastructure Technologies Program Multi-Year Research, Development and Demonstration Plan, U. S. Department of Energy (DOE), (2007) 14.

18. Wang, Y., Mukherjee, P.P., Mishler, J., Mukundan, R., and Borup, R.L. Cold start of polymer electrolyte fuel cells: three-stage startup characterization. *Electrochimica Acta* 55 (2010): 2636–44. doi: http://dx.doi.org/10.1016/j.electacta.2009.12.029

19. Springer, T.E., Zawodinski, T.A., and Gottesfeld, S. Polymer electrolyte fuel cell model. *J. Electrochem. Soc.* 138 (1991): 2334. doi: http://dx.doi.org/10.1149/1.2085971

20. Khandelwal, M., and Mench, M.M. Direct measurement of through-plane thermal conductivity and contact resistance in fuel cell materials. *J. Power Sources* 161 (2006): 1106. doi: http://dx.doi.org/10.1016/j.jpowsour.2006.06.092

21. Zamel, N., Litovsky, E., Li, X., and Kleiman, J. Measurement of the through-plane thermal conductivity of carbon paper diffusion media for the temperature range from −50 to +120°C. *Int. J. Hydrogen Energy* 36 (2011): 12618–25. doi: http://dx.doi.org/10.1016/j.ijhydene.2011.06.097

22. Burheim, O., Vie, P.J.S., Pharoah, J.G., and Kjelstrup, S. Ex situ measurements of through-plane thermal conductivities in a polymer electrolyte fuel cell *J. Power Sources* 195 (2010): 249–56. doi: http://dx.doi.org/10.1016/j.jpowsour.2009.06.077

23. Sadeghi, E., Djilali, N., and Bahrami, M. Effective thermal conductivity and thermal contact resistance of gas diffusion layers in proton exchange membrane fuel cells: Part 1. Effect of compressive load. *J. Power Sources* 196 (2011): 246–54. doi: http://dx.doi.org/10.1016/j.jpowsour.2010.06.039

24. Wang, Y., and Gundevia, M. Measurement of thermal conductivity and heat pipe effect in hydrophilic and hydrophobic carbon papers, in press. doi: http://dx.doi.org/10.1016/j.ijhydene.2011.11.010

25. Zhang, G., and Kandlikar, S.G. A critical review of cooling techniques in proton exchange membrane fuel cell stacks. *Int. J. Hydrogen Energy* 37 (2012): 2412–29. doi: http://dx.doi.org/10.1016/j.ijhydene.2011.11.010

26. Rogg, S., Hoglinger, M., Zwittig, E., Pfender, C., Kaiser, W., and Heckenberger, T. Cooling modules for vehicles with a fuel cell drive. *Fuel Cells* 3 (2003): 153–58. doi: http://dx.doi.org/10.1002/fuce.200332108

27. Burke, K.A., Jakupca, I., and Colozza, A. Development of passive fuel cell thermal management technology. AIAA 7th International Energy Conversion Engineering Conference, American Institute of Aeronautics and Astronautics Inc., (2009).

28. Fluckiger, R., Tiefenauer, A., Ruge, M., Aebi, C., Wokaun, A., and Buchi, F.N. Thermal analysis and optimization of a portable, edge-air-cooled PEFC stack. *J. Power Sources* 172 (2007): 324–33. doi: http://dx.doi.org/10.1016/j.jpowsour.2007.05.079

29. Wen, C.Y., and Huang, G.W. Application of a thermally conductive pyrolytic graphite sheet to thermal management of a PEM fuel cell. *J. Power Sources* 178 (2008): 132–40. doi: http://dx.doi.org/10.1016/j.jpowsour.2007.12.040

30. Vasiliev, L., and Vasiliev, Jr. L. Heat pipes in fuel cell technology. In Kakac, S., Pramuanjaroenkij, A., and Vasiliev, L. (Eds.). *Mini-Micro Fuel Cells Fundamentals and Applications*, Dordrecht, Springer Netherlands, (2008): 117–24. doi: http://dx.doi.org/10.1007/978-1-4020-8295-5_8

31. Faghri, A. Micro Heat Pipe Embedded Bipolar Plate for Fuel Cell Stacks, US Patent Appl Pub No. US 2005/002601 Appl. Pub. No.: US 2005/0026015, (2005).

32. Faghri, A. Integrated Bipolar Plate Heat Pipe for Fuel Cell Stacks, US Patent Appl. Pub. No.: US 2005/0037253, (2005).

33. Goebel, S.G. Evaporative Cooled Fuel Cell. US Patent 6960404, (2005).

34. Meyers, J.P., Darling, R.M., Evans, C., Balliet, R., and Perry, M.L. Evaporatively-cooled PEM fuel-cell stack and system. *ECS Trans.*, 3 (2006): 1207–14. doi: http://dx.doi.org/10.1149/1.2356240

35. Nolan, J., and Kolodziej, J. Modeling of an automotive fuel cell thermal system. *J. Power Sources* 195 (2010): 4743–52. doi: http://dx.doi.org/10.1016/j.jpowsour.2010.02.074

36. Seborg, D.E., Edgar, T.F., and Mellichamp, D.A. *Process Dynamics and Control*, Wiley, (1989).

37. Cengel, Y.A., and Boles, M.A. *Thermodynamics, An Engineering Approach*, 6th Ed. McGraw-Hill, (2007).

CHAPTER 9

COUPLED THERMAL-WATER MANAGEMENT: PHASE CHANGE

Phase change, such as liquid-water or ice formation, is frequently encountered in PEM fuel cells. Both liquid and solid phases of water affect fuel cell performance and durability and thus play an important role in PEM fuel cell operation. Water may "flood" electrodes, increasing resistances of transport for hydrogen fuel and oxygen reactant. Freeze and thaw cycles are common in PEM fuel cell operations during cold starts. Thus, the knowledge of vapor-to-liquid, liquid-to-ice, and vapor-to-ice phase changes as well as the consequences of these phase changes are urgently needed for developing efficient thermal and water management strategies and techniques in PEM fuel cells. In this chapter, the fundamentals of phase changes, their effects on fuel cell performance, their couplings with transport phenomena, and associated transient issues are outlined.

9.1 INTRODUCTION TO PHASE CHANGE

Phase change is involved in many power generation processes, for example, the Rankine-cycle based steam turbines and heat transfer units. In the Rankine cycle, liquid water is vaporized in boilers by heat addition, and high pressure steam drives turbines to produce mechanical work. At the low-temperature side, condensers change vapor to liquid water. During the phase change, pressure is usually assumed constant in thermodynamic analyses. The vapor-liquid mixture remains at constant temperature due to the latent heat absorption or release. In the mechanical vapor-compression refrigeration, heat withdrawn from an evaporator results from the refrigerant's phase change. Inside a freezer, ice forms on the chamber wall, resulting from vapor frosting. To defrost, heat can be added to melt the frost.

In PEM fuel cell operation, liquid water results from vapor condensation or water production by the ORR reaction directly, giving rise to complex two-phase phenomena, as explained in Chapter 6. Residual water stays in a PEM fuel cell after being shut down and can be removed by pumping in dry reactant gases to vaporize the liquid. The saturated-vapor pressure of water, $P_{w,sat}$, is important to the drying process and strongly dependent on temperature as shown in Eq. 9-1 [1]:

$$\log_{10} P_{w,sat} = -2.1794 + 0.02953\,(T{-}273.15) - 9.1837 \times 10^{-5}\,(T{-}273.15)^2 + 1.4454 \times 10^{-7}\,(T{-}273.15)^3 \tag{Eq. 9-1}$$

Using the ideal gas law, the vapor saturation concentration can be calculated, as displayed in Figure 9-1.

In Figure 9-1, both vapor saturation concentration ($C_{w,sat}$) and its derivative (dC_{sat}/dT or dC/dT in the figure) are shown as a function of temperature. At subfreezing temperatures, product water freezes. The saturated vapor pressure at subfreezing temperature can be evaluated by Eq. 9-2 with the coefficients for water and ice given below [2]:

$$P_{w,sat} = \sum_{n=0}^{6} C_n T^n. \tag{Eq. 9-2}$$

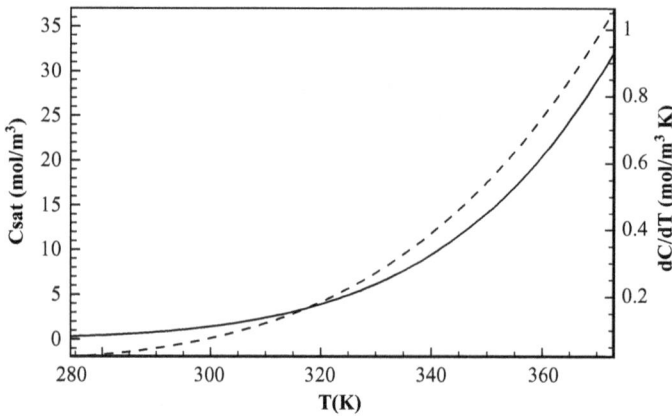

Figure 9-1. Water vapor saturation concentration ($C_{w,sat}$ or C_{sat}) and its derivative as a function of temperature.

Coefficient C_n	Water	Ice
C_0	6.1070422	6.1090668
C_1	4.4411566×10^{-1}	5.0249291×10^{-1}
C_2	1.4320982×10^{-2}	1.8684567×10^{-2}
C_3	2.6513961×10^{-4}	4.0559217×10^{-4}
C_4	3.0099985×10^{-6}	5.4323745×10^{-6}
C_5	2.0088796×10^{-8}	4.2255374×10^{-8}
C_6	$6.1926232 \times 10^{-11}$	$1.4687829 \times 10^{-10}$

In a dry environment, product ice dissimilates. When reaching the melting point, product ice melts by the waste heat generated within a PEM fuel cell. Because liquid water and ice strongly affect fuel cell operation, understanding the phase-change phenomena and their effects are critically important.

9.2 VAPOR–LIQUID PHASE CHANGE: EVAPORATION AND CONDENSATION

Condensed water is deposited on solid or liquid surfaces whereas evaporation occurs at gas–liquid interfaces and dissimilation at gas–ice interfaces. Condensation refers to the change of the physical state of matter from its vapor phase to liquid one, and evaporation is the reverse of condensation. In the microscopic view, when temperature becomes low, molecular movement slows down and intermolecular attraction prevails, bringing molecules closer to a distance comparable to their size. The latent heat is released, as a result of the potential energy change at the molecular level.

Using condensation in GDLs as an example, from the kinetic theory, the condensation rate (per unit interfacial area) is given by [3, 4]

$$\dot{m}_{g1,w} = \frac{1}{4} u_{\text{mol}} \alpha_m M_w \frac{P_{w,g} - P_{w,\text{sat}}}{R_g T},$$
(Eq. 9-3)

where u_{mol} is the mean molecular velocity $(8R_g T / \pi M)^{1/2}$, and α_m is the mass accommodation coefficient which is usually around 0.01–0.07.

The condensation rate is also limited by mass transport and heat transfer. The phase change occurs at the interface, where the concentration of the condensable vapor is lower when a large transport resistance is present. In addition, when efficient heat removal lacks, the latent heat release will raise local temperature, depressing the interfacial condensation. By defining an uptake coefficient Γ, one has

$$\dot{m}_{g1,w} = \frac{1}{4} u_{\text{mol}} \Gamma M_w \frac{P_{w,g} - P_{w,\text{sat}}}{R_g T},$$
(Eq. 9-4)

where

$$\frac{1}{\Gamma} = \frac{1}{\alpha_m} + \frac{u_{\text{mol}} l_D}{4 D_w^m} = \frac{1}{\alpha_m} + \frac{3}{4} \frac{1}{Kn}.$$
(Eq. 9-5)

l_D denotes the characteristic length for diffusion, and Kn is the Knudsen number defined as λ / l_D (the mean free path λ of water vapor molecules at 1 atm and 70°C is around 0.05 μm). In GDLs, l_D is ~10 μm, $Kn = 0.005$, and $\Gamma = 0.006$ [3].

The volumetric condensation rate $\dot{n}_{g1,w}$ is then derived as

$$\dot{n}_{g1,w} = \dot{m}_{g1,w} \frac{A_{1g}}{V}$$

$$= \frac{1}{4}\frac{A_{lg}}{V}u_{mol}\Gamma M_w \frac{P_{w,g} - P_{w,sat}}{R_g T} \qquad \text{(Eq. 9-6)}$$

$$= \gamma M_w \frac{P_{w,g} - P_{w,sat}}{R_g T}.$$

At 70°C and 1 atm, the volumetric condensation coefficient γ is estimated to be ~0.9 A_{lg}/V.

The liquid/gas specific interfacial area A_{lg}/V depends on the water saturation, and a maximum 20% of the solid/fluid specific interfacial area A_{sf}/V was obtained [5] for hydrophilic spheres packing at small water saturation. Due to hydrophobicity and fibrous medium, the liquid/gas specific interfacial area can be uniform (independent of local water saturation), since the microdroplets are assumed to be uniformly distributed in the diffusion medium. In addition, the phase equilibrium among water is frequently assumed to simplify two-phase flow models. Under this condition, the local water vapor becomes saturated as a result of sufficiently rapid phase change. In MPLs, GDLs, and catalyst layers, the equilibrium is generally satisfied due to the small pore dimensions and hence large interfacial areas.

9.2.1 VAPOR-PHASE WATER DIFFUSION AND HEAT PIPE EFFECT

In an isothermal two-phase region, vapor-phase water diffusion in GDLs vanishes because of the phase equilibrium. In the presence of temperature spatial variation or under nonisothermal conditions, the vapor saturation pressure differs from place to place, resulting in vapor-phase diffusion. To quantify the importance of this transport mechanism, the temperature gradient within a GDL must be understood first:

$$\frac{\Delta T}{\Delta x} \sim \frac{I(E' - V_{cell})}{k^{eff}}, \qquad \text{(Eq. 9-7)}$$

where the waste heat rate is approximated by $I(E' - V_{cell})$ with E' being the thermodynamic potential assuming all the enthalpy change of reactions is converted to electric energy. For 1 A/cm^2 at 0.6 V and k^{eff} of 3 W/m.K, the resulting temperature gradient is around 2000 K/m. Assuming phase equilibrium, the vapor-phase diffusion flux, driven by the temperature gradient, is expressed by

$$-D_w^{g,eff}(T,P)\frac{dC_w}{dx} = -D_w^{g,eff}(T,P)\frac{dC_{w,sat}(T)}{dx} = -D_w^{g,eff}(T,P)\frac{dC_{w,sat}}{dT}\frac{dT}{dx}. \text{ (Eq. 9-8)}$$

To evaluate the above flux, a dimensionless factor β is defined as the ratio of the vapor-phase diffusion flux to a PEM fuel cell's water production rate, namely,

$$\beta = \frac{D_w^{g,eff}(T,P)\dfrac{dC_{w,sat}(T)}{dT}\dfrac{dT}{dx}}{\dfrac{I}{2F}}, \qquad \text{(Eq. 9-9)}$$

where $C_{w,\text{sat}}(T)$ is the saturated water vapor concentration. The graph of β versus dT/dx is plotted in Figure 9-2 for three temperatures, showing that β varies strongly with temperature; it approximates 0.4 at 80°C with a temperature gradient of 2000 K/m. This means that this vapor-diffusion mechanism is able to transport ~40% of product water and hence its significance cannot be overlooked. The vapor-phase diffusion promotes phase-change heat transfer between the catalyst layer and land. This is realized through the following processes: water evaporation at hotter catalyst layer, vapor diffusion through the interstitial spaces, and subsequent vapor condensation on the cooler land surface. This mode of phase-change heat transfer is conventionally referred to as the heat pipe effect. Heat flux transported via heat pipe effect can be estimated by Eq. 9-10 [6]

$$h_{\text{lg}}\dot{m}_{\text{lg}} = h_{\text{lg}}M_wD_w^{g,\text{eff}}(T,P)\frac{dC_{w,\text{sat}}(T)}{dT}\frac{dT}{dx} = k_{\text{lg}}(T,P)\frac{dT}{dx}. \qquad \text{(Eq. 9-10)}$$

The above equation expresses the heat pipe effect using the apparent thermal conductivity $k_{\text{lg}}(T,P)$. From Figure 9-2, k_{lg} is estimated to be ~0.42 W/m.K at 80°C and 0.56 W/m.K at 90°C. The heat-pipe effect thus amounts to 15–18% of heat conduction through a GDL with a conductivity of 3 W/m.K.

In three dimensions, the above flux is written as

$$D_w^{g,\text{eff}}(T,P)\nabla C_w = D_w^{g,\text{eff}}(T,P)\nabla C_{w,\text{sat}}(T) = D_w^{g,\text{eff}}(T,P)\frac{dC_{w,\text{sat}}}{dT}\nabla T, \qquad \text{(Eq. 9-11)}$$

$$h_{\text{lg}}\dot{m}_{\text{lg}} = h_{\text{lg}}M_wD_w^{g,\text{eff}}(T,P)\frac{dC_{w,\text{sat}}(T)}{dT}\nabla T = k_{\text{lg}}(T,P)\nabla T. \qquad \text{(Eq. 9-12)}$$

As explained previously, the vapor-phase diffusion flux is driven by thermal gradients. Figure 9-3 shows the temperature-gradient vector at a cross-section of cathode GDL. Clearly, the vapor-phase diffusion, important to water management, takes place mostly in the in-plane

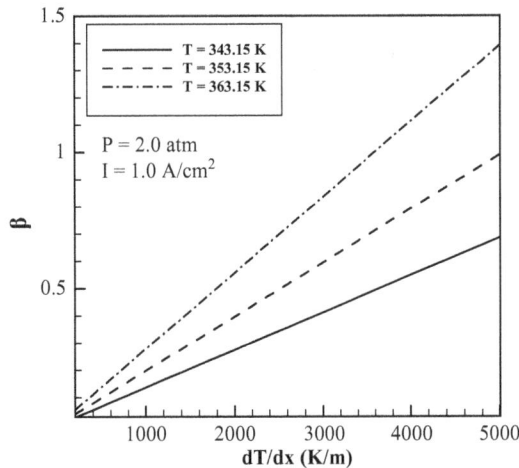

Figure 9-2. β as a function of the temperature gradient [6] (Courtesy of the Electrochemical Society).

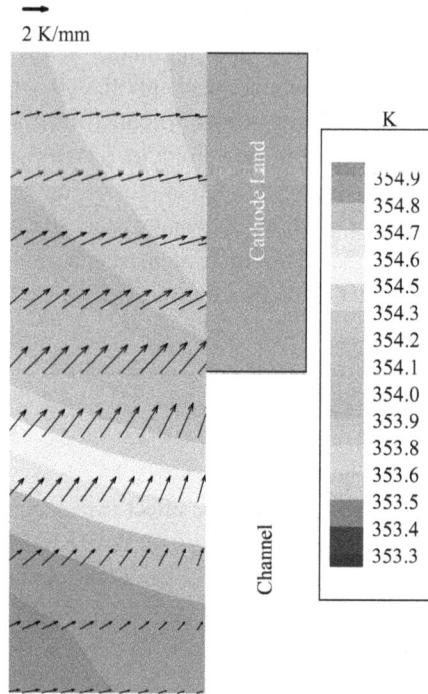

Figure 9-3. Thermal gradients and water vapor-phase diffusion flux in a cathode GDL [7] (Courtesy of the Electrochemical Society).

direction. Thus, a 2D geometry, that is, in-plane plus through-plane directions, is required to fully account for the effect of this vital mechanism on water transport.

Figure 9-4 presents the liquid saturation distributions in a cathode GDL to show the effect of vapor-phase water diffusion. In the in-plane direction between the channel and land, the vapor-phase diffusion driven by the temperature gradient is directed from the channel area to the land and thus opposes the capillarity-driven liquid water flow from the land to channel areas. The net result of the opposed water transport mechanisms is that the liquid saturation becomes higher under the land when there is vapor transport under a thermal gradient. The enlarged disparity in liquid saturation between the land and channel areas drives the transport of the additional liquid water to offset the vapor-phase diffusion along the opposite direction.

Suman *et al.* [8] investigated the phase change rate by assuming that product water exists in vapor phase. Figure 9-5 shows the distribution of the condensation/evaporation rate within GDLs with two different thermal conductivities. Positive value indicates condensation and negative value means evaporation. In addition to the heat pipe effect in GDLs, heat is also removed by the evaporation process in the catalyst layer region. The secondary heat pipe effect is observed between the region near the gas channel and current collecting land.

9.2.2 GDL DE-WETTING

As shown in Figure 9-6, the basic structure of PEM fuel cells — the 2D land-to-channel configuration –affects the two-phase distribution: due to the presence of impermeable land, liquid water is being removed via the in-plane direction first before entering a GFC. Therefore, high

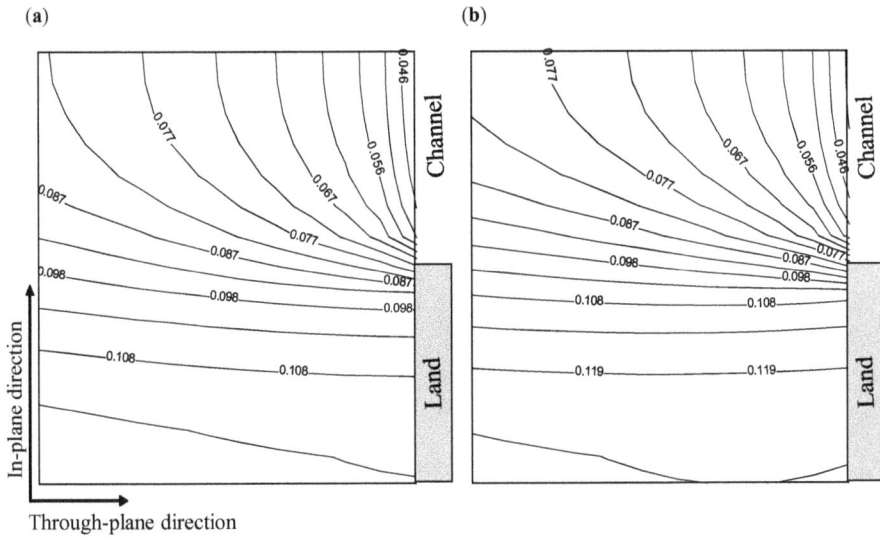

Figure 9-4. Liquid saturation contours in a cross-section of the cathode GDL: (a) without vapor-phase water diffusion, (b) with vapor-phase water diffusion (representing a full description of the two-phase, non-isothermal model). Note: the average current density at 0.61 V is predicted to be 1.33 and 1.35 A/cm^2 for the two cases, respectively [6] (Courtesy of the Electrochemical Society).

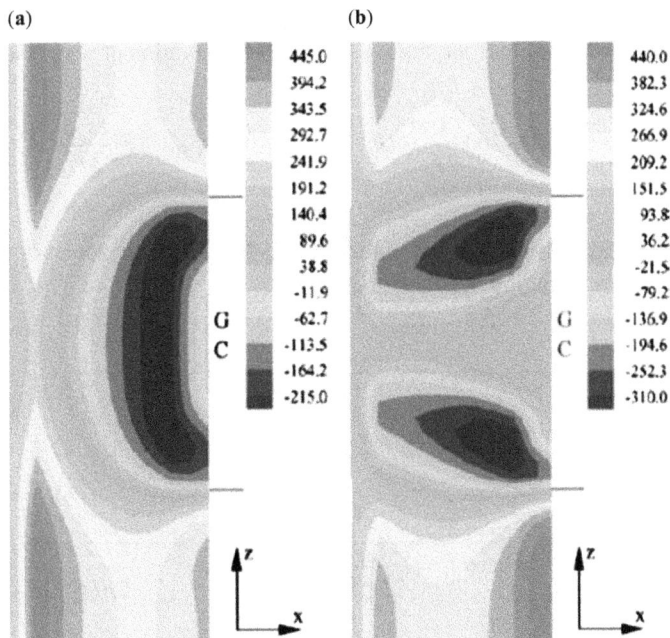

Figure 9-5. Volumetric condensation/evaporation rate (kg/m^{3}s) in GDLs with two GDL thermal conductivities: (a) 3 W/m·K , (b) 1.5 W/m·K [8] (Courtesy of the Electrochemical Society).

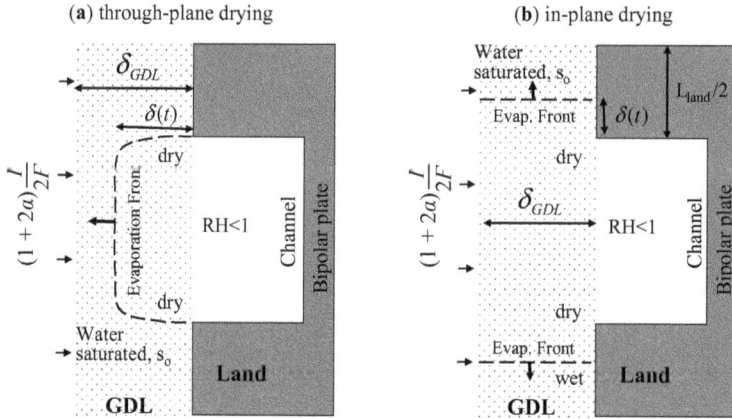

Figure 9-6. Schematics of GDL drying: (a) through-plane and (b) in-plane.

water content may appear under the land, as opposed to that under the GFC. A portion of liquid water resides inside GDLs after a PEM fuel cells shut down – this raises durability concerns. De-wetting or drying is thus important to removing residual water in PEM fuel cells. Because of the basic land-to-channel configuration, GDL drying takes place in two elementary processes in series: the through-plane drying followed by in-plane drying. In what follows, a general case is presented to show these two drying processes and their effects on cell voltage.

(a) Through-plane drying: Figure 9-6 (a) sketches the through-plane drying process occurring in the cathode GDL of a PEM fuel cell. For the GDL two-phase region with an average liquid saturation, assuming that there is water supply from fuel cell operation and water vapor removal from the evaporation front, the law of water conservation or water balance gives rise to

$$(1+2\alpha)\frac{I}{2F} - D_w^{g,\text{eff}}\frac{\Delta C}{\delta} = \frac{d}{dt}\left[(\delta_{\text{GDL}} - \delta)\frac{\varepsilon s_o \rho_l}{M_w}\right]. \qquad \text{(Eq. 9-13)}$$

Here, s_o is the average liquid water saturation in the two-phase zone prior to drying. Defining $Y = \delta/\delta_{\text{GDL}}$ and rearranging the above equation as

$$(1+2\alpha)\frac{I}{2F} - D_w^{g,\text{eff}}\frac{\Delta C}{\delta_{\text{GDL}}}\frac{1}{Y} = \frac{\varepsilon s_o \rho_l \delta_{\text{GDL}}}{M_w}\left(-\frac{dY}{dt}\right), \qquad \text{(Eq. 9-14)}$$

and defining

$$\tau = \frac{t}{t_1} \quad \text{where} \quad t_1 = \frac{\varepsilon s_o \rho_l (\delta_{\text{GDL}})^2}{2D_w^{g,\text{eff}} M_w \Delta C} \quad \text{and} \quad R = \frac{(1+2\alpha)I\delta_{\text{GDL}}}{2FD_w^{g,\text{eff}}\Delta C}, \qquad \text{(Eq. 9-15)}$$

one can obtain

$$R - \frac{1}{Y} = -2\frac{dY}{d\tau}. \qquad \text{(Eq. 9-16)}$$

The dimensionless parameter R measures the ratio of the ORR's water addition rate to removal rate from the liquid region. When $R < 1$, the de-wetting process can dry up a flooded GDL completely, while when $R > 1$, only part of the GDL can be de-wetted. The solution to Eq. 9-16 at RH < 1 can be derived analytically as

$$\tau = -\frac{2}{R}Y - \frac{2}{R^2}\ln(1-RY) \qquad \text{(Eq. 9-17)}$$

when $R \rightarrow 0$,

$$\ln(1-RY) = -RY - \frac{(RY)^2}{2} + O(R^3). \qquad \text{(Eq. 9-18)}$$

Therefore,

$$\tau = Y^2 + O(R) \quad \text{when } R \rightarrow 0 \qquad \text{(Eq. 9-19)}$$

When $R = 0$, for example, at open circuit, one arrives at

$$\tau = Y^2 \text{ or } Y = \sqrt{\tau} \qquad \text{(Eq. 9-20)}$$

When $Y = 1$, that is, the through-plane de-wetting is completed, $\tau = 1$. Assuming the vapor-phase concentration at the evaporation front is equal to the saturation value and channel flow is at the infinite rate, it follows that

$$\Delta C = C_{w,\text{sat}}(1-RH). \qquad \text{(Eq. 9-21)}$$

Consequently,

$$t_1 = \frac{s_o \rho_l (\delta_{\text{GDL}})^2}{2D_w^g \varepsilon^{n-1} M_w C_{w,\text{sat}}(1-RH)},$$

$$R = \frac{(1+2\alpha)I\delta_{\text{GDL}}}{2FD_w^{g,\text{eff}} C_{w,\text{sat}}(1-RH)}, \qquad \text{(Eq. 9-22)}$$

where n is the Bruggeman factor accounting for torturosity of the diffusive paths in GDLs, and RH is the relative humidity of the channel gas. In addition, t_1 can be regarded as the time constant for the through-plane drying upon fuel cell shutdown or $R = 0$, which is a function of the GDL thickness, porosity, gas RH in the channel, and $C_{w,\text{sat}}(T)$ or local temperature.

Graphs of Y versus τ are plotted in Figure 9-7 for different values of R. The Y–τ curve is highly dependent on R, especially as R approaches unity. Under the condition of the numerical simulation case, $R \sim 0.2$; therefore the time for through-plane drying, that is, $Y = 1$ or $\tau \sim 1.2$, is ~ 1.5 s, which is of the same order of magnitude as the duration of Stage 1 shown in Figure 9-8.

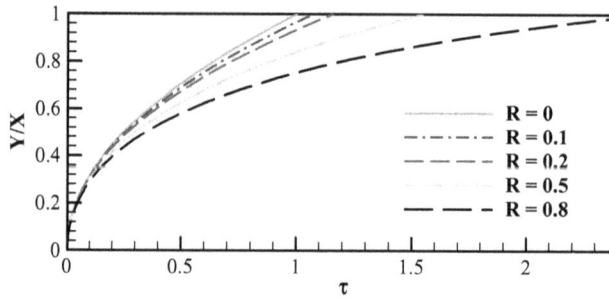

Figure 9-7. Graphs of Y or X versus τ [7] (Courtesy of the Electrochemical Society).

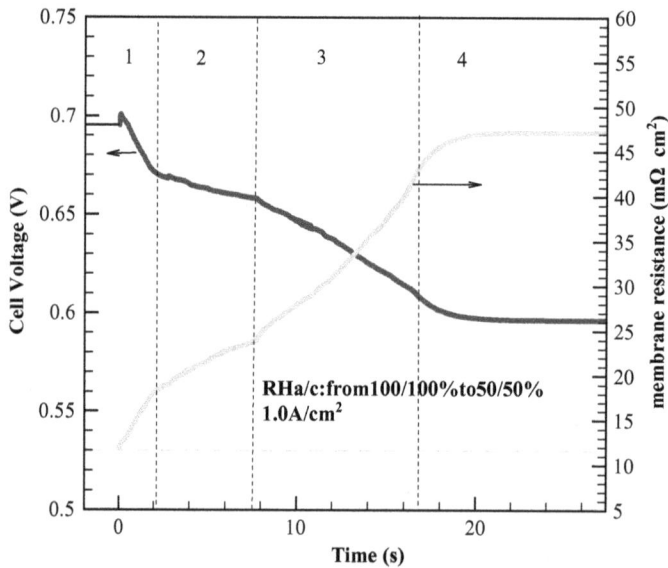

Figure 9-8. Evolutions of cell voltage and membrane resistance during GDL dewetting [7] (Courtesy of the Electrochemical Society).

(b) In-plane drying: Figure 9-6 (b) schematically shows the in-plane drying process on the cathode side. Following the same approach as for the through-plane drying, one derives

$$(1+2\alpha)\frac{I}{2F}\frac{L_{\text{land}}}{2} - D_w^{g,\text{eff}}\frac{\Delta C}{\delta}\delta_{\text{GDL}} = \frac{d}{dt}\left[\left(\frac{L_{\text{land}}}{2}-\delta\right)\delta_{\text{GDL}}\frac{\varepsilon s_o \rho_l}{M_w}\right]. \quad \text{(Eq. 9-23)}$$

Defining $X = \dfrac{\delta}{\dfrac{L_{\text{land}}}{2}}$ and rearranging the above equation as

$$(1+2\alpha)\frac{I}{2F}\frac{L_{\text{land}}}{2} - \frac{D_w^{g,\text{eff}}\Delta C\delta_{\text{GDL}}}{\dfrac{L_{\text{land}}}{2}}\frac{1}{X} = -\frac{L_{\text{land}}}{2}\frac{\delta_{\text{GDL}}\varepsilon s_o \rho_l}{M_w}\frac{dX}{dt} \quad \text{(Eq. 9-24)}$$

and further defining

$$\tau = \frac{t}{t_2} \quad \text{where } t_2 = \frac{\varepsilon s_o \rho_l \left(\dfrac{L_{\text{land}}}{2}\right)^2}{2D_w^{g,\text{eff}} M_w \Delta C} = \frac{s_o \rho_l \left(\dfrac{L_{\text{land}}}{2}\right)^2}{2D_w^g \varepsilon^{n-1} M_w C_{w,\text{sat}}(1-RH)} \quad \text{and} \quad \text{(Eq. 9-25)}$$

$$R = \frac{\dfrac{(1+2\alpha)I}{2F}\dfrac{L_{\text{land}}}{2}}{\dfrac{D_w^{g,\text{eff}}\Delta C}{\dfrac{L_{\text{land}}}{2}}\delta_{\text{GDL}}} = \frac{\dfrac{(1+2\alpha)I}{2F}\left(\dfrac{L_{\text{land}}}{2}\right)^2}{D_w^{g,\text{eff}}\Delta C \delta_{\text{GDL}}}.$$

One can obtain

$$R - \frac{1}{X} = -2\frac{dX}{d\tau}. \qquad \text{(Eq. 9-26)}$$

Comparing Eq. 9-16 with Eq. 9-26, it is clear that the dynamics of in-plane and through-plane dryings share the same mathematical characteristics. Thus, Eq. 9-17 and Figure 9-7 also describe the in-plane drying process, provided that the location of the evaporation front is normalized by their respective length scales. In addition, one can obtain the following relationship between t_1 and t_2:

$$t_2 = t_1 \left(\frac{L_{\text{land}}/2}{\delta_{\text{GDL}}}\right)^2. \qquad \text{(Eq. 9-27)}$$

In the simulation case study to be presented in the following section, $t_2 = 6.25t_1$ for the same value of s_o. In addition, t_2 is inversely proportional to D_w^g as described in Eq. 9-17. Thus, the time constant of in-plane drying in the anode is roughly three times shorter than that in the cathode based on the anode and cathode water diffusivities.

9.2.3 GDL DE-WETTING AND VOLTAGE LOSS

To understand the effects of the GDL de-wetting phenomena, the voltage loss is monitored during the process. To precisely describe the nonisothermal phenomena, the partial differential equation of energy conservation is solved, which accounts for rates of energy storage, convective/conductive heat transfer, and heating sources. In addition to the waste heat by PEM fuel cells as discussed in Chapter 8, the heat release/absorption also arises from water condensation/evaporation, and the rate of heat release/absorption, S_{lg}, is given by

$$S_{\text{lg}} = h_{\text{lg}} \dot{m}_{\text{lg}}, \qquad \text{(Eq. 9-28)}$$

where h_{1g} and $\dot{m}_{1g}$ are, respectively, the latent heat and rate of vapor-liquid phase change. The latter is readily calculated through the liquid continuity equation, namely:

$$\dot{m}_{1g} = \rho_l \frac{\partial s}{\partial t} + \nabla \cdot (\rho_l \vec{u}_l), \qquad \text{(Eq. 9-29)}$$

where the liquid-phase velocity in the multiphase mixture (M^2) model is related to its mobility λ_l:

$$\rho_l \vec{u}_l = \vec{j}_l + \lambda_l \rho \vec{u}. \qquad \text{(Eq. 9-30)}$$

Here, $\vec{j}_l$ is the capillary diffusion flux determined by the gradient of capillary pressure, ∇p_c:

$$\vec{j}_l = \frac{\lambda_l \lambda_g}{\nu} K[\nabla p_c + (\rho_l - \rho_g)\vec{g}]. \qquad \text{(Eq. 9-31)}$$

To shed light on the phasechange process of drying, the following operation is considered: a PEM fuel cell initially operates at steady state, 1.0 A/cm², and full humidification. Of interest is the transient process after switching to dry operation, that is, the RH on both the anode and cathode side becomes 50% (or RHa/c = 50%/50%). The gas flow rates (or stoichiometric flow ratios) are set sufficiently large so that channel flooding is absent. Figure 9-8 displays the time responses of cell voltage and membrane resistance upon the change on relative humidity. The voltage prediction clearly experiences four major stages of reduction as marked in the figure. The voltage drop can be explained by the membrane resistance evolution plotted in the same figure. Note that the cell performance in this range of the current is controlled by the ohmic loss. The membrane resistance increases due to the feed with a lower-humidity gas, which gradually dries up water inside the fuel cell.

Figure 9-9 shows the liquid water saturation distributions during Stage 1. Figure 9-9(a) presents the initial condition (liquid water saturation) at a steady state. Initially liquid exists in both the anode and cathode. Upon the change to the 50% RH reactant gas, the GDL's liquid water under the channel is first evaporated and then removed by the dry channel stream. This first stage of de-wetting is characterized as through-plane drying under the channel, that is, the evaporation front moving inwards in the through-plane direction. The land's water changes little though the level under the channel experiences a significant decrease.

The GDL de-wetting under the channel is followed by in-plane drying under the land. Since the water diffusivity in the anode hydrogen gas is several times larger than that in the cathode, the in-plane drying in the anode occurs in a faster pace. The evolution of liquid water saturation in Stage 2 of the anode in-plane drying is displayed in Figure 9-10. This figure shows that liquid water evaporates away from the anode GDL under the land and the evaporation front propagates from the channel towards the land area. The process in Stage 2 takes ~5 s.

Obviously, water loss under the land occurs simultaneously on both cathode and anode sides. The cathode in-plane drying takes longer due to the lower water diffusivity on this side. Note that the duration of the cathode in-plane drying encompasses both Stages 2 and 3. In Stage 2, liquid water remains under the lands of both sides; thus, the resulting increase in the

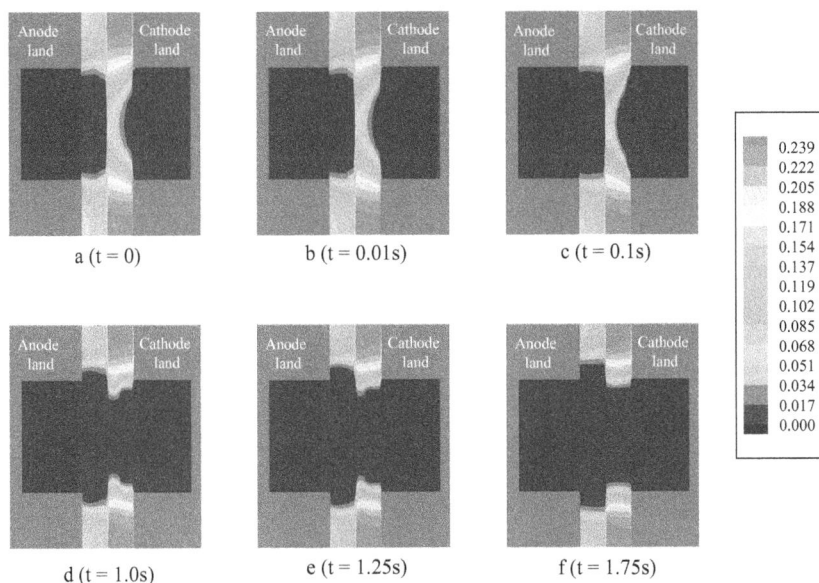

Figure 9-9. Liquid water saturation distributions at time instants of 0, 0.01, 0.1, 1.0, 1.25, and 1.75 s, respectively [7] (Courtesy of the Electrochemical Society).

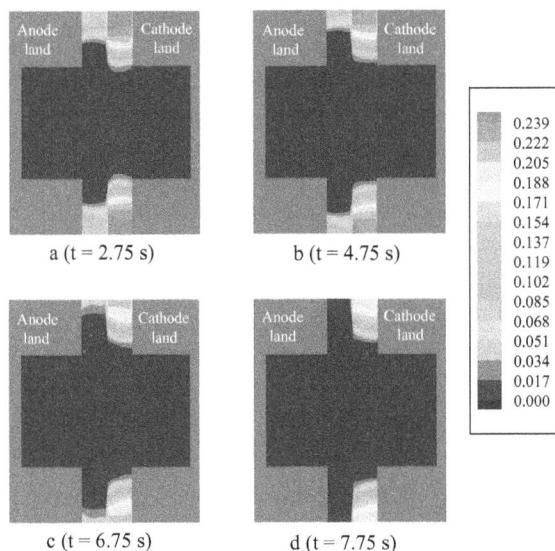

Figure 9-10. Liquid water saturation distributions at time instants of 2.75, 4.75, 6.75, and 7.75 s, respectively [7] (Courtesy of the Electrochemical Society).

average membrane resistance and loss in the cell voltage are relatively small. Once liquid water disappears in the anode under both channel and land after Stage 2, the rise in the membrane resistance accelerates as shown in Figure 9-8 during Stage 3. Figure 9-11 displays the water saturation contours in Stage 3, indicative of cathode in-plane drying. At this stage, the cathode liquid water underneath the land is removed through two mechanisms: one is evaporation at the evaporation front and the other is water back-diffusion through the membrane to the dry anode.

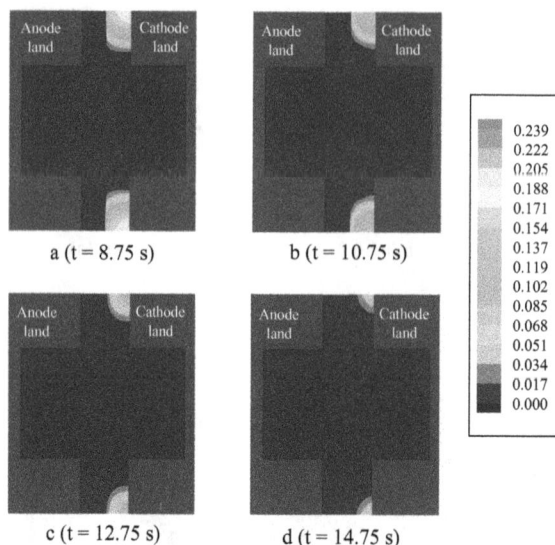

Figure 9-11. Liquid water saturation distributions at time instants of 8.75, 10.75, 12.75, and 14.75 s, respectively [7] (Courtesy of the Electrochemical Society).

Figure 9-12 displays the water vapor concentrations in Stage 3. Though the cathode liquid water area shrinks significantly, the water vapor concentration remains high in the GDL. Consequently, the membrane maintains adequate hydration. Once all liquid water disappears and there is no water vapor supply from evaporation, membrane dehydration accelerates, resulting in the final stage of cell voltage or membrane resistance evolution during the de-wetting process, as shown in Figure 9-8. The membrane dehydration process lasts a time scale of seconds. Figure 9-13 displays the water vapor concentration distributions in the GDL in Stage 4. It is seen that water vapor concentrations throughout the PEM fuel cell are below the saturation value at 80°C, that is, 15.9 mol/m^3. In addition, the GDL maintains a relatively higher value of water vapor concentration on both anode and cathode at 16.75 s and the water vapor concentration falls quickly over the subsequent seconds until another steady state at 22.75 s.

Figure 9-14 shows the membrane water content during the four stages. The de-wetting process of GDLs and catalyst layers is followed by membrane dehydration. The latter is the main reason for the increase of the membrane resistance and cell voltage drop as shown in Figure 9-8. In addition, the under-land region always maintains a higher water content than the under-channel region. At 5.75 s in Stage 2 when the GDL de-wetting under the channel is completed, there is a large difference between the under-land and the under-channel water content. From 16.75 to 22.75 s when there is no liquid water in the fuel cell, that is, in Stage 4, the membrane content decreases, indicative of dehydration.

Figure 9-15 displays the current density contours. Initially high current density appears under the channel. This is due to the full-humidity operation initially at $t = 0$, which maintains a high water content in the membrane, rendering the ohmic resistance relatively small. Consequently, oxygen transport determines the local fuel cell performance. As the GDL de-wets under the channel, the local membrane resistance increases, leading to a shift of the current density peak from the under-channel to the under-land areas. The difference between the current

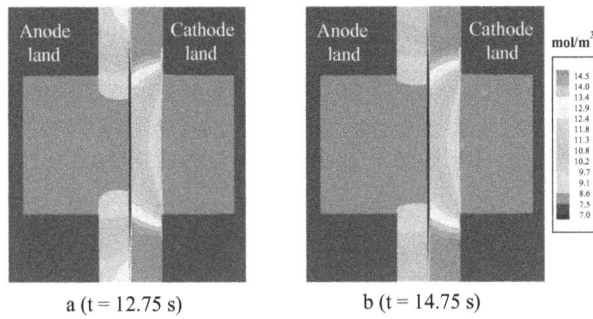

a (t = 12.75 s)   b (t = 14.75 s)

Figure 9-12. Water vapor concentration contours at time instants of 12.75 and 14.75 s, respectively [7] (Courtesy of the Electrochemical Society).

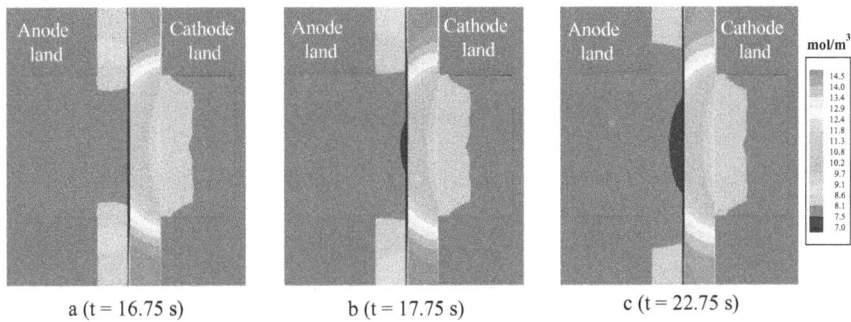

a (t = 16.75 s)   b (t = 17.75 s)   c (t = 22.75 s)

Figure 9-13. Water vapor concentration contours at time instants of 16.75, 17.75, and 22.75 s, respectively [7] (Courtesy of the Electrochemical Society).

Through-plane direction

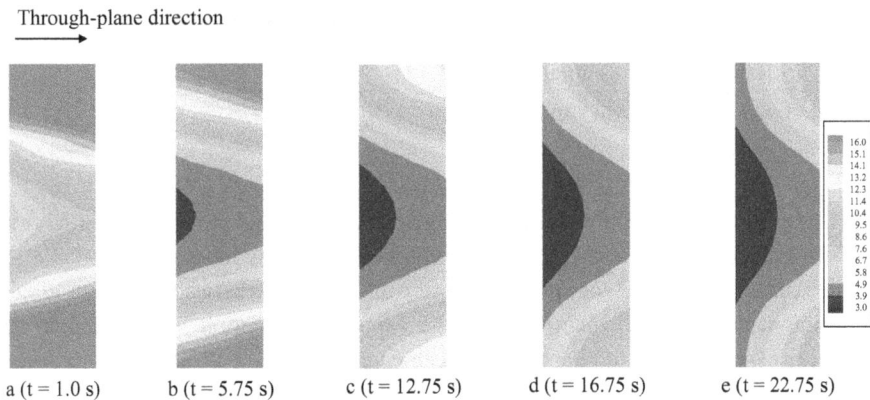

a (t = 1.0 s)   b (t = 5.75 s)   c (t = 12.75 s)   d (t = 16.75 s)   e (t = 22.75 s)

Figure 9-14. Water content contours in the membrane at time instants of 1.0, 5.75, 12.75, 16.75, and 22.75 s, respectively [7] (Courtesy of the Electrochemical Society).

densities under the land and channel is further increased until 5.75 s, consistent with the occurrence of the maximum non-uniformity in the membrane water content. After 5.75 s, in-plane drying toward the land area increases the local membrane resistance and hence the difference in local current density between the channel and land areas diminishes.

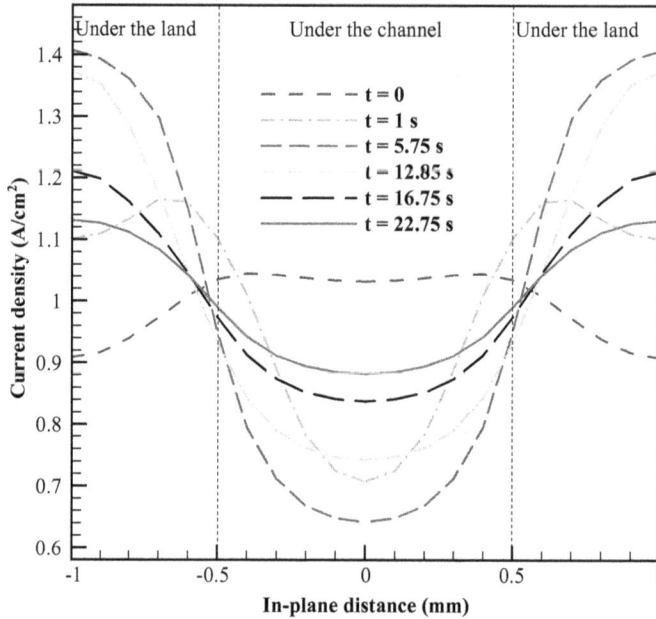

Figure 9-15. Evolution of current density profiles [7] (Courtesy of the Electrochemical Society).

9.2.4 A GENERAL DEFINITION OF THE DAMKOHLER NUMBER, DA

Two-phase transport is important to PEM fuel cell operation. Under sufficiently dry operating conditions, liquid water presence will be completely avoided. Dry operation, however, exhibits a large ohmic voltage loss. To measure the relative importance of water production to water removal via vapor diffusion, a dimensionless group Da (the Damkohler number) is defined as follows [9]:

$$Da = \frac{\text{Rate of water production}}{\text{Rate of water removal via vapor diffusion}} = \frac{I/2F}{D_w^{g,\text{eff}} \Delta C_w / H_{\text{GDL}}} = \frac{I H_{\text{GDL}}}{2F D_w^{g,\text{eff}} \Delta C_w},$$

(Eq. 9-32)

where F is the Faraday's constant, I is the average current density output by the PEM fuel cell, H_{GDL} is the GDL thickness, and $D_w^{g,\text{eff}}$ is the effective diffusivity for water vapor. ΔC_w is the water vapor capacity in the gaseous phase, denoting the difference between the saturation vapor concentration at the GDL–CL interface and the vapor concentration in the channel stream. Under the extreme condition of completely dry channel stream, ΔC_w can be approximated by the saturated water vapor concentration $C_{w,\text{sat}}$ at the average cell temperature. In defining Da as given by Eq. 9-32, water addition or source due to electro-osmatic drag from the anode to the cathode is neglected. A more precise definition of Da will be presented in the next section. With the physical meaning of Da, the following statements can be made:

- When $Da \to 0$, we have single-phase operation. For example, at dry operation and high operating temperature, ΔC_w is sufficiently large such that the Da is small and the PEM fuel cell is subject to low-humidity operation.

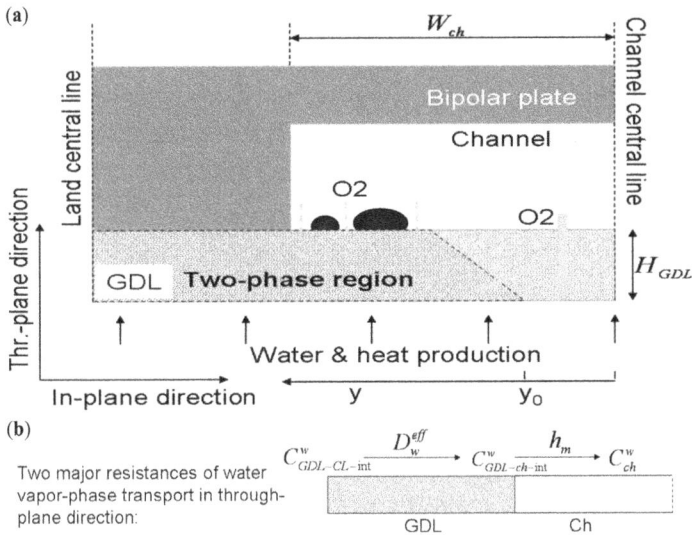

Figure 9-16. (a) Schematic of the channel-land structure of a PEM fuel cell cathode and the concept of single-phase passage in GDLs for oxygen transport, and (b) vapor-phase water transport resistance in the through-plane direction.

- When $Da \to \infty$, we have full two-phase operation (which means liquid water is present everywhere in the cathode). For example, under isothermal condition and fully humidified channel condition, ΔC_w is equal to zero, leading to an infinite Da. The fuel cell cathode is thus subject to two-phase flow regime.
- When $Da = o(1)$ (i.e., Da is on the order of unity), we may have partially two-phase operation, which means liquid water is present only in some regions of the fuel cell.

Both water and waste heat are produced during PEM fuel cell operation. Without waste heat or external heating sources, under full-humidification operation the entire cathode will be subjected to two-phase flow. The waste heat generation raises local temperature, evaporating liquid water and diminishing the two-phase region. Figure 9-16 schematically shows the basic land-channel structure of a PEM fuel cell and water transport.

9.2.4.1 Local Heating and Vapor-Phase Removal

During PEM fuel cell operation, both water and waste heat are produced and then removed via GDLs: water is removed by channel gas flow, whereas most heat is taken away via the land. The portion of heat removed by the channel stream is small under normal conditions and can be neglected in analysis. For the part of GDL under channels, the maximum in-plane variation of temperature can be evaluated by

$$\Delta T_{max}^{\text{in-plane}} \approx \frac{\frac{1}{2} I(E' - V_{cell}) W_{ch}^2}{2 H_{GDL} k_{GDL,W}^{eff}}, \qquad \text{(Eq. 9-33)}$$

where W_{ch} is half of the channel in-plane length, H_{GDL} is the GDL thickness, and $k_{GDL,W}^{eff}$ is the effective in-plane conductivity. E' is the thermodynamic potential assuming all the enthalpy change of reactions is converted to electric energy. We also assume a uniform in-plane current density and equal amount of heat removed by the anode and cathode. The through-plane change $\Delta T_{max}^{thr\text{-}plane}$ is approximated by

$$\Delta T_{max}^{thr\text{-}plane} = \frac{\frac{1}{2} I(E'-V_{cell})}{\dfrac{k_{GDL,H}^{eff}}{H_{GDL}}}.$$

(Eq. 9-34)

$k_{GDL,H}^{eff}$ and $k_{GDL,W}^{eff}$ denote the through- and in-plane conductivities, respectively. Assuming that the two elementary thermal resistors, that is, the through- and in-plane ones, are arranged in series, the total temperature change becomes the sum of the above two variations:

$$\Delta T_{max} = \Delta T_{max}^{in\text{-}plane} + \Delta T_{max}^{thr\text{-}plane} = \Delta T_{max}^{thr\text{-}plane} f_{max}(\frac{W_{ch}}{H_{GDL}}, \frac{k_{GDL,H}^{eff}}{k_{GDL,W}^{eff}}),$$

(Eq. 9-35)

where $f_{max} = 1 + \dfrac{W_{ch}^2 k_{GDL,H}^{eff}}{2 H_{GDL}^2 k_{GDL,W}^{eff}}$. Note that the above assumes negligible heat removal by channel flow, which is a good approximation under normal operating conditions, as discussed in Chapter 8.

Equation 9-35 presents a simplified equation to evaluate the maximum temperature variation in a GDL due to the waste heat generation. Upon this temperature change, the change in the local-vapor saturation concentration can be determined as follows:

$$\Delta C_{w,sat} = C_{w,sat}(T_0 + \Delta T_{max}) - C_{w,sat}(T_0),$$

(Eq. 9-36)

where T_0 is the temperature of the bipolar plates (note that the bipolar plate temperature can be assumed uniform due to their large thermal conductivity) and $C_{w,sat}$ is given by the ideal gas law:

$$C_{w,sat}(T) = \frac{P_{w,sat}(T)}{RT}.$$

(Eq. 9-37)

The linear approximation of Eq. 9-36 yields

$$\Delta C_{w,sat} \approx \frac{dC_{w,sat}(T_0)}{dT} \Delta T_{max} = f_{max} \frac{I(E'-V_{cell})H_{GDL}}{2k_{GDL,H}^{eff}} \frac{dC_{w,sat}(T_0)}{dT}.$$

(Eq. 9-38)

Gas flow channels are grooved in the bipolar plates with a dimension around 1 mm. The channel stream temperature can be assumed to be equal to that of the local bipolar plate. In GDLs, local equilibrium holds true because the interfacial area is large between the two phases.

Then, ΔC_{sat}^w in Eq. 9-38 yields the maximum difference in the saturated vapor concentration between local GDL sites and channel stream. Assuming that diffusion dominates vapor transport, the capability or maximum rate of vapor removal is estimated by

$$G_{w,\text{diff},\max} = \frac{\Delta C_{w,\text{sat}} + C_{w,\text{sat}}(T_0) - C_{w,\text{ch}}}{\dfrac{H_{\text{GDL}}}{D_w^{g,\text{eff}}} + \dfrac{1}{h_m}} = \frac{\Delta C_{w,\text{sat}} + C_{w,\text{sat}}(T_0) - C_{w,\text{ch}}}{\dfrac{H_{\text{GDL}}}{D_w^{g,\text{eff}}}\left(1 + \dfrac{D_w}{h_m H_{\text{ch}}}\dfrac{H_{\text{ch}} D_w^{g,\text{eff}}}{H_{\text{GDL}} D_w^g}\right)},\ \text{(Eq. 9-39)}$$

where $C_{w,\text{ch}}$ is the vapor concentration in the channel. The Sherwood number then appears in the denominator on the right-hand side of Eq. 9-39: $Sh = \dfrac{h_m H_{\text{ch}}}{D_w^g}$ where h_m is the mass transfer coefficient.

9.2.4.2 A Specific Damkohler Number

Under the channel central line, the lateral water transport is negligible. The through-plane flux equals to the net water addition rate, $\dfrac{I(1+2\alpha)}{2F}$, where α is the net water transport coefficient. The Damkohler number at the CL–GDL interface under the channel centerline Da_0 can be expressed through comparing water addition with vapor removal capability:

$$Da_0 = \frac{I(1+2\alpha)H_{\text{GDL}}}{2FD_w^{g,\text{eff}}(\Delta C_{w,\text{sat}} + C_{w,\text{sat}}(T_0) - C_{w,\text{ch}})}\left(1 + \frac{1}{Sh}\frac{H_{\text{ch}} D_w^{g,\text{eff}}}{H_{\text{GDL}} D_w^g}\right)$$

$$= \frac{\dfrac{I(1+2\alpha)}{2F}}{D_w^{g,\text{eff}}\dfrac{(\Delta C_{w,\text{sat}} + C_{w,\text{sat}}(T_0) - C_{w,\text{ch}})}{H_{\text{GDL}}}}\left(1 + \frac{1}{Sh}\frac{H_{\text{ch}} D_w^{g,\text{eff}}}{H_{\text{GDL}} D_w^g}\right),\qquad \text{(Eq. 9-40)}$$

where the suffix 0 represents the location at the channel centerline. This location is the starting site where liquid water evaporates as a result of the presence of the maximum temperature. Da_0 represents a parameter that indicates the two-phase regime in the cathode GDL – this will be detailed in the next section. The above expression is further rearranged as follows:

$$Da_0 = \frac{\dfrac{I(1+2\alpha)}{2F}}{D_w^{g,\text{eff}}\dfrac{f_{\max}\dfrac{I(E'-V_{\text{cell}})H_{\text{GDL}}}{2k_{\text{GDL},H}^{\text{eff}}}\dfrac{dC_{w,\text{sat}}(T_0)}{dT} + C_{w,\text{sat}}(T_0)(1-RH_{\text{local}})}{H_{\text{GDL}}}}\left(1 + \frac{1}{Sh}\frac{H_{\text{ch}} D_w^{g,\text{eff}}}{H_{\text{GDL}} D_w^g}\right)$$

$$= \frac{I(1+2\alpha)H_{\text{GDL}}}{2FD_w^{g,\text{eff}}\left[f_{\max}\dfrac{I(E'-V_{\text{cell}})H_{\text{GDL}}}{2k_{\text{GDL},H}^{\text{eff}}}\dfrac{dC_{w,\text{sat}}(T_0)}{dT} + C_{w,\text{sat}}(T_0)(1-RH_{\text{local}})\right]}\left(1 + \frac{1}{Sh}\frac{H_{\text{ch}} D_w^{g,\text{eff}}}{H_{\text{GDL}} D_w^g}\right).$$

$$\text{(Eq. 9-41)}$$

In the absence of supersaturated state, $Da_0 > 1$ means the entire cathode GDL–CL interface is subjected to two-phase flow, whereas for $Da_0 < 1$ a portion of the interface becomes free of liquid water.

Several limiting cases can be readily obtained from the above equation:

1. For sufficiently fast channel stream such that $1/Sh \to 0$, $\dfrac{1}{Sh} \dfrac{H_{ch} D_w^{g,\text{eff}}}{H_{GDL} D_w^g}$ vanishes; consequently,

$$Da_0 = \frac{I(1+2\alpha)H_{GDL}}{2FD_w^{g,\text{eff}}\left[f_{\max} \dfrac{I(E'-V_{\text{cell}})H_{GDL}}{2k_{\text{GDL},H}^{\text{eff}}} \dfrac{dC_{w,\text{sat}}(T_0)}{dT} + C_{w,\text{sat}}(T_0)(1-RH_{\text{local}}) \right]} \cdot \quad \text{(Eq. 9-42)}$$

2. When $\dfrac{H_{GDL}}{D_w^{g,\text{eff}}} \to 0$, either the GDL thickness is small or the water diffusivity is large, $Da_0 \to 0$,

   that is, the cathode GDL–CL interface under the channel centerline is always dry. From Eq. 9-32, a conclusion can be drawn under this extreme condition that the entire cathode GDL is in the single-phase regime.

3. When $C_{w,\text{sat}}(T_0)(1-RH_{\text{local}})$ is sufficiently large, for example, for sufficiently dry operation, or high GDL conductivity, Da_0 can be rewritten as

$$Da_0 = \frac{I(1+2\alpha)H_{GDL}}{2FD_w^{g,\text{eff}} C_{w,\text{sat}}(T_0)(1-RH_{\text{local}})} \left(1 + \frac{1}{Sh} \frac{H_{ch} D_w^{g,\text{eff}}}{H_{GDL} D_w^g} \right). \quad \text{(Eq. 9-43)}$$

With sufficiently fast channel flows, a simpler expression is reached:

$$Da_0 = \frac{I(1+2\alpha)H_{GDL}}{2FD_w^{g,\text{eff}} C_{w,\text{sat}}(T_0)(1-RH_{\text{local}})}. \quad \text{(Eq. 9-44)}$$

4. when $f_{\max} \dfrac{I(E'-V_{\text{cell}})H_{GDL}}{2k_{\text{GDL},H}^{\text{eff}}} \dfrac{dC_{w,\text{sat}}(T_0)}{dT} \gg C_{w,\text{sat}}(T_0)(1-RH_{\text{local}})$, Eq. 9-41 reduces to

$$Da_0 = \frac{(1+2\alpha)k_{\text{GDL},H}^{\text{eff}}}{FD_w^{g,\text{eff}} f_{\max}(E'-V_{\text{cell}}) \dfrac{dC_{w,\text{sat}}(T_0)}{dT}} \left(1 + \frac{1}{Sh} \frac{H_{ch} D_w^{g,\text{eff}}}{H_{GDL} D_w^g} \right). \quad \text{(Eq. 9-45)}$$

This scenario occurs when the gas flow channel is subjected to two-phase flow, in which the local RH is unity. Two-phase channel stream emerges usually downstream because of water production from the ORR. The following discussion focuses on this scenario only to explore the complex interaction between heat transfer and two-phase flow. In addition, current density I disappears in the Da_0 expression; however, it affects Da_0 through other parameters such as cell voltage. Figure 9-17 plots the Da_0 profile as a function of temperature. The gray region represents the area that the single-phase flow appears at the GDL–CL interface, whereas the upper area

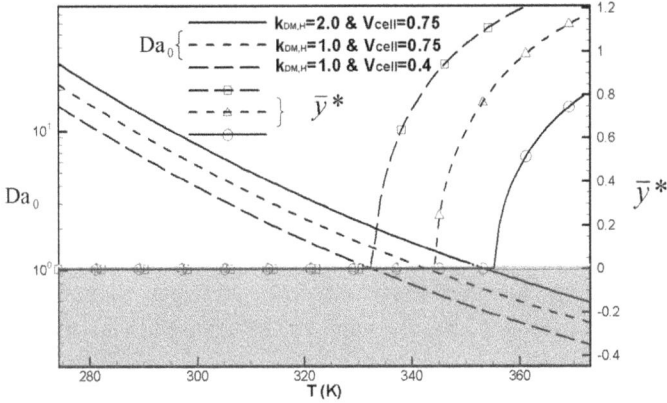

Figure 9-17. Da_0 and $\bar{y}*$ as a function of temperature. The thermal conductivity and cell voltage units are in W m^{-1} K^{-1} and V, respectively [9]. $k_{DM,H}$ refers to $k_{GDL,H}^{eff}$.

represents the condition that the entire interface is subjected to two-phase regime. Da_0 changes inversely with operation temperature as indicated in this figure. At low temperature, the cathode GDL–CL interface is more prone to two-phase formation. Figure 9-17 indicates that both the cell voltage and GDL thermal property affect Da_0.

$k_{GDL,H}^{eff}$ is the effective through-plane conductivity. This coefficient takes into accounts the heat-pipe effect in the two-phase flow regime. Assuming $k_{GDL,H}^{eff}$ is the sum of the GDL intrinsic conductivity $k_{DM,H}$ and heat-pipe apparent conductance $k_{lg} = \left(h_{lg} M_w D_w^{g,eff} \dfrac{dC_{w,sat}(T)}{dT} \right)$, Da_0 can be rewritten as

$$Da_0 = \frac{k_{lg} + k_{GDL,H}}{f_{max} F D_w^{g,eff} (E'-V_{cell}) \dfrac{dC_{w,sat}(T_0)}{dT}} (1+2\alpha) \left(1+\frac{1}{Sh}\frac{H_{ch} D_w^{g,eff}}{H_{GDL} D_w^g}\right)$$

$$= \left(h_{lg} M_w + \frac{k_{GDL,H}}{D_w^{g,eff} \dfrac{dC_{w,sat}(T_0)}{dT}} \right) \frac{1+2\alpha}{f_{max} F(E'-V_{cell})} \left(1+\frac{1}{Sh}\frac{H_{ch} D_w^{g,eff}}{H_{GDL} D_w^g}\right).$$

(Eq. 9-46)

In the above, $\dfrac{dC_{w,sat}(T)}{dT}$ is approximated by using $\dfrac{dC_{w,sat}(T_0)}{dT}$.

9.2.4.3 Liquid-Free Passages

In a PEM fuel cell, the presence of liquid water in the cathode increases the oxygen transport resistance. Equation 9-46 implies that a single-phase zone may exist in GDLs even though the channel flow is two-phase. This liquid-free area will be effective passages for transporting reactants without any concern of "flooding." A schematic of this transport path is sketched in Figure 9-16. Again from Eq. 9-46, the following qualitative conclusions can be drawn directly:

a) Under full humidification, it is possible that a portion of the cathode GDL–CL interface is free of liquid (i.e., $Da_0 < 1$).
b) Occurrence of the free-liquid area at the cathode CL–GDL interface depends on a number of parameters, not limited to the water vapor diffusivity.
c) Operating temperature can significantly alter the liquid-water region in GDLs.
d) The GDL anisotropy, local net water transport coefficient, and geometrical parameters such as GDL thickness and channel width can be adjusted to modify the flow regime in GDLs.

For $Da_0 < 1$, no liquid water appears at the GDL–CL interface under the channel center-line. In the in-plane direction from channels to lands, temperature decreases, reducing the local vapor saturation pressure. Liquid emerges when the vapor phase is unable to remove all the added water. To assess the onset of the two-phase flow, one can evaluate the in-plane tempera-ture profile as follows (see the y-direction in Fig. 9-16):

$$\Delta T^{\text{in-plane}}(\bar{y}) \approx \tfrac{1}{2}I(E'-V_{\text{cell}})\frac{W_{\text{ch}}^2}{2H_{\text{GDL}}^2 k_{\text{GDL},W}^{\text{eff}}}(1-\bar{y}^2), \qquad \text{(Eq. 9-47)}$$

where the dimensionless location $\bar{y}$ is defined as $\dfrac{y}{W_{\text{ch}}}$. Following the same procedure, a local Da number can be defined as

$$Da(\bar{y}) = \frac{k_{\text{GDL},H}^{\text{eff}}(1+2\alpha)}{f(\bar{y})FD_w^{g,\text{eff}}(E'-V_{\text{cell}})\dfrac{dC_{w,\text{sat}}(T_0)}{dT}}\left(1+\frac{1}{Sh}\frac{H_{\text{ch}}D_w^{g,\text{eff}}}{H_{\text{GDL}}D_w^g}\right), \quad \text{(Eq. 9-48)}$$

where $f(\bar{y}) = 1 + \dfrac{W_{\text{ch}}^2 k_{\text{GDL},H}^{\text{eff}}}{2H_{\text{GDL}}^2 k_{\text{GDL},W}^{\text{eff}}}(1-\bar{y}^2)$. $Da(\bar{y}^*) = 1$ gives the onset location of liquid $\bar{y}^*$:

$$\bar{y}^* = \sqrt{\frac{f_{\max}(1-Da_0)}{f_{\max}-1}}. \qquad \text{(Eq. 9-49)}$$

Figure 9-17 also plots $\bar{y}_{Da<1}$ as a function of temperature. It should be noted that the above derivation neglects the effect of lands. As $\bar{y}^*$ approaches to unity, that is, at the land-channel edge, the above solution yields a large error in evaluation due to the effect of lateral water transport.

9.2.4.4   2D Numerical Simulation

As an example to show the single-phase passage in GDLs, 2D numerical results from three case studies are presented below, pertaining to the flow regimes in the cathode GDL. Figure 9-18 shows the geometry of a PEM fuel cell cathode and the computational domain. The PEM fuel cell is taken to operate under full humidification without irreducible liquid water. The governing equations consist of those of two-phase flow and heat transfer in the M^2 model (see Appendix VI.B). Details of the governing equations and boundary conditions as well as numerical procedure can be found in Ref. [9]. Results from three case studies with the following operating conditions are presented.

Case 1: operating temperature 80°C with a GDL thermal conductivity 3.0 W/m·K.
Case 2: operating temperature 80°C with a GDL thermal conductivity 2.0 W/m·K.
Case 3: operating temperature 90°C with a GDL thermal conductivity 2.0 W/m·K.

The fuel cell performance (measured by polarization data) is set to be the same for the three cases. The polarization data are listed in Table 9-1. For all the current densities selected, a constant gas flow rate is assumed in the channel, yielding a Sherwood number of ~3.66. Table 9-2 summarizes all the cases and corresponding computed Da_0. The first case has all the Da_0 s over 1.0 except for the largest current density. Reducing Da_0 can be achieved

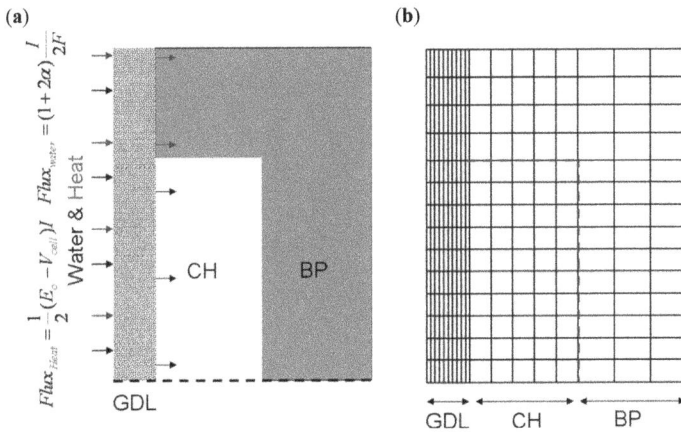

Figure 9-18. (a). PEM fuel cell domain used in the 2D numerical case studies. (b) The computational domain. E_0 refers to E' .

Table 9-1. Polarization data employed in the 2D numerical simulation case studies

Current density (A/cm²)	Cell voltage (V)
0.1	0.85
0.5	0.72
0.8	0.67
1.0	0.62
1.3	0.54
1.5	0.43

Table 9-2. Polarization data employed in the 2D numerical simulation case studies

Temperature & conductivity	Current density (A/cm2)	By Eq. 9-46 Da_0	By Eq. 9-49 $\bar{y}*$	2D Simulation $\bar{y}*$
	0.1	2.22	0	0
	0.5	1.59	0	0
	0.8	1.43	0	0
Case 1: 80°C & 3 W/m·K	1.0	1.30	0	0
	1.3	1.14	0	0
	1.5	0.97	18%	20%
	0.1	1.53	0	0
	0.5	1.09	0	0
	0.8	0.98	12%	5%
Case 2: 80°C & 2 W/m·K	1.0	0.90	33%	15%
	1.3	0.78	48%	35%
	1.5	0.67	60%	55%
	0.1	1.13	0	0
	0.5	0.81	23%	40%
	0.8	0.73	42%	45%
	1.0	0.66	53%	50%
	1.3	0.58	63%	60%
Case 3: 90°C & 2 W/m·K	1.5	0.49	73%	70%

through several ways: one is to use low thermal conductivity GDLs as in Case 2. As seen from Table 9-2, a wider operating range as in Case 2 exhibits a lower Da_0. Another is to operate the PEM fuel cell at a higher temperature (hence a higher value of $\frac{dp_{w,sat}}{dT}$) as in Case 3, which exhibits a smaller Da_0 than Case 2 does.

Figure 9-19 displays the temperature contours for Case 1. Peak temperature clearly appears at the channel centerline for all the current densities. A small temperature variation appears in the bipolar plate due to its large conductivity. In the channel flow, only near the GDL surface

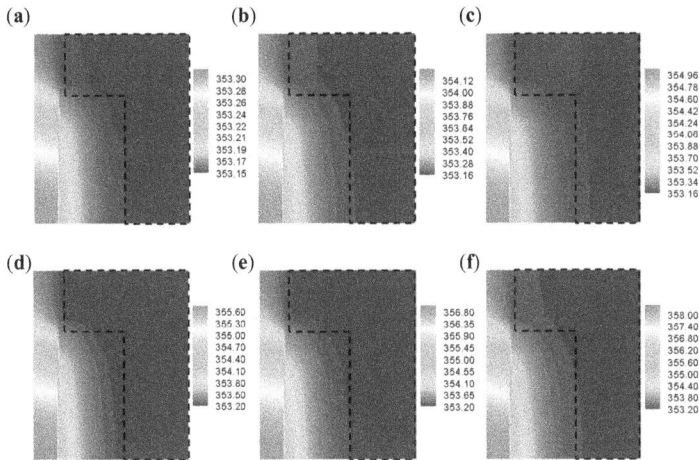

Figure 9-19. Temperature (in K) contours of (a) 0.1 A/cm²; (b) 0.5 A/cm²; (c) 0.8 A/cm²; (d) 1.0 A/cm²; (e) 1.3 A/cm²; and (f) 1.5 A/cm². The BP outer surface is set at 353.15 K and GDL thermal conductivity is 3 W/m·K, that is, Case 1 [9].

show remarkable temperature contours, which is the thermal boundary layer. Figure 9-20 presents results from Case 2, showing a similar trend as Figure 9-19. However, the magnitudes of peak temperatures are different: Case 2 has a larger variation. The locally higher temperature improves the water removal capability through the vapor phase. For $Da_0<1$, the vapor phase is able to remove all the product water; thus liquid water near the centerline disappears. Case 3 exhibits similar temperature contours as Case 2 does and is therefore not displayed.

Figures 9-21 to 9-23 display liquid water distributions for the three cases. Figure 9-21 shows that in almost all the operation conditions the entire GDL–CL interface (the left side) is subject to two-phase flow, consistent with the Damkohler number analysis using Da_0; see Table 9-2. Only the highest current density, that is, Figure 9-21(f), has a value of Da_0 below one, which has a "dryout" centerline region. The predicted dry region is about 20% portion of the GDL, close to the analytical value of ~18% presented in Table 9-2. Figure 9-22 displays liquid contours for Case 2, in which the centerline GDL–CL interface is free of liquid for the latter four current densities, consistent with the Da_0 analysis. Some cases, however, show discrepancies between the simulation and analytical results. For example, for 1.0 A/cm² the 2D model predicts a 15% portion of "dry" GDL–CL interface, whereas the analytical solution gives a value of about 33%. Analytical $\bar{y}*$ is sensitive to Da_0 when the latter approaches unity, as shown in Figure 9-17; thus a large discrepancy can yield. Because of this sensitivity, $\bar{y}*$ is more meaningful for the purpose of qualitative analysis. One major reason for the discrepancy is due to the temperature estimate being an approximation, not an exact solution. Another is that the channel stream provides some cooling, which is excluded in the analysis. Furthermore, the heat-pipe effect only occurs in the two-phase region, the analysis assumes that it is active in the entire GDL. Figure 9-23 presents the contours for Case 3. Again, consistent with Table 9-2, Da_0 < 1, indicating a centerline CL–GDL interface being free of liquid water. The predicted $\bar{y}*$ is close to the analytical result.

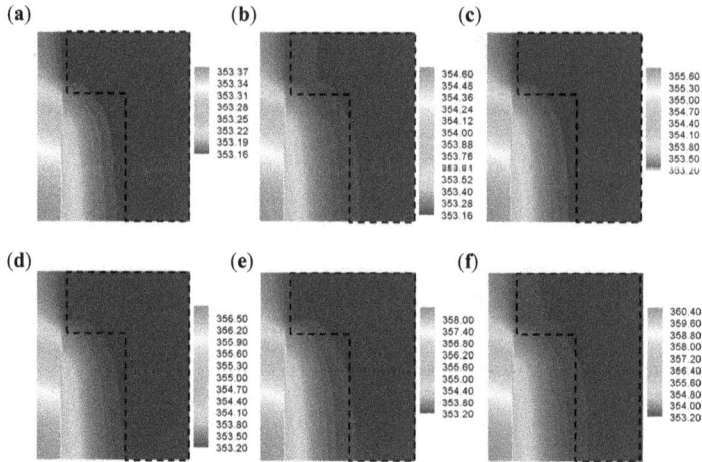

Figure 9-20. Temperature (in K) contours of (a) 0.1 A/cm²; (b) 0.5 A/cm²; (c) 0.8 A/cm²; (d) 1.0 A/cm²; (e) 1.3 A/cm²; and (f) 1.5 A/cm². The BP outer surface is set at 353.15 K and GDL thermal conductivity is 2 W/m.K, that is, Case 2 [9].

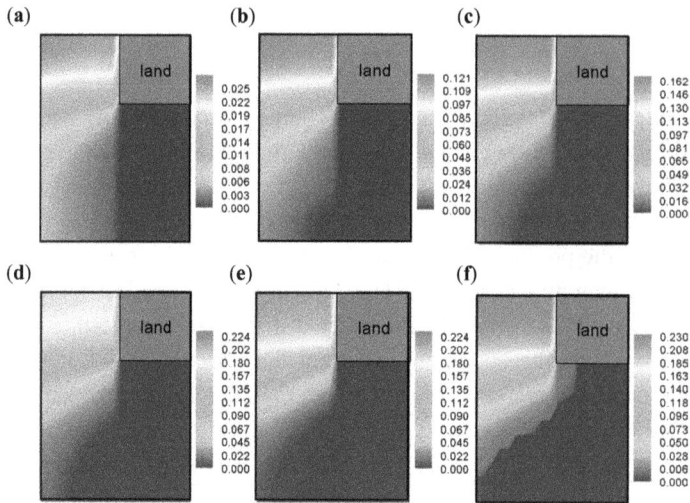

Figure 9-21. Liquid saturation contours of (a) 0.1 A/cm²; (b) 0.5 A/cm²; (c) 0.8 A/cm²; (d) 1.0 A/cm²; (e) 1.3 A/cm²; and (f) 1.5 A/cm². The bipolar plate (BP) outer surface is set at 353.15 K and GDL thermal conductivity is 3 W/m.K, that is, Case 1 [9].

For $Da_0 > 1$, two-phase flow is present at the entire GDL–catalyst layer interface. A high volume fraction of residual liquid water can be present in the two-phase region, leading to "flooding." It is difficult to quantify residual liquid as it depends on several factors, such as the pore-level structure and local wettability, and can vary greatly in the spatial dimension. Hickner *et al.* [10] used neutron radiography imaging to probe liquid water content across GDLs at vanishingly small current density and the results they obtained show a considerable spatial variation. In addition to a large variation, the neutron images obtained indicate that the residual liquid content is comparable with the liquid saturation level at 1 A/cm². However, the residual water can be evaporated in dry or partially humidified environment.

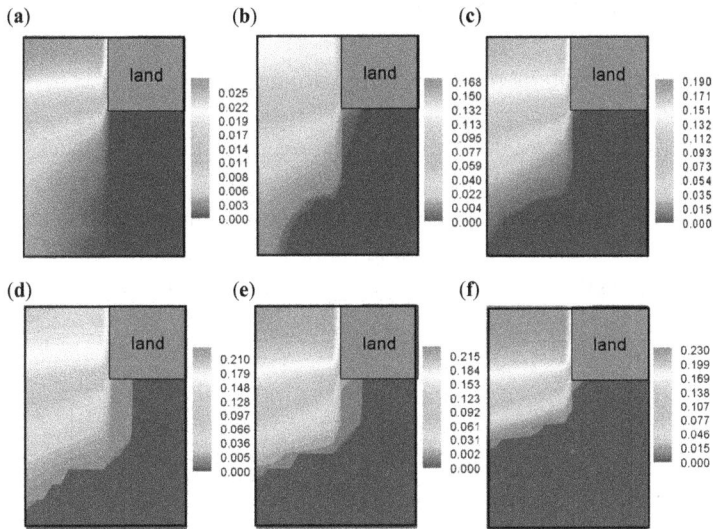

Figure 9-22. Liquid saturation contours of (a) 0.1 A/cm²; (b) 0.5 A/cm²; (c) 0.8 A/cm²; (d) 1.0 A/cm²; (e) 1.3 A/cm²; and (f) 1.5 A/cm². The BP outer surface is set at 353.15 K and GDL thermal conductivity is 2 W/m·K, that is, Case 2 [9].

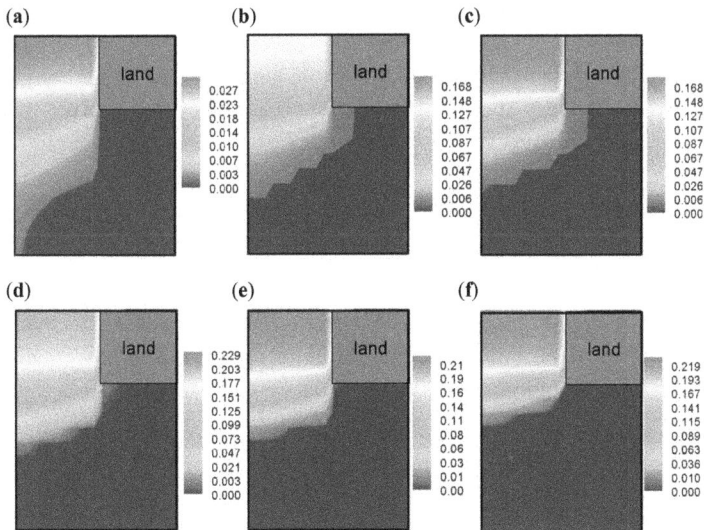

Figure 9-23. Liquid saturation contours of (a) 0.1 A/cm²; (b) 0.5 A/cm²; (c) 0.8 A/cm²; (d) 1.0 A/cm²; (e) 1.3 A/cm²; and (f) 1.5 A/cm². The BP outer surface is set at 363.15 K and GDL thermal conductivity is 2 W/m·K, that is, Case 3 [9].

For $Da_0 < 1$, the CL–GDL interface is free of liquid and hence residual liquid water at the centerline. Thus, creating such a region with $Da_0 < 1$ is beneficial to avoiding severe flooding. Due to the adjacency to a two-phase zone, the single-phase region is highly humidified, avoiding local dryness occurrence. Both extremes, either flooding or complete dryout, are detrimental to PEM fuel cell operation.

9.3 FREEZING/THAWING

9.3.1 Temperature Spatial and Temporal Variation

Three stages of cold start are defined and discussed in Chapter 7: product water freezes and ice forms in the second stage. The waste heat generation by the PEM fuel cell keeps increasing its temperature till the barrier of $0°C$, at which the heat melts ice. This phase change of ice melting is defined as the third stage of cold start. The time constant of the ice melting $\tau_{T,2}$ is evaluated by equating the amount of heat generation to the amount of latent heat absorption by the produced ice:

$$\int_{\tau_{T,1}}^{\tau_{T,1}+\tau_{T,2}} I(E'-V_{cell})dt = \rho_{ice}h_{sl}\delta_{CL}\varepsilon_{CL}s_{ice}^{max} = \rho_{ice}h_{sl}\delta_{CL}\varepsilon_{CL}\left(\frac{\tau_{T,1}}{\tau_{ice,2}}-k_\tau\right),$$

$$\text{where } \tau_{T,1} = \frac{m\overline{C}_p}{I(E'-V_{cell})A_m}(273.15-T_o) \text{ and } \tau_{ice,2} = \frac{2F\varepsilon_{CL}\rho_{ice}\delta_{CL}}{(1+2\alpha)M_w I}. \quad \text{(Eq. 9-50)}$$

The time constants $\tau_{T,1}$ and $\tau_{ice,2}$ are derived in Chapter 8 and Chapter 7, respectively, representing the timescales of fuel cell self-heating up to $0°C$ and ice occupying the void of the cathode catalyst layer. Given the typical values of the parameters, $s_{ice}^{max}=1$ yields $\tau_{T,2} \sim 2.0$ s at 0.1 A/cm^2, which is fairly short and indicates a relatively rapid process of ice melting. For comparison, the ratio of $\tau_{T,2}$ to $\tau_{ice,2}$ is given by

$$\frac{\tau_{T,2}}{\tau_{ice,2}} = \frac{h_{sl}(1+2\alpha)M_w}{2F(E'-V_{cell})}s_{ice}^{max}. \quad \text{(Eq. 9-51)}$$

Note that the electrode properties and current density vanish in the above equation. The ratio varies from 0 to 0.04, indicating that the third region is short. We can then obtain s_{ice} through the following equation:

$$s_{ice} = s_{ice}^{max} - \frac{\int_{\tau_{T,1}}^{t} I(E'-V_{cell})dt}{\rho_{ice}h_{sl}\delta_{CL}\varepsilon_{CL}} \quad \tau_{T,1} < t \le \tau_{T,1}+\tau_{T,2}. \quad \text{(Eq. 9-52)}$$

For a constant rate of heat generation, one can simplify the above expression as

$$s_{ice} = s_{ice}^{max}\frac{\tau_{T,2}+\tau_{T,1}-t}{\tau_{T,2}} = \left(\frac{\tau_{T,1}}{\tau_{ice,2}}-k_\tau\right)\frac{\tau_{T,2}+\tau_{T,1}-t}{\tau_{T,2}} \quad \tau_{T,1} < t \le \tau_{T,1}+\tau_{T,2}. \quad \text{(Eq. 9-53)}$$

Ice and liquid water coexist in the period of $\tau_{T,1} < t \le \tau_{T,1}+\tau_{T,2}$. For $t > \tau_{T,1}+\tau_{T,2}$, all the water exists in liquid form, which can be drained out of the cathode catalyst layer by taking advantage of the capillary pressure gradients that are present.

9.3.2 NONISOTHERMAL COLD START

Cold start is essentially a nonisothermal process: specifically, temperature increases from the starting subfreezing temperature; the process overcomes the energy barrier at the freezing point, and the fuel cell eventually reaches the operating temperature, for example, around 80°C. Thus, fundamental understanding of a practical cold-start process requires the knowledge of the heating mechanisms, thermal properties such asspecific heat capacity, heat transfer, and phase change. Three heating mechanisms are covered in Chapter 8 in detail. In addition, the latent heat is released or absorbed during freezing or melting. External heating, an option to improve the cold-start capability, can be added to a PEM fuel cell by taking advantage of its existing cooling units.

In the nonisothermal process of cold start, a PEM fuel cell increases its temperature from subzero to operating temperature. A cold-start model needs to account for ice formation mechanism and ice effects, species transport, energy conservation, and electrochemical reaction kinetics, most of which have been discussed in previous chapters. In what follows, a brief discussion on the modeling of ice formation and its effects as well as the associated heat source is presented.

The conservation equations of mass and species are expressed as [11–13]

$$\text{mass balance: } \frac{\partial\left[\varepsilon\left(1-s_{\text{ice}}\right)\rho\right]}{\partial t}+\nabla\left(\rho\vec{u}\right)=0, \tag{Eq. 9-54}$$

$$\text{species transport: } \frac{\partial\left[\varepsilon\left(1-s_{\text{ice}}\right)C_i\right]}{\partial t}+\nabla\left(\vec{u}C_i\right)=\nabla\left(D_i^{g,\text{eff}}\nabla C_i\right)+S_i, \tag{Eq. 9-55}$$

where s_{ice} is the ice volume fraction. The effective diffusion coefficient and catalyst surface area are modified as follows to accounting for the effect of ice presence:

$$D_i^{g,\text{eff}}=[\varepsilon(1-s_{\text{ice}})]^{\tau d}\,D_i^g \text{ and } a=(1-s_{\text{ice}})^{\tau a}\,a_0. \tag{Eq. 9-56}$$

Ice is produced under the condition that the water vapor pressure reaches its saturation value. Ice can be produced directly at the reaction surface by the ORR or deposit from the oversaturated vapor phase. Assuming that the latter is the only mechanism, the mass conservation of ice can be developed by considering the storage rate and phase change rate limited by mass transfer:

$$\frac{\partial\left[\varepsilon\rho_{\text{ice}}s_{\text{ice}}\right]}{\partial t}=S_{\text{ice}}M_w, \tag{Eq. 9-57}$$

where the volumetric desublimation rate S_{ice} is expressed as

$$S_{\text{ice}}=h_m\left(p_w-p_{w,\text{sat}}\right) \text{ where } h_m=\frac{k_c\varepsilon\left(1-s_{\text{ice}}\right)x_v}{2RT}\left[1+\frac{\left|p_w-p_{w,\text{sat}}\right|}{p_w-p_{w,\text{sat}}}\right]. \tag{Eq. 9-58}$$

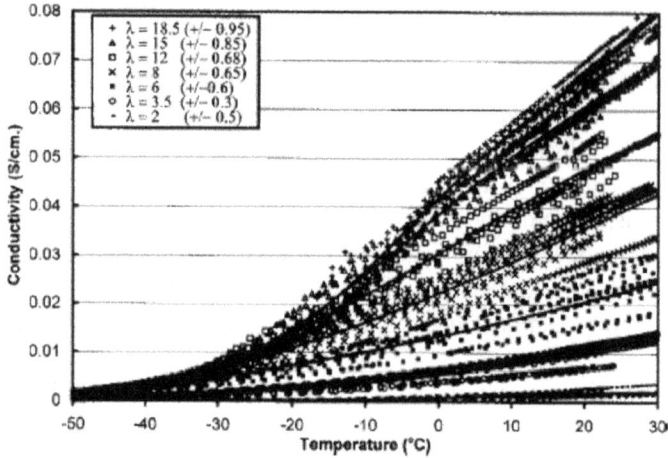

Figure 9-24. Ionic conductivity of Nafion 117 in a wide range of temperature [14] (Courtesy of the Electrochemical Society).

The energy conservation equation follows the general framework for heat transfer with the heating source expressed as follows:

$$S_T = j\left(\eta + T\frac{dU_0}{dT}\right) + \frac{i_e^2}{\kappa} + \frac{i_s^2}{\sigma} + h_{sg}M_wS_{ice} + S_{external}, \qquad \text{(Eq. 9-59)}$$

where $S_{external}$ accounts for any externally added heat sources on a volumetric basis. Surface heating can be added to the boundary condition of the energy conservation equation. The ionic conductivity of the Nafion membrane varies with the water content and temperature and may follow different trends between the normal (> 0°C) and subzero conditions, as discussed in Chapter 4. Figure 9-24 presents experimental data of the Nafion ionic conductivity for a wide range of temperature and water content.

9.3.3 FREEZING/THAWING AND DEGRADATION

Freezing/thawing operation is frequently encountered in fuel cell startups under subfreezing environment or conditions. Ice source comes from either water production at the reaction sites or residual water remained from the previous operation cycle. In startup from a freezing temperature, residual ice or newly produced ice melts when local temperature reaches 0°C. The water density changes during phase change (e.g., 0.9998 g/cm^3 at 0°C and 0.9168 g/cm^3 at −30°C, respectively) may cause physical damage to the electrodes during ice formation [15]. In addition, thermal cycling between a subzero temperature and the optimal operating temperature becomes part of the freezing/thawing process and can be detrimental. Figure 9-25 shows changes in the OCV (open circuit voltage) and current density after a thermal cycle. Before the thermal cycle, OCV was 1.0 V and the current density at 0.6 V is 880 mA/cm^2. It is seen that the thermal cycling has no obvious effect on the OCV, whereas the current density

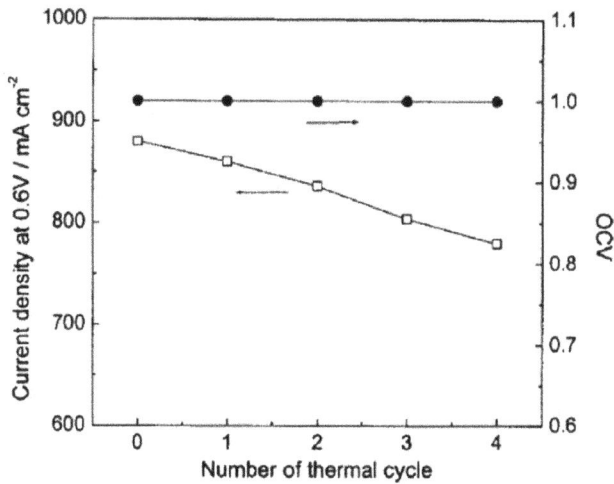

Figure 9-25. Effects of thermal cycling from 80°C to −10°C on the OCV and current density at 0.6 V [16] (Courtesy of the Electrochemical Society).

at 0.6 V decreases to 860, 836, 804, and 780 mA/cm^2, respectively, after the first four cycles. The degradation rate, defined as the ratio of change in current density to the current density measured before the cycle, is 2.8% per cycle on average. Gas purging is a popular method to alleviate degradation caused by the freezing/thawing cycling, as shown in Figure 9-26. The OCV is almost constant at 0.97 V in this case. The current density changes slightly from 880 to 872 mA/cm^2 after 15 cycles. The degradation rate with gas purging is only 0.06% per thermal cycle, indicating the significant benefit of gas purging in freeze/thaw cycling.

1. *Layer delamination*: ice formation and melting inside the catalyst layer causes physical expansion or shrinkage. Thus, repetitive freezing/thawing cycles may delaminate the catalyst layers from both the membrane and GDLs. This mechanical delamination increases the thermal and electrical contact resistances at these interfaces since the components are no longer in seamless contacts. Membrane cracks may form as a result of the delamination, leading to gas crossover and further direct mixing of hydrogen and oxygen. Mukundan *et al.* [17] showed that fuel cell performance decreased with each freezing/thawing cycle from −80°C to 80°C. After nine cycles the cell potential dropped from the initial 0.6 to 0.5 V (at a constant current density of 1.0 A/cm^2). Figure 9-27 presents the performance of a PEM fuel cell subject to freeze/thaw cycling and clearly shows a loss in performance. It is observed that the electrode peeled off from the electrolyte at several locations. Yan *et al.* [18] showed the morphology of fuel cell components of fresh samples and after normal operation and operation at subzero conditions. Massive damage was observed in the MEA and water freezing caused delamination of catalyst layers from both the membrane and GDL under subzero conditions.

The effects of thermal cycles on the ohmic and charge-transfer resistance can be evaluated through the AC impedance as shown in Figure 9-28, which presents the Nyquist plots for thermal cycling with and without gas purge. In the Nyquist plot, the left point of intersection with the *x*-axis indicates the ohmic resistance, and the diameter of the semicircle gives the

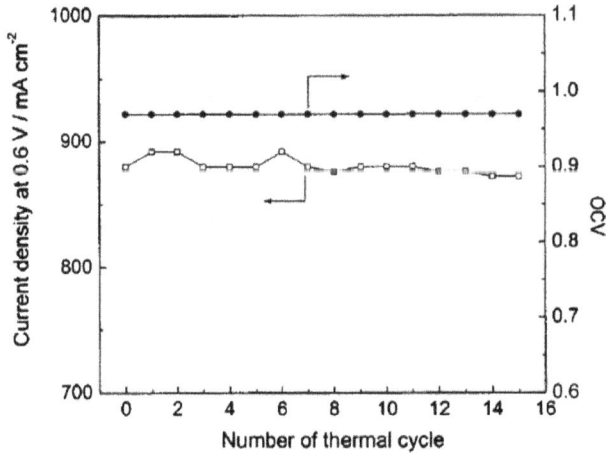

Figure 9-26. Effects of thermal cycling from 80°C to −10°C with gas purge on the OCV and current density at 0.6 V [16] (Courtesy of the Electrochemical Society).

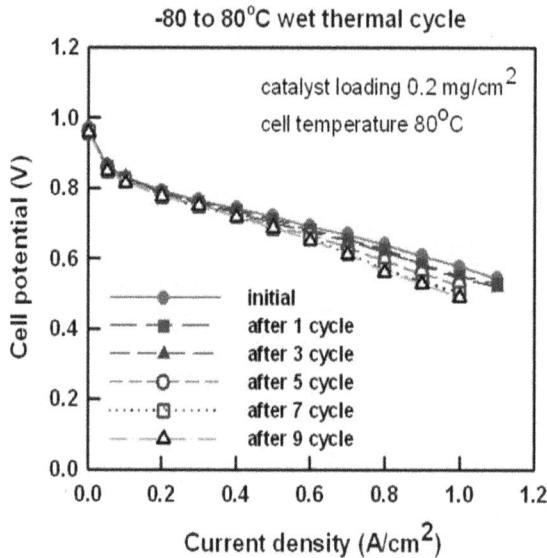

Figure 9-27. Effect of freeze/thaw cycling of a 5 cm^2 fuel cell from −80°C to +80°C on fuel cell performance [17] (Courtesy of the Electrochemical Society).

charge-transfer resistance. In the thermal cycle without gas purging, the semicircles move right and become larger in diameter, that is, both the ohmic and the charge-transfer resistance are increased. In four cycles, the ohmic and the charge-transfer resistances increase from 0.26 to 0.65 Vcm^2 and from 0.16 to 0.28 Vcm^2, respectively. In contrast, with gas-purging, the ohmic resistance and the charge-transfer resistance remain almost constant around 0.25 and 0.22–0.25 V cm^2, respectively. These results confirm that using gas purging alleviates degradation.

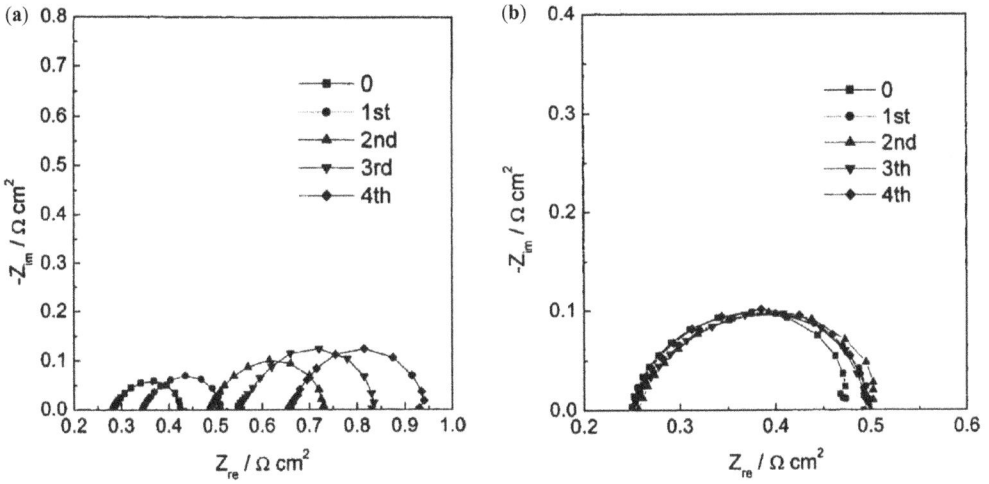

Figure 9-28. Effects of thermal cycling from 80°C to −10°C (a) without and (b) with gas purge on Nyquist plots measured at 0.7 V. The applied frequency varies from 1 mHz to 10 kHz with an excitation voltage of 5 mV [16] (Courtesy of the Electrochemical Society).

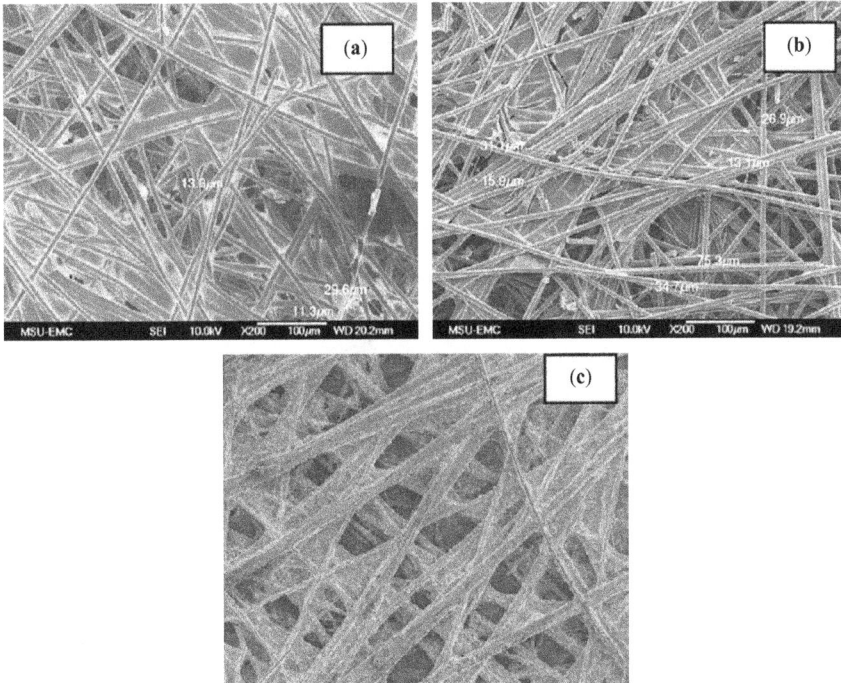

Figure 9-29. GDL images: (a) fresh carbon paper, (b) carbon paper after operation at room temperature, and (c) carbon paper after operation at −15°C [18] (Courtesy of Elsevier).

2. *Material damage*: deep cycles of freezing/thawing may cause damage on the GDL structure. In GDLs, the pores are not uniform, and water tends to be trapped in large pores as a result of capillary action. In freezing/thawing cycles, the bulk of liquid water constantly freezes under subzero temperature and melts during the thawing operation, which may damage the GDL structure due to phase change. Figure 9-29 compares the GDL images for three cases, clearly showing damage to the GDL fibers and binder in a failed MEA for the freezing operation (see Fig. 9-29(c)). The damage may appear in the GDL coating, Teflon, and the binder structure on the GDL surface. In Figure 9-30, damage is shown in the sample of the failed MEA as evidenced by the rough membrane surface after exposure to subzero freezing (Fig. 9-30(c)). The

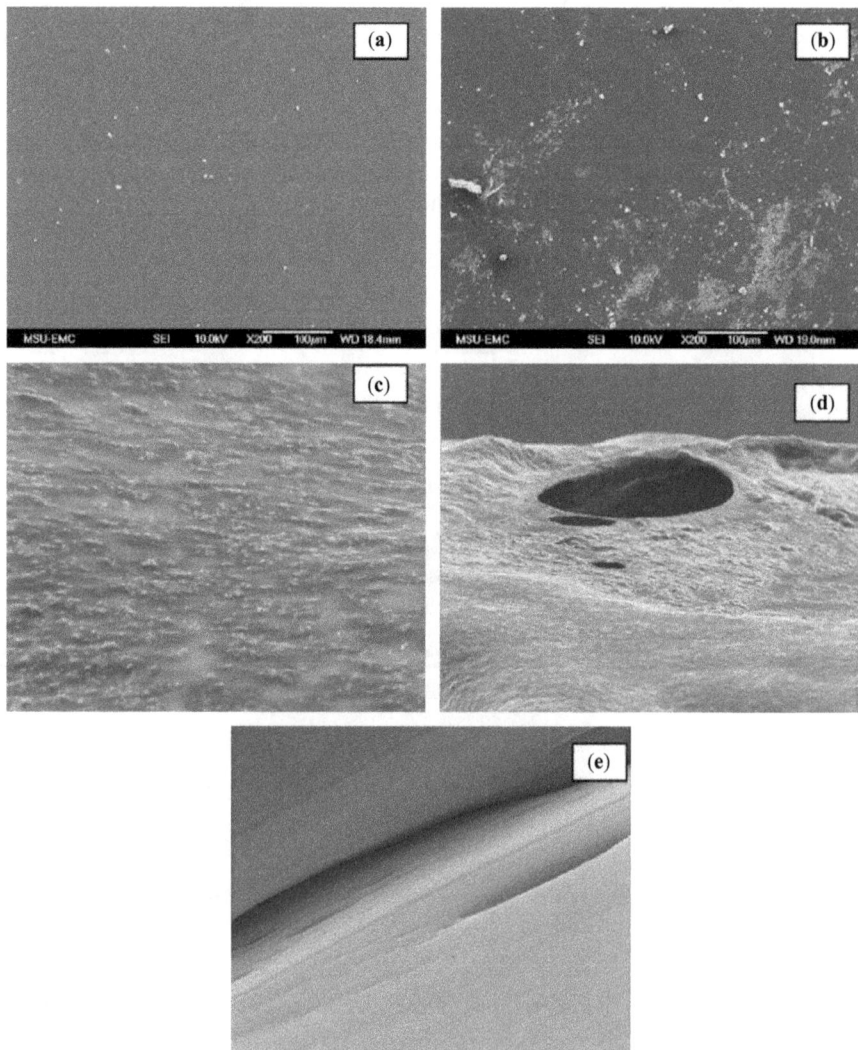

Figure 9-30. Membrane images: (a) fresh Nafion membrane, (b) membrane after operation at room temperature, (c) membrane after operation at −15°C, (d) membrane from cathode outlet regions after operation at −15°C, and (e) membrane from cathode outlet regions after operation at −15°C [18] (Courtesy of Elsevier).

membrane shows signs of pinhole formation (Fig. 9-30(d)) and microcracking (Fig. 9-30(e)) after subzero operation. Furthermore, freezing/thawing may cause change in the surface properties of materials. After subfreezing operation, it was found that the flow-field channel pattern is visible on the cathode GDL as dark lines (microscope image in Fig. 9-31(a)). A hydrophobicity test shows that the channel pattern is more hydrophilic (Fig. 9-31(b)) so that water droplets agglomerated.

3. *The ECSA loss*: the electrochemical surface area or ECSA is a key property of the electrodes, characterizing the surface area available for electrochemical reactions. The electrochemically active surface is usually referred to as the triple-phase boundaries (TPB) of the ionomer, carbon-supported catalyst, and gas phase. This reaction surface can be permanently damaged during freeze/thaw cycles. One possible explanation is that the associated solid–liquid phase change may alter the CL micro structure. Wang *et al.* [20] indicated that the phase change occurs relatively fast within a time constant about 1 s. Furthermore, most ice exists in the cathode CL, and ice may nearly fill the entire void space before operation failure. Another cause of degradation may be the oxygen starvation caused by ice formation, which may damage the carbon support. The cyclic voltammetry (CV) is frequently used to evaluate the ECSA of Pt/C catalysts by adsorption of atomic hydrogen in acidic media. Figure 9-32 shows the curves of the voltage and current at a voltage sweeping rate of 60 mV/s. Table 9-3 summarizes the ECSA losses, respectively, after 4 cold starts and 10–11 cold starts at $-10°C$. Figure 9-33 displays the cell performance (in terms of I–V curves) before and after 10–11 cold starts for different PTFE loadings, indicating performance decays. The figure also indicates that higher PTFE loading yields a smaller ECSA loss for the experimental fuel cell, which may be due to liquid water drainage when ice melts: the higher loading improves liquid water removal.

Table 9-3 lists the values of the ECSA loss [21].

9.4 SYSTEM-LEVEL ANALYSIS OF COUPLED THERMAL AND WATER MANAGEMENT

The open system of a PEM fuel cell is drawn in Figure 9-34, showing the inputs and outputs of interest. The analysis is established based on the balance of mass and energy. Anode reformate (hydrogen gas, water, and CO_2), air, and cooling water are independent streams and not mixed. The following assumptions are made for the analysis:

1. Product water generated in the cathode is assumed to be in the liquid state.
2. Equilibrium between vapor phase and liquid water holds true.
3. The fuel cell is treated as a lumped system without any spatial variation in temperature.

Table 9-3. ECSA loss after cold-start cycling

PEM fuel cell description	ECSA loss after 4 cold-start cycles	ECSA loss after 10 or 11 cold-start cycles
Gore MEA/Low PTFE paper GDL	46%	53% (11 starts)
Gore MEA/High PTFE paper GDL	21%	32% (10 starts)

Figure 9-31. Flow-field channel pattern on the cathode GDL after subzero operation (spray test with water). Hydrophilic channel pattern is visible [19] (Courtesy of Elsevier).

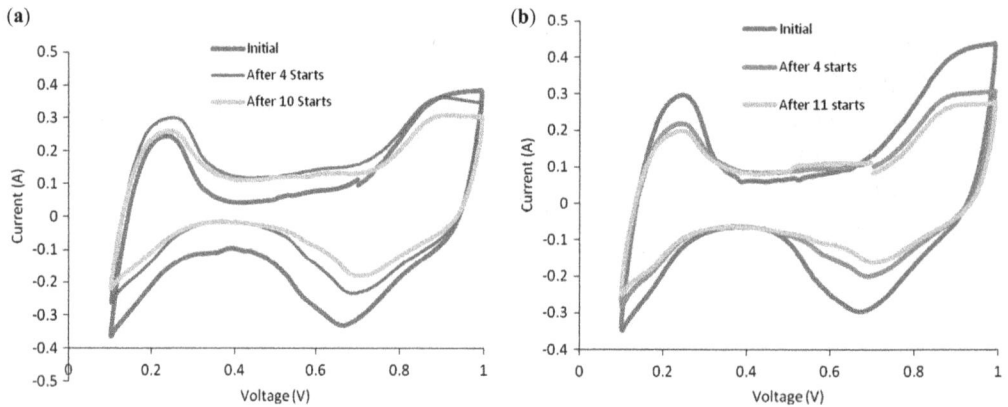

Figure 9-32. Cyclic voltammetry (CV) testing at the voltage sweeping rate of 60 mV/s for PEM fuel cells with: (a) high-PTFE carbon paper and (b) low-PTFE carbon paper. The corresponding ECSA losses are listed in Table 9-3 [21].

To monitor whether liquid water emerges from the streams, the relative humidity (RH) is used, as the ratio of the water content to the saturated vapor content. Assuming that no super-cooled vapor exists, the stream is free of liquid water when RH < 1; otherwise liquid water may emerge. The following discussion is based on Ref. [22].

9.4.1 FLOW RATES OF SPECIES AND TWO-PHASE FLOWS

The molar flow rate for hydrogen gas at the anode inlet and air at the cathode inlet can be obtained through the operating current I and stoichiometric flow ratios ξ:

$$N_{H_2,a,in} = \frac{I\xi_a}{2F},$$

(Eq. 9-60)

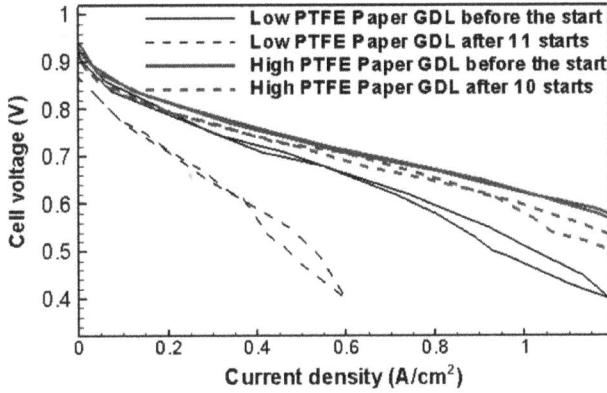

Figure 9-33. PEM fuel cell performance before and after cold starts for Gore MEAs at $-10°C$.

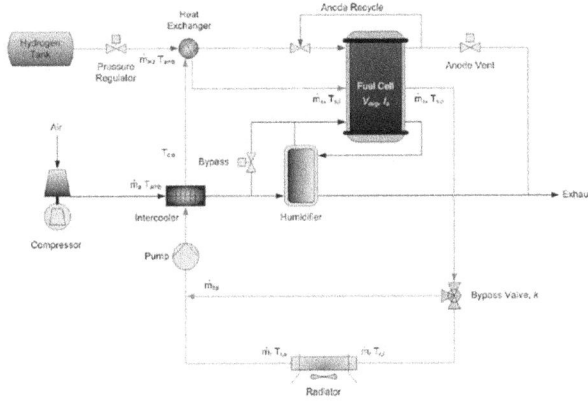

Figure 9-34. Open-system schematic of a PEM fuel cell [22].

$$N_{\text{air},c,\text{in}} = \frac{I\xi_c}{4F\beta_{O_2}}, \tag{Eq. 9-61}$$

where β_{O_2} is the molar fraction of oxygen in air stream. The maximum rate of water vapor fed into the anode inlet is evaluated as

$$N_{w,g,a,\text{in,max}} = \left(N_{H_2,a,\text{in}} + N_{CO_2,a,\text{in}}\right)\frac{P_{w,\text{sat},a,\text{in}}}{P_{g,a,\text{in}} - P_{w,\text{sat},a,\text{in}}}. \tag{Eq. 9-62}$$

The vapor saturation pressure is a function of temperature, as shown in Eq. 9-1 or Eq. 9-2. When the water feeding rate during operation is over this maximum rate, that is, $N_{w,a,\text{in}} \geq N_{w,g,a,\text{in,max}}$, liquid water emerges with the vapor and liquid phase flow rates calculated by

$$N_{w,g,a,\text{in}} = N_{w,g,a,\text{in,max}}, \qquad (\text{Eq. 9-63})$$

$$N_{w,l,a,\text{in}} = N_{w,a,\text{in}} - N_{w,g,a,\text{in}}. \qquad (\text{Eq. 9-64})$$

For the case of $N_{w,a,\text{in}} < N_{w,g,a,\text{in,max}}$, no liquid water will be present at the inlet flow:

$$N_{w,g,a,\text{in}} = \left(N_{\text{H}_2,a,\text{in}} + N_{\text{CO}_2,a,\text{in}}\right)\frac{P_{w,\text{sat},a,\text{in}}RH_{a,\text{in}}}{P_{g,a,\text{in}} - P_{w,\text{sat},a,\text{in}}RH_{a,\text{in}}}, \qquad (\text{Eq. 9-65})$$

$$N_{w,l,a,\text{in}} = 0. \qquad (\text{Eq. 9-66})$$

The maximum flow rate of water vapor at the anode outlet is calculated as follows:

$$N_{w,g,a,\text{out,max}} = \left(N_{\text{H}_2,a,\text{out}} + N_{\text{CO}_2,a,\text{out}}\right)\frac{P_{w,\text{sat},a,\text{out}}}{P_{g,a,\text{out}} - P_{w,\text{sat},a,\text{out}}}. \qquad (\text{Eq. 9-67})$$

Similarly, when $N_{w,a,\text{in}} - N_{\text{trans}} \geq N_{w,g,a,\text{out,max}}$, liquid emerges at the anode outlet:

$$N_{w,g,a,\text{out}} = N_{w,g,a,\text{out,max}}, \qquad (\text{Eq. 9-68})$$

$$N_{w,l,a,\text{out}} = N_{w,a,\text{in}} - N_{\text{trans}} - N_{w,g,a,\text{out}}. \qquad (\text{Eq. 9-69})$$

Otherwise, for $N_{w,a,\text{in}} - N_{\text{trans}} < N_{w,g,a,\text{out,max}}$, the anode outlet is free of liquid water:

$$N_{w,g,a,\text{out}} = N_{w,a,\text{in}} - N_{\text{trans}} \qquad (\text{Eq. 9-70})$$

$$N_{w,l,a,\text{out}} = 0. \qquad (\text{Eq. 9-71})$$

For the cathode inlet, the maximum rate of water vapor fed into the cathode inlet is

$$N_{w,g,c,\text{in,max}} = \left(N_{\text{O}_2,c,\text{in}} + N_{\text{N}_2,c,\text{in}}\right)\frac{P_{w,\text{sat},c,\text{in}}}{P_{g,c,\text{in}} - P_{w,\text{sat},c,\text{in}}}. \qquad (\text{Eq. 9-72})$$

For $N_{w,c,\text{in}} \geq N_{w,g,c,\text{in,max}}$, the gas phase becomes saturated with water vapor. The vapor and liquid water flow rates are derived as

$$N_{w,g,c,\text{in}} = N_{w,g,c,\text{in,max}} \qquad (\text{Eq. 9-73})$$

$$N_{w,l,c,\text{in}} = N_{w,c,\text{in}} - N_{w,g,c,\text{in}}. \qquad (\text{Eq. 9-74})$$

For $N_{w,c,\text{in}} < N_{w,g,c,\text{in,max}}$, the cathode inlet stream is relatively dry with no liquid water so the vapor and liquid water flow rates are given by

$$N_{w,g,c,\text{in}} = \left(N_{O_2,c,in} + N_{N_2,c,in}\right) \frac{p_{w,\text{sat},c,\text{in}} RH_{c,\text{in}}}{p_{g,c,\text{in}} - p_{w,\text{sat},c,\text{in}} RH_{c,\text{in}}}, \qquad \text{(Eq. 9-75)}$$

$$N_{w,l,c,\text{in}} = 0. \qquad \text{(Eq. 9-76)}$$

For the cathode outlet, the maximum rate of water vapor at the cathode outlet is evaluated as

$$N_{w,g,c,\text{out,max}} = \left(N_{O_2,c,\text{out}} + N_{N_2,c,\text{out}}\right) = \frac{p_{w,\text{sat},c,\text{out}}}{p_{g,c,\text{out}} - p_{w,\text{sat},c,\text{out}}}. \qquad \text{(Eq. 9-77)}$$

Assuming that product water is in the liquid phase, then the water production rate is equal to hydrogen gas consumption rate:

$$N_{w,l,\text{prod}} = N_{H_2,\text{cons}} = N_{H_2,a,\text{in}} - N_{H_2,a,\text{out}}, \qquad \text{(Eq. 9-78)}$$

when $\left(N_{w,c,\text{in}} + N_{w,l,\text{prod}} + N_{\text{trans}}\right) \geq N_{w,g,c,\text{out,max}}$, water exists in both vapor and liquid phases so that

$$N_{w,g,c,\text{out}} = N_{w,g,c,\text{out,max}} \qquad \text{(Eq. 9-79)}$$

$$N_{w,l,c,\text{out}} = N_{w,c,\text{in}} + N_{w,l,\text{prod}} + N_{\text{trans}} - N_{w,g,c,\text{out}}. \qquad \text{(Eq. 9-80)}$$

Otherwise, for $\left(N_{w,c,\text{in}} + N_{w,l,\text{prod}} + N_{\text{trans}}\right) < N_{w,g,c,\text{out,max}}$, no liquid emerges, and all the cathode outlet water is in the vapor phase such that

$$N_{w,g,c,\text{out}} = N_{w,c,\text{in}} + N_{w,l,\text{prod}} + N_{\text{trans}}, \qquad \text{(Eq. 9-81)}$$

$$N_{w,l,c,\text{out}} = 0. \qquad \text{(Eq. 9-82)}$$

9.4.2 ENERGY BALANCE

The waste heat generation by a PEM fuel cell is given by (as shown in Fig. 8-1)

$$q_{\text{theo}} = n(E' - E_{\text{cell}})I. \qquad \text{(Eq. 9-83)}$$

The specific heat capacity is assumed to be constant in the given temperature range. The sensible heat of the anode stream can be derived by accounting for all the possible species in the anode (i.e., water vapor, liquid water, and carbon dioxide from reformate):

$$q_{\text{sens},a} = N_{H_2,a,\text{out}} c_{p,H_2,g} \left(T_{a,\text{out}} - T_0 \right)$$
$$+ N_{w,g,a,\text{out}} c_{p,H_2O,g} \left(T_{a,\text{out}} - T_0 \right)$$
$$+ N_{CO_2,a,\text{out}} c_{p,CO_2,g} \left(T_{a,\text{out}} - T_0 \right)$$
$$+ N_{w,l,a,\text{out}} c_{p,w,l} \left(T_{a,\text{out}} - T_0 \right)$$
$$- N_{w,l,a,\text{in}} c_{p,w,l} \left(T_{a,\text{in}} - T_0 \right) \qquad \text{(Eq. 9-84)}$$
$$- N_{w,g,a,\text{in}} c_{p,H_2O,g} \left(T_{a,\text{in}} - T_0 \right)$$
$$- N_{H_2,a,\text{in}} c_{p,H_2,g} \left(T_{a,\text{in}} - T_0 \right)$$
$$- N_{CO_2,a,\text{in}} c_{p,CO_2,g} \left(T_{a,\text{in}} - T_0 \right)$$

The latent heat release/absorption in the anode is proportional to the phase-change rate:

$$q_{\text{latent},a} = \left(N_{w,g,a,\text{out}} - N_{w,g,a,\text{in}} + N_{\text{trans}} \right) h_{lg} \qquad \text{(Eq. 9-85)}$$

Similarly, the sensible heat in the cathode stream (oxygen, nitrogen, water vapor, and water liquid) is written as

$$q_{\text{sens},c} = N_{O_2,c,\text{out}} c_{p,O_2,g} \left(T_{c,\text{out}} - T_0 \right)$$
$$+ N_{w,g,c,\text{out}} c_{p,H_2O,g} \left(T_{c,\text{out}} - T_0 \right)$$
$$+ N_{N_2,c,\text{out}} c_{p,N_2,g} \left(T_{c,\text{out}} - T_0 \right)$$
$$+ N_{w,l,c,\text{out}} c_{p,w,l} \left(T_{c,\text{out}} - T_0 \right)$$
$$- N_{w,l,c,\text{in}} c_{p,w,l} \left(T_{c,\text{in}} - T_0 \right) \qquad \text{(Eq. 9-86)}$$
$$- N_{w,g,c,\text{in}} c_{p,H_2O,g} \left(T_{c,\text{in}} - T_0 \right)$$
$$- N_{O_2,c,\text{in}} c_{p,O_2,g} \left(T_{c,\text{in}} - T_0 \right)$$
$$- N_{N_2,c,\text{in}} c_{p,N_2,g} \left(T_{c,\text{in}} - T_0 \right).$$

The latent heat in the cathode is evaluated in a different way due to the ORR's water generation. For

$$N_{w,l,c,\text{in}} \geq \left(N_{w,g,c,\text{out}} - N_{\text{trans}} - N_{w,g,c,\text{in}} \right),$$

$$q_{\text{latent},c} = \left(N_{w,g,c,\text{out}} - N_{\text{trans}} - N_{w,g,c,\text{in}} \right) h_{lg}. \qquad \text{(Eq. 9-87)}$$

Otherwise, all the liquid water into the cathode inlet is evaporated, along with a portion of the product water, yielding

$$q_{\text{latent},c} = N_{w,l,c,\text{in}} H_{\text{vaporization},c1} + (N_{w,g,c,\text{out}} - N_{\text{trans}} - N_{w,g,c,\text{in}} - N_{w,l,c,\text{in}}) h_{lg}. \qquad \text{(Eq. 9-88)}$$

The sensible heat in water coolant stream is evaluated using the following equation:

$$q_{\text{sens},w} = N_{w,\text{out}} c_{p,w,l} \left(T_{w.\text{out}} - T_0 \right) - N_{w,\text{in}} c_{p,w,l} \left(T_{w.\text{in}} - T_0 \right). \qquad \text{(Eq. 9-89)}$$

The heat loss from the fuel cell to the ambient is then calculated through the energy balance of the system:

$$q_{\text{loss}} = q_{\text{theo}} - q_{\text{elec}} - q_{\text{sens}} - q_{\text{latent}}, \qquad \cdot \text{(Eq. 9-90)}$$

where

$$q_{\text{sens}} = q_{\text{sens},a} + q_{\text{sens},c} + q_{\text{sens},w} \qquad \text{(Eq. 9-91)}$$
$$q_{\text{latent}} = q_{\text{latent},a} + q_{\text{latent},c}.$$

During transient, the storage rate adds to the energy balance equation, yielding

$$m_{\text{stack}} C_{p,\text{stack}} \frac{dT_{\text{stack}}}{dt} = q_{\text{theo}} - q_{\text{elec}} - q_{\text{sens}} - q_{\text{latent}} - q_{\text{loss}},$$

or

$$\frac{dT_{\text{stack}}}{dt} = \frac{q_{\text{theo}} - q_{\text{elec}} - q_{\text{sens}} - q_{\text{latent}} - q_{\text{loss}}}{m_{\text{stack}} C_{p,\text{stack}}}, \qquad \text{(Eq. 9-92)}$$

where m_{stack} is the total mass of the fuel cell stack, $C_{p,\text{stack}}$ is the average specific heat of the stack, and dT_{stack}/dt is the temperature change with respect to time.

9.5   CHAPTER SUMMARY

This chapter discusses the phase-change phenomena in PEM fuel cells, including the liquid–vapor (evaporation) and liquid–solid (ice formation or icing) phase changes. The key points in this chapter are summarized as follows:

- Heat transfer and water transport are highly coupled in PEM fuel cell operation. Water vapor-phase diffusion is significant and is able to transport ~40% or higher of water produced at 80°C, and heat pipe effect and water vapor-phase diffusion promote each other; the vapor-phase diffusion driven by temperature gradients occurs from the catalyst layer (CL) to the gas flow channel (GFC), and from the region under channel to that under land, and vapor-phase diffusion increases liquid water saturation level under land.
- At steady state, a dimensionless group or number Da is defined and employed to characterize the single- and two-phase regions in PEM fuel cells. Under full humidification (i.e., $Da_0 < 1$), it is possible that a portion of the cathode gas-diffusion-layer (GDL)/CL interface is free of liquid; occurrence of the liquid-free area along the cathode CL–GDL

interface depends on water vapor diffusivity and other parameters. Operating temperature can significantly alter the presence and sizes of the liquid-water regions in GDLs, and GDL material and geometric parameters, GDL/GFC interfacial properties, and GFC geometry and flow conditions can be optimized to create the desired single-phase flow regions in GDLs.

- During transient, GDL de-wetting (which is due to dry reactant flows and caused by the through-plane and in-plane drying) and its coupling with electrochemical reaction is examined and expressions for time constants of the through-plane and in-plane drying are presented; by proper scaling, the in-plane and through-plane transport processes are found to follow the same mathematical equation, and different stages of the GDL drying exhibit different slopes of voltage drop, as a result of the ohmic voltage loss.

- In the cold start of a PEM fuel cell, three stages exist with the third stage being signified by ice melting. The time constant for the melting of product ice during cold start is relatively small, on the order of a second; modeling of non-isothermal cold start by employing an isothermal cold-start model plus an approach for describing the heating mechanisms for a PEM fuel cell is demonstrated; and freezing/thawing operation during cold start may cause electrode degradation, including the physical structure damage of GDLs and CLs, and the loss of electrochemical surface areas.

REFERENCES

1. Springer, T.E., Zawodinski, T.A., and Gottesfeld, S. Polymer electrolyte fuel cell model. *J. Electrochem. Soc. 138* (1991): 2334. doi: http://dx.doi.org/10.1149/1.2085971

2. Rasmussen, L.A. On the approximation of saturation vapor pressure. *J. Appl. Meteorol. 17* (1978): 1564. doi: http://dx.doi.org/10.1175/1520-0450(1978)017<1564:OTAOSV>2.0.CO;2

3. Nam, J.H., and Kaviany, M. Effective diffusivity and water-saturation distribution in single- and two-layer PEMFC diffusion medium. *Int. J. Heat Mass Transf 46* (2003): 4595–4611.

4. Schwartz, S.E. Historical Perspective on Heterogeneous Gas-Particle Interaction, Mass Accommodation Workshop, Aerodyne Research Incorporated, (2002).

5. Kim, H., Rao, P.S.C., and Annable, M.D. Determination of effective air–water interfacial area in partially saturated porous media using surfactant adsorption. *Water Resour. Res. 33* (1997): 2705–11. doi: http://dx.doi.org/10.1029/97WR02227

6. Wang, Y., and Wang, C.Y. A non-isothermal two-phase model for polymer electrolyte fuel cells. *J. Electrochem. Soc. 153* (2006): A1193–1200. doi: http://dx.doi.org/10.1149/1.2193403

7. Wang, Y., and Wang, C.Y. Two-phase transients of polymer electrolyte fuel cells. *J. Electrochem. Soc. 154* (2007): B636–643. doi: http://dx.doi.org/10.1149/1.2734076

8. Basu, S., Wang, C.Y., and Chen, K.S. Phase change in a polymer electrolyte fuel cell. *J. Electrochem. Soc. 156* (2009): B748. doi: http://dx.doi.org/10.1149/1.3115470

9. Wang, Y., and Chen, K.S. Elucidating two-phase transport in a polymer electrolyte fuel cell: Part 1. Characterizing flow regimes with a dimensionless group. *Chem. Eng. Sci. 66* (2011): 3557–67. doi: http://dx.doi.org/10.1016/j.ces.2011.04.016

10. Hickner, M.A., Siegel, N.P., Chen, K.S., Hussey, D.S., Jacobson, D.L., and Arif, M. Understanding liquid–water distribution and removal phenomena in an operating PEMFC via neutron radiography. *J. Electrochem. Soc. 155* (2008): B294. doi: http://dx.doi.org/10.1149/1.2825298

11. Jiang, F., Fang, W., and Wang, C.Y. Non-isothermal cold start of polymer electrolyte fuel cells. *Electrochimica Acta 53* (2007): 610. doi: http://dx.doi.org/10.1016/j.electacta.2007.07.032

12. Jiao, K., and Li, X. Three-dimensional multiphase modeling of cold start processes in polymer electrolyte membrane fuel cells. *Electrochimica Acta 54* (2009): 6876. doi: http://dx.doi.org/10.1016/j.electacta.2009.06.072

13. Meng, H., and Ruan, B. Numerical studies of cold-start phenomena in PEM fuel cells: a review. *Int. J. Energy Res. 35* (2011): 2. doi: http://dx.doi.org/10.1002/er.1730

14. Thompson, E.L., Capehart, T.W., Fuller, T.J., and Jorne, J. Investigation of low-temperature proton transport in Nafion using direct current conductivity and differential scanning calorimetry. *J. Electrochem. Soc. 153* (2006): A2351. doi: http://dx.doi.org/10.1149/1.2359699

15. Schmittinger, W., and Vahidi, A. A review of the main parameters influencing long-term performance and durability of PEM fuel cells. *J. Power Sources 180* (2008): 1–14. doi: http://dx.doi.org/10.1016/j.jpowsour.2008.01.070

16. Cho, E.A., Ko, J.J., Ha, H.Y., Hong, S.A., Lee, K.Y., Lim, T.W., and Oh, I.H. Effects of water removal on the performance degradation of PEMFCs repetitively brought to <0°C. *J. Electrochem. Soc. 151* (2004): A661–65. doi: http://dx.doi.org/10.1149/1.1683580

17. Mukundan, R., Kim, Y.S., Garzon, F.H., and Pivovar, B. Freeze/thaw effects in PEM fuel cells. *ECS Trans. 1* (2006): 403. doi: http://dx.doi.org/10.1149/1.2214572

18. Yan, Q., Toghiani, H., Lee, Y.W., Liang, K., and Causey, H. Effect of sub-freezing temperatures on a PEM fuel cell performance, startup and fuel cell components. *J. Power Sources 160* (2006): 1242–50. doi: http://dx.doi.org/10.1016/j.jpowsour.2006.02.075

19. Oszcipok, M., Riemann, D., Kronenwett, U., Kreideweis, M., and Zedda, M. Statistic analysis of operational influences on the cold start behaviour of PEM fuel cells. *J. Power Sources 145* (2005): 407–15. doi: http://dx.doi.org/10.1016/j.jpowsour.2005.02.058

20. Wang, Y. Analysis of the key parameters in the cold start of polymer electrolyte fuel cells. *J. Electrochem. Soc. 154* (2007): B1041. doi: http://dx.doi.org/10.1149/1.2767849

21. Mishler, J., Wang, Y., Mukherjee, P.P., Mukundan, R., and Borup, R.L. Subfreezing operation of polymer electrolyte fuel cells: ice formation and cell performance loss. *Electrochimica Acta 65* (2012): 127–33. doi: http://dx.doi.org/10.1016/j.electacta.2012.01.020

22. Yu, X., Zhou, B., and Sobiesiak, A. Water and thermal management for Ballard PEM fuel cell stack. *J. Power Sources 147* (2005): 184–95. http://dx.doi.org/10.1016/j.jpowsour.2005.01.030

THERMODYNAMIC PROPERTIES OF AIR, HYDROGEN GAS, AND WATER VAPOR*

Table A.2-a1. Ideal-gas properties of air (k: 1.406 – 1.392)

T (K)	$H^\dagger$ (kJ/kg)	$U^\dagger$ (kJ/kg)	S^{**} (kJ/(kg K))	C_p^* (kJ/(kg K))	C_v^* (kJ/(kg K))
230	232.0409	165.5811	1.776346	0.9939	0.7069
240	242.1921	172.9011	1.789524	0.9952	0.7082
243	245.2402	175.0744	1.793309	0.9956	0.7086
250	251.7154	180.1442	1.802048	0.9966	0.7096
260	262.4987	187.4031	1.814533	0.9980	0.7110
270	272.6462	194.6583	1.826454	0.9994	0.7124
273	275.6964	196.8337	1.830082	0.9998	0.7128
280	282.8037	201.9056	1.838426	1.0008	0.7138
290	292.9554	209.1629	1.849896	1.0023	0.7153
300	303.1092	216.4244	1.861485	1.0038	0.7168
310	313.2751	223.688	1.872625	1.0053	0.7183
320	323.4352	230.9558	1.883931	1.0068	0.7198
330	333.6074	238.2278	1.894828	1.0083	0.7213
340	343.7838	245.5098	1.905932	1.0099	0.7229
350	353.9644	252.7902	1.916853	1.0115	0.7245
353	357.023	254.9819	1.920122	1.0120	0.7250
360	364.1492	260.0906	1.927815	1.0131	0.7261
370	382.2482	267.3873	1.938616	1.0147	0.7277
380	384.5593	274.6982	1.949677	1.0163	0.7293

(Continued)

Table A.2-a1. Ideal-gas properties of air (k: $1.406 - 1.392$) (*Continued*)

T (K)	$H^{\dagger}$ (kJ/kg)	$U^{\dagger}$ (kJ/kg)	S^{**} (kJ/(kg K))	C_p^{*} (kJ/(kg K))	C_v^{*} (kJ/(kg K))
390	394.7746	282.0212	1.960601	1.0180	0.7310
393	397.8395	284.2192	1.963892	1.0185	0.7315
400	404.9983	289.3584	1.971593	1.0197	0.7327

*Based on a third-degree polynomial equation in Y. A. Cengel, M. A. Boles. *Thermodynamics, An Engineering Approach*, 6th Ed. McGraw-Hill (2007).

†Based on data for O_2 (21%) + N_2 (79%) from NIST Chemistry WebBook.

**Based on $S = \sum_k n_k \left(c_{vk} \log T + R \log \dfrac{V}{n_k} \right)$ using data from NIST Chemistry WebBook.

Table A.2-a2. Ideal-gas properties of hydrogen gas (k: $1.404 - 1.403$)

T (K)	H (kJ/kg)	U (kJ/kg)	S (kJ/(kg K))	$C_p^{\dagger}$ (kJ/(kg K))	$C_v^{\dagger}$ (kJ/(kg K))
260	3655.9716	2583.9832	62.8192	14.1329	10.0089
270	3798.3412	2684.6836	63.3564	14.1883	10.0643
273†	3842.3887	2715.7560	63.5690	14.2033	10.0793
280	3941.2068	2786.3762	63.8753	14.2357	10.1117
290	4084.0725	2888.0687	64.3764	14.2763	10.1523
300	4227.4342	2989.7613	64.8620	14.3111	10.1871
310	4370.7959	3092.1979	65.3251	14.3410	10.2170
320	4514.1576	3194.6346	65.7881	14.3666	10.2426
330	4658.0154	3297.0713	66.2239	14.3886	10.2646
340	4801.8731	3399.5079	66.6597	14.4074	10.2834
350	4946.2270	3502.4406	67.0717	14.4236	10.2996
353†	4989.4351	3532.7869	67.2526	14.4280	10.3040
360	5090.5808	3605.3733	67.4837	14.4375	10.3135
370	5234.6866	3708.5541	67.8738	14.4494	10.3254
380	5378.7924	3811.7348	68.2640	14.4596	10.3356
390	5523.3942	3914.9156	68.6345	14.4683	10.3443
393†	5567.4961	3945.7895	68.8038	14.4707	10.3467
400	5667.9961	4018.0963	69.0051	14.4759	10.3519

Table A.2-a3. Ideal-gas properties of oxygen (k: 1.409 – 1.375)

T (K)	H (kJ/kg)	U (kJ/kg)	S (kJ/(kg K))	$C_p^{\dagger}$ (kJ/(kg K))	$C_v^{\dagger}$ (kJ/(kg K))
230	209.1953	149.4431	6.1709	0.9108	0.6510
240	218.2582	155.9121	6.2095	0.9115	0.6517
243[†]	220.8715	157.8291	6.2240	0.9117	0.6519
250	227.3523	162.4123	6.2466	0.9123	0.6525
260	236.4464	168.9126	6.2823	0.9132	0.6534
270	245.5717	175.4128	6.3167	0.9143	0.6545
273[†]	248.2640	177.6587	6.3303	0.9147	0.6549
280	254.6971	181.9443	6.3500	0.9156	0.6558
290	263.8536	188.5071	6.3821	0.9170	0.6572
300	273.0102	195.0698	6.4131	0.9185	0.6587
310	282.1981	201.6638	6.4433	0.9202	0.6604
320	291.4172	208.2578	6.4725	0.9220	0.6622
330	300.6363	214.9143	6.5009	0.9240	0.6642
340	309.8866	221.5708	6.5285	0.9261	0.6663
350	319.1682	228.2273	6.5554	0.9283	0.6685
353[†]	321.9528	230.2516	6.5670	0.9289	0.6691
360	328.4811	234.9463	6.5816	0.9306	0.6708
370	337.7939	241.6653	6.6072	0.9330	0.6732
380	347.1693	248.4156	6.6322	0.9355	0.6757
390	356.5446	255.1971	6.6566	0.9381	0.6783
393[†]	359.3043	257.1991	6.6672	0.9389	0.6791
400	365.9825	262.0098	6.6804	0.9408	0.6810

*Source: Y. A. Cengel, M. A. Boles. *Thermodynamics, An Engineering Approach*, 6th Ed. McGraw-Hill (2007).
[†] NIST Chemistry WebBook.

APPENDIX II.B

CALCULATION OF THE ENTHALPY, ENTROPY, AND GIBBS FREE ENERGY FOR A SUBSTANCE AND THE OVERALL PEM FUEL CELL REACTION

This appendix, following Larminie and Dicks (2003), provides an example of calculating Δh, Δg, and Δs, as discussed in Chapter 2, for the overall reaction of PEM fuel cells:

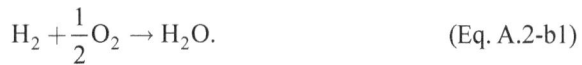

$$H_2 + \frac{1}{2}O_2 \rightarrow H_2O. \qquad \text{(Eq. A.2-b1)}$$

The Gibbs free energy of a system is defined in terms of the entropy and the enthalpy:

$$G = H - TS \qquad \text{(Eq. A.2-b2)}$$

Using the molar-based quantities, the above equation changes to

$$g = h - Ts. \qquad \text{(Eq. A.2-b3)}$$

We focus on the change in the free energy, which determines the thermodynamics reversible potential for PEM fuel cells. Under constant temperature, we obtain

$$\Delta g = \Delta h - T\Delta s. \qquad \text{(Eq. A.2-b4)}$$

The value of Δh is the difference between the products and the reactants for the overall fuel cell reaction:

$$\Delta h = h_{H_2O} - h_{H_2} - \frac{1}{2}h_{O_2}. \qquad \text{(Eq. A.2-b5)}$$

Similarly, Δs is the difference between the products and reactants:

$$\Delta s = s_{H_2O} - s_{H_2} - \frac{1}{2}s_{O_2}.$$ (Eq. A.2-b6)

h and s vary with temperature, determined through the specific heat at constant pressure c_p:

$$h(T) = h(T_o) + \int_{T_o}^{T} c_p dT$$ (Eq. A.2-b7)

$$s(T) = s(T_o) + \int_{T_o}^{T} \frac{1}{T} c_p dT$$ (Eq. A.2-b8)

The value for the molar entropy and enthalpy of formation at $T_o = 298.15$ K are available in thermodynamics tables (e.g., Keenan and Kaye, 1948), given in Table A.2-b1 at standard conditions.

Empirical correlations for c_p are available in many thermodynamics texts (e.g., VanWylen and Sonntag, 1986), and the correlations given below are accurate within 0.6% over the range of 300–3500 K.

$$c_{p,H_2O(v)} = 143.05 - 58.040T^{0.25} + 8.2751T^{0.5} - 0.036989T$$

$$c_{p,H_2} = 56.505 - 22222.6T^{-0.75} + 1166500T^{-1} - 560700T^{-1.5}$$ (Eq. A.2-b9)

$$c_{p,O_2} = 37.432 + 2.0102 \times 10^{-5} T^{1.5} - 178570T^{-1.5} + 2368800T^{-2},$$

where c_p is in (J/mol K). The above correlations, along with Eqs. A.2-b7 and A.2-b8, are substituted into Eqs. A.2-b4, A.2-b5, and A.2-b6 for calculating Δh, Δs, and Δg. The results are listed in Table A.2-b2.

For the case of liquid water, Table A.2-b1 provides standard values for h and s at 25°C. At 80°C, Eqs. A.2-b7 and A.2-b8 can be used to find h and s, where c_p of liquid water can be assumed constant for the given temperature range.

Table A.2-b1. h and s at 298.15 K for PEM fuel cells (Courtesy of John Wiley & Sons)

	h (J mol^{-1})	s (J mol^{-1}K^{-1})
H$_2$O (liquid)	−285838	70.05
H$_2$O (steam)	−241827	188.83
H$_2$	0	130.59
O$_2$	0	205.14

Table A.2-b2. Δh, Δs, and Δg under various temperatures for $H_2 + \frac{1}{2}O_2 \rightarrow H_2O$ (Courtesy of John Wiley & Sons)

Temperature (°C)	Δh (kJ mol^{-1})	Δs (kJ K^{-1} mol^{-1})	Δg (kJ mol^{-1})
100	−242.6	−0.0466	−225.2
300	−244.5	−0.0507	−215.4
500	−246.2	−0.0533	−205.0
700	−247.6	−0.0549	−194.2
900	−248.8	−0.0561	−183.1

*SOURCES

1. Keenan, J.H., and Kaye, J. *Gas Table*, Wiley, 1948.
2. Larminie, J., and Dicks, A. *Fuel Cell Systems Explained*, 2nd edn. Wiley, 2003.
3. Van Wylen, G.J., and Sonntag, R.E. *Fundamental of Classical Thermodynamics*, 3rd edn. Wiley, 1986.

Mass, Momentum, and Energy Conservation Equations in the Cartesian, Cylindrical, and Spherical Coordinates

Table A.3-a1. Mass conservation equations in the cartesian, cylindrical, and spherical coordinates

Cartesian coordinates (x, y, z):

$$\frac{\partial \rho}{\partial t} + \frac{\partial}{\partial x}(\rho u_x) + \frac{\partial}{\partial y}(\rho u_y) + \frac{\partial}{\partial z}(\rho u_z) = 0$$

Cylindrical coordinates (r, θ, z):

$$\frac{\partial \rho}{\partial t} + \frac{1}{r}\frac{\partial}{\partial r}(\rho r u_r) + \frac{1}{r}\frac{\partial}{\partial \theta}(\rho u_\theta) + \frac{\partial}{\partial z}(\rho u_z) = 0$$

Spherical coordinates (r, θ, ϕ):

$$\frac{\partial \rho}{\partial t} + \frac{1}{r^2}\frac{\partial}{\partial r}(\rho r^2 u_r) + \frac{1}{r \sin \theta}\frac{\partial}{\partial \theta}(\rho u_\theta \sin \theta) + \frac{1}{r \sin \theta}\frac{\partial}{\partial \phi}(\rho u_\phi) = 0$$

Table A.3-a2. Momentum equations in the Cartesian coordinate

Cartesian coordinates (x, y, z) x-direction:

$$\rho\left(\frac{\partial u_x}{\partial t} + u_x\frac{\partial u_x}{\partial x} + u_y\frac{\partial u_x}{\partial y} + u_z\frac{\partial u_x}{\partial z}\right) = -\frac{\partial p}{\partial x} + \mu\left(\frac{\partial^2 u_x}{\partial x^2} + \frac{\partial^2 u_x}{\partial y^2} + \frac{\partial^2 u_x}{\partial z^2}\right) + \rho g_x.$$

Cartesian coordinates (x, y, z) y-direction:

$$\rho\left(\frac{\partial u_y}{\partial t} + u_x\frac{\partial u_y}{\partial x} + u_y\frac{\partial u_y}{\partial y} + u_z\frac{\partial u_y}{\partial z}\right) = -\frac{\partial p}{\partial y} + \mu\left(\frac{\partial^2 u_y}{\partial x^2} + \frac{\partial^2 u_y}{\partial y^2} + \frac{\partial^2 u_y}{\partial z^2}\right) + \rho g_y.$$

Cartesian coordinates (x, y, z) z-direction:

$$\rho\left(\frac{\partial u_z}{\partial t} + u_x\frac{\partial u_z}{\partial x} + u_y\frac{\partial u_z}{\partial y} + u_z\frac{\partial u_z}{\partial z}\right) = -\frac{\partial p}{\partial z} + \mu\left(\frac{\partial^2 u_z}{\partial x^2} + \frac{\partial^2 u_z}{\partial y^2} + \frac{\partial^2 u_z}{\partial z^2}\right) + \rho g_z.$$

Table A.3-a3. Momentum equations in the cylindrical coordinate

Cylindrical coordinates (r, θ, z) r-direction:

$$\rho\left(\frac{\partial u_r}{\partial t} + u_r\frac{\partial u_r}{\partial r} + \frac{u_\theta}{r}\frac{\partial u_r}{\partial \theta} - \frac{u_\theta^2}{r} + u_z\frac{\partial u_r}{\partial z}\right)$$

$$= -\frac{\partial p}{\partial r} + \mu\left[\frac{\partial}{\partial r}\left(\frac{1}{r}\frac{\partial}{\partial r}(ru_r)\right) + \frac{1}{r^2}\frac{\partial^2 u_r}{\partial \theta^2} - \frac{2}{r^2}\frac{\partial u_\theta}{\partial \theta} + \frac{\partial^2 u_r}{\partial z^2}\right] + \rho g_r.$$

Cylindrical coordinates (r, θ, z) θ-direction:

$$\rho\left(\frac{\partial u_\theta}{\partial t} + u_r\frac{\partial u_\theta}{\partial r} + \frac{u_\theta}{r}\frac{\partial u_\theta}{\partial \theta} - \frac{u_r u_\theta}{r} + u_z\frac{\partial u_\theta}{\partial z}\right)$$

$$= -\frac{1}{r}\frac{\partial p}{\partial \theta} + \mu\left[\frac{\partial}{\partial r}\left(\frac{1}{r}\frac{\partial}{\partial r}(ru_\theta)\right) + \frac{1}{r^2}\frac{\partial^2 u_\theta}{\partial \theta^2} + \frac{2}{r^2}\frac{\partial u_r}{\partial \theta} + \frac{\partial^2 u_\theta}{\partial z^2}\right] + \rho g_\theta.$$

Cylindrical coordinates (r, θ, z) z-direction:

$$\rho\left(\frac{\partial u_z}{\partial t} + u_r\frac{\partial u_z}{\partial r} + \frac{u_\theta}{r}\frac{\partial u_z}{\partial \theta} + u_z\frac{\partial u_z}{\partial z}\right)$$

$$= -\frac{\partial p}{\partial z} + \mu\left[\frac{1}{r}\frac{\partial}{\partial r}\left(r\frac{\partial u_z}{\partial r}\right) + \frac{1}{r^2}\frac{\partial^2 u_z}{\partial \theta^2} + \frac{\partial^2 u_z}{\partial z^2}\right] + \rho g_z.$$

Table A.3-a4. Momentum equations in the spherical coordinate

Spherical coordinates (r, θ, ϕ) r-direction:

$$\rho\left(\frac{\partial u_r}{\partial t}+u_r\frac{\partial u_r}{\partial r}+\frac{u_\theta}{r}\frac{\partial u_r}{\partial \theta}+\frac{u_\phi}{r\sin\theta}\frac{\partial u_r}{\partial \phi}-\frac{u_\theta^2+u_\phi^2}{r}\right)$$

$$=-\frac{\partial p}{\partial r}+\mu\left[\nabla^2 u_r-\frac{2}{r^2}u_r-\frac{2}{r^2}\frac{\partial u_\theta}{\partial \theta}-\frac{2}{r^2}u_\theta\cot\theta-\frac{2}{r^2\sin\theta}\frac{\partial u_\phi}{\partial \phi}\right]+\rho g_r.$$

Spherical coordinates (r, θ, ϕ) θ-direction:

$$\rho\left(\frac{\partial u_\theta}{\partial t}+u_r\frac{\partial u_\theta}{\partial r}+\frac{u_\theta}{r}\frac{\partial u_\theta}{\partial \theta}+\frac{u_\phi}{r\sin\theta}\frac{\partial u_\theta}{\partial \phi}+\frac{u_r u_\theta}{r}-\frac{u_\phi^2\cot\theta}{r}\right)$$

$$=-\frac{1}{r}\frac{\partial p}{\partial \theta}+\mu\left[\nabla^2 u_\theta+\frac{2}{r^2}\frac{\partial u_r}{\partial \theta}-\frac{u_\theta}{r^2\sin^2\theta}-\frac{2\cos\theta}{r^2\sin^2\theta}\frac{\partial u_\phi}{\partial \phi}\right]+\rho g_\theta.$$

Spherical coordinates (r, θ, ϕ) ϕ-direction:

$$\rho\left(\frac{\partial u_\varphi}{\partial t}+u_r\frac{\partial u_\varphi}{\partial r}+\frac{u_\theta}{r}\frac{\partial u_\varphi}{\partial \theta}+\frac{u_\varphi}{r\sin\theta}\frac{\partial u_\varphi}{\partial \varphi}+\frac{u_\varphi u_r}{r}+\frac{u_\theta u_\varphi}{r}\cot\theta\right)$$

$$=-\frac{1}{r\sin\theta}\frac{\partial p}{\partial \varphi}+\mu\left[\nabla^2 u_\varphi-\frac{u_\varphi}{r^2\sin^2\theta}+\frac{2}{r^2\sin\theta}\frac{\partial u_r}{\partial \varphi}+\frac{2\cos\theta}{r^2\sin^2\theta}\frac{\partial u_\theta}{\partial \varphi}\right]+\rho g_\varphi.$$

Table A.3-a5. Energy equations in the Cartesian, cylindrical, and spherical coordinates

Cartesian coordinates (x, y, z):

$$\rho c_p \left(\frac{\partial T}{\partial t} + u_x \frac{\partial T}{\partial x} + u_y \frac{\partial T}{\partial y} + u_z \frac{\partial T}{\partial z} \right) - k \left(\frac{\partial^2 T}{\partial x^2} + \frac{\partial^2 T}{\partial y^2} + \frac{\partial^2 T}{\partial z^2} \right)$$

$$- \beta T \left(\frac{\partial p}{\partial t} + u_x \frac{\partial p}{\partial x} + u_y \frac{\partial p}{\partial y} + u_z \frac{\partial p}{\partial z} \right) + \mu \Phi + \dot{q}$$

$$\Phi = 2 \left[\left(\frac{\partial u_x}{\partial x} \right)^2 + \left(\frac{\partial u_y}{\partial y} \right)^2 + \left(\frac{\partial u_z}{\partial z} \right)^2 \right] + \left[\left(\frac{\partial u_x}{\partial y} + \frac{\partial u_y}{\partial x} \right)^2 + \left(\frac{\partial u_y}{\partial z} + \frac{\partial u_z}{\partial y} \right)^2 + \left(\frac{\partial u_z}{\partial x} + \frac{\partial u_x}{\partial z} \right)^2 \right]$$

$$- \frac{2}{3} \left(\frac{\partial u_x}{\partial x} + \frac{\partial u_y}{\partial y} + \frac{\partial u_z}{\partial z} \right)^2.$$

Cylindrical coordinates (r, θ, z):

$$\rho c_p \left(\frac{\partial T}{\partial t} + u_r \frac{\partial T}{\partial r} + \frac{u_\theta}{r} \frac{\partial T}{\partial \theta} + u_z \frac{\partial T}{\partial z} \right) = \frac{k}{r} \frac{\partial}{\partial r} \left(r \frac{\partial T}{\partial r} \right) + \frac{k}{r^2} \frac{\partial^2 T}{\partial \theta^2} + k \frac{\partial^2 T}{\partial z^2}$$

$$- \beta T \left(\frac{\partial p}{\partial t} + u_r \frac{\partial p}{\partial r} + \frac{u_\theta}{r} \frac{\partial p}{\partial \theta} + u_z \frac{\partial p}{\partial z} \right) + \mu \Phi + \dot{q}$$

$$\Phi = 2 \left[\left(\frac{\partial u_r}{\partial r} \right)^2 + \left(\frac{1}{r} \frac{\partial u_\theta}{\partial \theta} + \frac{u_r}{r} \right)^2 + \left(\frac{\partial u_z}{\partial z} \right)^2 \right] + \left[r \frac{\partial}{\partial r} \left(\frac{u_\theta}{r} \right) + \frac{1}{r} \left(\frac{\partial u_r}{\partial \theta} \right) \right]^2$$

$$+ \left[\frac{1}{r} \frac{\partial u_z}{\partial \theta} + \frac{\partial u_x}{\partial z} \right]^2 + \left[\frac{\partial u_r}{\partial z} + \frac{\partial u_z}{\partial r} \right]^2 - \frac{2}{3} \left[\frac{1}{r} \frac{\partial}{\partial r} (r u_r) + \frac{1}{r} \frac{\partial u_\theta}{\partial \theta} + \frac{\partial u_z}{\partial z} \right]^2.$$

Spherical coordinates (r, θ, ϕ):

$$\rho c_p \left(\frac{\partial T}{\partial t} + u_r \frac{\partial T}{\partial r} + \frac{u_\theta}{r} \frac{\partial T}{\partial \theta} + \frac{u_\varphi}{r \sin \theta} \frac{\partial T}{\partial \varphi} \right) = \frac{k}{r^2} \frac{\partial}{\partial r} \left(r^2 \frac{\partial T}{\partial r} \right) + \frac{k}{r^2 \sin \theta} \frac{\partial}{\partial \theta} \left(\sin \theta \frac{\partial T}{\partial \theta} \right) + \frac{k}{r^2 \sin^2 \theta} \frac{\partial^2 T}{\partial \varphi^2}$$

$$- \beta T \left(\frac{\partial p}{\partial t} + u_r \frac{\partial p}{\partial r} + \frac{u_\theta}{r} \frac{\partial p}{\partial \theta} + \frac{u_\varphi}{r \sin \theta} \frac{\partial p}{\partial \varphi} \right) + \mu \Phi + \dot{q}$$

$$\Phi = 2 \left[\left(\frac{\partial u_r}{\partial r} \right)^2 + \left(\frac{1}{r} \frac{\partial u_\theta}{\partial \theta} + \frac{u_r}{r} \right)^2 + \left(\frac{1}{r \sin \theta} \frac{\partial u_\varphi}{\partial \varphi} + \frac{u_r}{r} + \frac{u_\theta \cot \theta}{r} \right)^2 \right] + \left[r \frac{\partial}{\partial r} \left(\frac{u_\theta}{r} \right) + \frac{1}{r} \frac{\partial u_r}{\partial \theta} \right]^2$$

$$+ \left[\frac{\sin \theta}{r} \frac{\partial}{\partial \theta} \left(\frac{u_\varphi}{\sin \theta} \right) + \frac{1}{r \sin \theta} \frac{\partial u_\theta}{\partial \varphi} \right]^2 + \left[\frac{1}{r \sin \theta} \frac{\partial u_r}{\partial \varphi} + r \frac{\partial}{\partial \theta} \left(\frac{u_\varphi}{r} \right) \right]^2$$

$$- \frac{2}{3} \left[\frac{1}{r^2} \frac{\partial}{\partial r} (r^2 u_r) + \frac{1}{r \sin \theta} \frac{\partial}{\partial \theta} (u_\theta \sin \theta) + \frac{1}{r \sin \theta} \frac{\partial u_\varphi}{\partial \varphi} \right]^2.$$

Table A.3-a6. Species equations in the Cartesian, cylindrical, and spherical coordinates

Cartesian coordinates (x, y, z):

$$\frac{\partial C_i}{\partial t} + u_x \frac{\partial C_i}{\partial x} + u_y \frac{\partial C_i}{\partial y} + u_z \frac{\partial C_i}{\partial z} = D_i \left[\frac{\partial^2 C_i}{\partial x^2} + \frac{\partial^2 C_i}{\partial y^2} + \frac{\partial^2 C_i}{\partial z^2} \right] + R_{Vi}.$$

Cylindrical coordinates (r, θ, z):

$$\frac{\partial C_i}{\partial t} + u_r \frac{\partial C_i}{\partial r} + \frac{u_\theta}{r} \frac{\partial C_i}{\partial \theta} + u_z \frac{\partial C_i}{\partial z} = D_i \left[\frac{1}{r} \frac{\partial}{\partial r} \left(r \frac{\partial C_i}{\partial r} \right) + \frac{1}{r^2} \frac{\partial^2 C_i}{\partial \theta^2} + \frac{\partial^2 C_i}{\partial z^2} \right] + R_{Vi}.$$

Spherical coordinates (r, θ, ϕ):

$$\frac{\partial C_i}{\partial t} + u_r \frac{\partial C_i}{\partial r} + \frac{u_\theta}{r} \frac{\partial C_i}{\partial \theta} + \frac{u_\varphi}{r \sin \theta} \frac{\partial C_i}{\partial \varphi}$$

$$= D_i \left[\frac{1}{r^2} \frac{\partial}{\partial r} \left(r^2 \frac{\partial C_i}{\partial r} \right) + \frac{1}{r^2 \sin \theta} \frac{\partial}{\partial \theta} \left(\sin \theta \frac{\partial C_i}{\partial \theta} \right) + \frac{1}{r^2 \sin^2 \theta} \frac{\partial^2 C_i}{\partial \varphi^2} \right] + R_{Vi}.$$

MATHEMATICAL BASICS AND RELATIONS* (COURTESY OF JOHN WILEY AND SONS)

1. Definitions

(a) Dyadic product. $\left(\vec{a}\,\vec{c}\right)_{ij} = a_i c_j$. ($\vec{a}\,\vec{c}$ is a tensor.)

(b) Double dot product.

$$\vec{\sigma} : \vec{\tau} = \sum_i \sum_j \sigma_{ij} \tau_{ji}.$$

(c) A tensor operating on a vector from the right yields a vector.

$$\vec{a} \cdot \vec{\tau} = \sum_i \sum_j \vec{e}_i a_j \tau_{ji}.$$

(d) Transpose of a tensor.

$$\left(\vec{\tau}^*\right)_{ij} = \tau_{ji} \text{ or } \vec{\tau} \cdot \vec{a} = \vec{a} \cdot \vec{\tau}^*.$$

(e) Product of two tensors.

$$\left(\vec{\tau} \cdot \vec{\sigma}\right) \cdot \vec{v} = \vec{\tau} \cdot \left(\vec{\sigma} \cdot \vec{v}\right) \text{ or } \left(\vec{\tau} \cdot \vec{\sigma}\right)_{ij} = \sum_k \tau_{ik} \sigma_{kj}.$$

(f) The divergence of a tensor is a vector.

$$\nabla \cdot \vec{\tau} = \sum_i \sum_j \vec{e}_i \left(\frac{\partial \tau_{ji}}{\partial x_j}\right).$$

(g) Laplacian of a scalar.

$$\nabla^2 \Phi = \nabla \cdot \nabla \Phi = \sum_i \left(\frac{\partial^2 \Phi}{\partial x_i^2}\right)$$

(h) Gradient of a vector. $\left(\nabla \vec{v}\right)_{ij} = \partial v_j / \partial x_i$.

(i) Laplacian of a vector. $\nabla^2 \vec{v} = \nabla \cdot \nabla \vec{v} = \nabla(\nabla \cdot \vec{v}) - \nabla \times \nabla \times \vec{v}$.

2. Algebra

(a) $\vec{r} : (\vec{a}\vec{b}) - \vec{b} \cdot (\vec{r} \cdot \vec{a})$.

(b) $(\vec{u}\vec{v}):(\vec{w}\vec{z}) = (\vec{u}\vec{w}):(\vec{v}\vec{z}) = (\vec{u} \cdot \vec{z})(\vec{v} \cdot \vec{w})$.

(c) $\vec{a} \cdot (\vec{b}\vec{c}) = (\vec{a} \cdot \vec{b})\vec{c}$.

(d) $(\vec{a}\vec{b}) \cdot \vec{c} = \vec{a}(\vec{b} \cdot \vec{c})$.

(e) $\vec{a} \times (\vec{b} \times \vec{c}) = \vec{b}(\vec{a} \cdot \vec{c}) - \vec{c}(\vec{a} \cdot \vec{b})$.

(f) $\vec{u} \cdot (\vec{v} \times \vec{w}) = \vec{v} \cdot (\vec{w} \times \vec{u})$.

(g) $(\vec{u} \times \vec{v}) \cdot (\vec{w} \times \vec{z}) = (\vec{u} \cdot \vec{w})(\vec{v} \cdot \vec{z}) - (\vec{u} \cdot \vec{z})(\vec{v} \cdot \vec{w})$.

(h) $\vec{v} \cdot (\tau^* \cdot \vec{w}) = \vec{w} \cdot (\vec{\tau} \cdot \vec{v})$.

3. Differentiation of products

(a) $\nabla \phi \psi = \phi \nabla \psi + \psi \nabla \phi$ (a vector).

(b) $\nabla \phi \vec{v} = \phi \nabla \vec{v} + (\nabla \phi) \vec{v}$ (a tensor).

(c) $\nabla(\vec{a} \cdot \vec{c}) = \vec{a} \cdot \nabla \vec{c} + \vec{c} \cdot \nabla \vec{a} + \vec{a} \times \nabla \times \vec{c} + \vec{c} \times \nabla \times \vec{a}$

(d) $= (\nabla c) \cdot \vec{a} + (\nabla a) \cdot \vec{c}$ (a vector).

(e) $\nabla \cdot (\phi \vec{v}) = \phi \nabla \cdot \vec{v} + \vec{v} \cdot \nabla \phi$ (a scalar).

(f) $\nabla \cdot (\vec{v} \times \vec{w}) = \vec{w} \cdot (\nabla \times \vec{v}) - \vec{v} \cdot (\nabla \times \vec{w})$ (a scalar).

(g) $\nabla \times (\phi \vec{v}) = \phi \nabla \times \vec{v} + (\nabla \phi) \times \vec{v}$ (a vector).

(h) $\nabla \times (\vec{b} \times \vec{c}) = \vec{b}(\nabla \cdot \vec{c}) - \vec{c}(\nabla \cdot \vec{b}) + \vec{c} \cdot \nabla \vec{b} - \vec{b} \cdot \nabla \vec{c}$ (a vector).

(i) $\nabla \cdot (\vec{a}\vec{b}) = (\nabla \cdot \vec{a})\vec{b} + \vec{a} \cdot \nabla b$ (a vector).

(j) $\nabla \cdot (\phi \vec{\tau}) = \phi \nabla \cdot \vec{\tau} + (\nabla \phi) \cdot \vec{\tau}$ (a vector).

(k) $\nabla \cdot (\vec{u} \cdot \vec{\tau}) = \vec{\tau} : \nabla \vec{u} + \vec{u} \cdot \nabla \cdot \vec{\tau}^*$ (a scalar).

4. Various forms of Gauss's law (divergence theorem) and Stoke's law (dS =area element, dl = line element, dv =volume element. Integration over a closed surface or a closed curve is denoted by a circle through the integral sign. In the first case, $\overrightarrow{dS}$ is normally outward from the surface;

in the second case, $\overrightarrow{dl}$ and $\overrightarrow{dS}$ are related by a right-hand screw rule, that is, a right-hand screw turned in the direction of $\overrightarrow{dl}$ advances in the direction of $\overrightarrow{dS}$.)

(a) $\oint \overrightarrow{dS} \cdot \vec{F} = \int dv \nabla \cdot \vec{F}.$

(b) $\oint \overrightarrow{dS} \phi = \int dv \nabla \phi.$

(c) $\oint (\overrightarrow{dS} \cdot \vec{G}) \vec{F} = \int dv \vec{F} \nabla \cdot \vec{G} + \int dv \vec{G} \cdot \nabla \vec{F}.$

(d) $\oint \overrightarrow{dS} \times \vec{F} = \int dv \nabla \times \vec{F}.$

(e) $\oint \overrightarrow{dS} \cdot \vec{\tau} = \int dv \nabla \cdot \vec{\tau}.$

(f) $\oint \overrightarrow{dS} \cdot (\Psi \nabla \phi - \phi \nabla \Psi) = \int dv (\Psi \nabla^2 \phi - \phi \nabla^2 \Psi).$

(g) $\oint \overrightarrow{dl} \cdot \vec{F} = \int \overrightarrow{dS} \cdot \nabla \times \vec{F}.$

(h) $\oint \overrightarrow{dl} \phi = \int \overrightarrow{dS} \times \nabla \phi.$

5. Miscellaneous

(a) $\nabla \cdot \nabla \times \vec{E} = 0.$

(b) $\nabla \times \nabla \phi = 0.$

(c) $\vec{w} \cdot \nabla \vec{v} = \sum_i \sum_j \vec{e}_i w_j \frac{\partial v_i}{\partial x_j}.$

(d) $D/Dt = \partial / \partial t + \vec{v} \cdot \nabla.$

(e) $D\vec{v}/Dt = \partial \vec{v} / \partial t + \frac{1}{2} \nabla v^2 - \vec{v} \times \nabla \times \vec{v}$ where $\vec{v}$ is the mass − average velocity.

6. Scalar product of vectors

$$\vec{a} \cdot \vec{b} = \left(\sum_i a_i \vec{e}_i \right) \cdot \left(\sum_j b_j \vec{e}_j \right) = \sum_i \sum_j a_i b_j (\vec{e}_i \cdot \vec{e}_j) = \sum_i \sum_j a_i b_j \delta_{ij} = \sum_i a_i b_i.$$

7. Vector product of vectors

$$\vec{a} \times \vec{b} = \begin{vmatrix} \vec{e}_1 & \vec{e}_2 & \vec{e}_3 \\ a_1 & a_2 & a_3 \\ b_1 & b_2 & b_3 \end{vmatrix}.$$

8. Multiple products

(a) $\vec{a} \cdot (\vec{b} \times \vec{c}) = \vec{b} \cdot (\vec{c} \times \vec{a}) = \vec{c} \cdot (\vec{a} \times \vec{b}).$

(b) $\vec{a} \times (\vec{b} \times \vec{c}) = (\vec{a} \cdot \vec{c}) \vec{b} - (\vec{a} \cdot \vec{b}) \vec{c}.$

(c) $\left(\vec{a} \times \vec{b}\right) \times \vec{c} = \left(\vec{a} \cdot \vec{c}\right)\vec{b} - \left(\vec{b} \cdot \vec{c}\right)\vec{a}.$

(d) $\left(\vec{a} \times \vec{b}\right) \cdot \left(\vec{c} \times \vec{d}\right) = \left(\vec{a} \cdot \vec{c}\right)\left(\vec{b} \cdot \vec{d}\right) - \left(\vec{a} \cdot \vec{d}\right)\left(\vec{b} \cdot \vec{c}\right).$

9. Identity tensor

$$\vec{\delta} = \sum_i \sum_j \delta_{ij} \vec{e}_i \vec{e}_j = \sum_i \vec{e}_i \vec{e}_i = \begin{bmatrix} 1 & 0 & 0 \\ 0 & 1 & 0 \\ 0 & 0 & 1 \end{bmatrix}.$$

10. Vector-differential operators

(a) Gradient

$$\nabla = \vec{e}_1 \frac{\partial}{\partial x_1} + \vec{e}_2 \frac{\partial}{\partial x_2} + \vec{e}_3 \frac{\partial}{\partial x_3} = \sum_i \vec{e}_i \frac{\partial}{\partial x_i}.$$

(b) Divergence

$$\nabla \cdot \vec{v} = \frac{\partial v_1}{\partial x_1} + \frac{\partial v_2}{\partial x_2} + \frac{\partial v_3}{\partial x_3} = \sum_i \frac{\partial v_i}{\partial x_i}.$$

(c) Curl

$$\nabla \times \vec{v} = \left(\frac{\partial v_3}{\partial x_2} - \frac{\partial v_2}{\partial x_3}\right)\vec{e}_1 + \left(\frac{\partial v_1}{\partial x_3} - \frac{\partial v_3}{\partial x_1}\right)\vec{e}_2 + \left(\frac{\partial v_2}{\partial x_1} - \frac{\partial v_1}{\partial x_2}\right)\vec{e}_3.$$

(d) Laplacian

$$\nabla \cdot \nabla = \nabla^2 = \frac{\partial^2}{\partial x_1^2} + \frac{\partial^2}{\partial x_2^2} + \frac{\partial^2}{\partial x_3^2} = \sum_i \frac{\partial^2}{\partial x_i^2}.$$

(e) Material derivative

$$\frac{D}{Dt} = \frac{\partial}{\partial t} + \vec{v} \cdot \nabla = \frac{\partial}{\partial t} + \sum_i v_i \frac{\partial}{\partial x_i}$$

*SOURCE

J. Newman and K. E. Thomas-Alyea. *Electrochemical Systems*, Wiley-Interscience, 3rd edn (May 27, 2004).

R. B. Bird, W. E. Stewart, and E. N. Lightfoot. *Transport Phenomena*, Wiley, 2nd edn (2001).

APPENDIX III.C

HENRY'S CONSTANT FOR SELECTED GASES IN WATER AT MODERATE PRESSURE*

	H or $k_i = p_{A,i}/x_{A,i}$ (bars)			
T(K)	CO_2	CH_4	O_2	H_2
273	753	24,092	26,761	60,748
280	960	27,800	30,500	61,500
290	1,300	35,200	37,600	66,500
300	1,660	40,829	43,879	71,634
310	2,175	50,000	52,500	76,000
320	2,650	56,300	56,800	78,600
323	2,870	58,000	58,000	79,000
333	3,669	69,263	72,019	84,499
353	5,519	90,935	92,957	92,002
363	6,656	103,029	104,502	95,663

*SOURCES

1. F.P. Incropera, D.P. DeWitt, T.L. Bergman, and A.S. Lavine. *Fundamentals of Heat and Mass Transfer*, 6th Edn, John Wiley & Sons, 2006.
2. NIST Chemistry WebBook.

APPENDIX IV.A

MEMBRANE MATERIALS

Table A.4-a1. PEM fuel cells with modified SPFA membranes (Zhang *et al.* 2006) (Courtesy of Elsevier)

Membranes	Fuel cell conditions ($T_{anode}/T_{cell}/T_{cathode}$/pressure)	Performance V (V) / I (mA/cm^2)	Reference
Nafion 112	90/100/90°C, 1 bar, H$_2$/air	0.69/400	Song *et al.* 2008
	90/120/90°C, 1 bar, H$_2$/air	0.58/400	
Nafion 115/SiO$_2$ (6%)	130/130/130°C, 3 bar, H$_2$/O$_2$	0.4/1000	Adjemian *et al.* 2002
	130/140/130°C, 3 bar, H$_2$/O$_2$	0.4/400	
SDF-F/SiO$_2$ (4%)	108/110/108°C, 2 bar, H$_2$/O$_2$	0.6/800	Kim *et al.* 2004
	113/120/113°C, 2.5 bar, H$_2$/O$_2$	0.6/700	
Nafion/SiO$_2$/PWA[a]		0.4/540	
Nifion/SiO$_2$	100/110/100°C, 1.4 bar, H$_2$/O$_2$	0.4/320	Shao *et al.* 2006
Nafion/WO$_3$		0.4/300	
Nafion/TiO$_2$		0.4/185	
Nafion 115/ZrP	130/130/130°C, 3 bar, H$_2$/O$_2$	0.45/1000	Costamagna *et al.* 2002
Recast Nafion 115/ZrP	130/130/130°C, 3 bar, H$_2$/O$_2$	0.45/1500	
ZrP/Nafion 115 composite	130/120/130°C, 3 bar, H$_2$/O$_2$	0.6/610	
	130/130/130°C, 3 bar, H$_2$/O$_2$	0.6/530	Yang *et al.* 2004
	130/140/130°C, 3 bar, H$_2$/O$_2$	0.6/200	
Nafion/PTA-I		0.6/120	Ramani *et al.* 2004
Nafion/PTA-II	90/120/90°C, 1 bar, H$_2$/air	0.6/60	(1), Ramani *et al.* 2004 (2)
Mordenite/Nafion	90/110/90°C, H$_2$/O$_2$, 1 bar	0.6/280	Kwak *et al.* 2003,
	90/120/90°C, H$_2$/O$_2$, 1 bar	0.6/110	Kwak *et al.* 2004
	90/130/90°C, H$_2$/O$_2$, 1 bar	0.6/80	
Nafion-Teflon-Zr(HPO$_4$)	87/105/91°C, H$_2$/O$_2$	0.6/690	
	87/105/91°C, H$_2$/air	0.6/350	Si *et al.* 2004
	87/120/91°C, H$_2$/O$_2$	0.6/400	
	87/120/91°C, H$_2$/air	0.6/200	
15% SO$_4^{2-}$-ZrO$_2$/ Nafion	110/120/110°C, 3 bar, H$_2$/O$_2$	0.6/1500	Zhai *et al.* 2006

[a]Phosphotungstic acid

Table A.4-a2. Inorganic-organic composite membranes (Li et al. 2003) (Courtesy of American Chemical Society)

Organic component	Inorganic component	Ionic conductivity	Reference
SPEEK	SiO_2, ZrP, Zr-SPP	0.09 S cm^{-1} at 100°C, 100% RH, 95°C	Bonnet et al. 2000
	HPA	10^{-1} S cm^{-1} above 100°C	Zaidi et al. 2000
	BPO_4	5×10^{-1} S cm^{-1} at 160°C, 100% RH	Mikhailenko et al. 2001
	SiO_2	$0(3-4) \times 10^{-2}$ S cm^{-1} at 100°C, 100% RH	Roziére et al. 2000
SPSF	PWA	0.15 S cm^{-1} at 130°C, 100% RH	Hickner et al. 2001
	PAA	0.135 S cm^{-1} at 50°C, 100% RH	Poinsignon et al. 2000, 2001
	PAA	2×10^{-2} S cm^{-1} at 80°C, 98% RH	Genova-Dimitrova et al. 2001
PBI	ZrP + H_3PO_4	9×10^{-2} S cm^{-1} at 200°C, 5% RH	He et al. 2003
	PWA/SiWA + H_3PO_4	$(3-4) \times 10^{-2}$ S cm^{-1} at 200°C, 5% RH	Staiti and Minutoli 2001
	SiWA + SiO_2	2.2×10^{-3} S cm^{-1} at 160°C, 100% RH	Staiti et al. 2000
	PWA + SiO_2 + H_3PO_4	1.5×10^{-3} S cm^{-1} at 150°C, 100% RH	
PVDF	SiO_2, TiO_2:Al_2O_3, doping acids	>0.2 S cm^{-1} at 25°C	Peled et al. 1998
		>0.45 S cm^{-1} at 25°C and 80°C	Peled et al. 2000
	CsHSO4	10^{-2} S cm^{-1} at >150°C, 80% RH	Boysen et al. 2000
GPTS	SiWA + SiO_2; SiWA + ZrP	1.9×10^{-2} S cm^{-1} at 150°C, 100% RH	Park and Nagai 2001
GPTMS	PVA + PWA + P_2O_5 + Gl	10^{-3} S cm^{-1} at 60–90°C, 75–100% RH	Thanganathan et al. 2012.
Polysilsesquloxanes	PWA	3×10^{-2} S cm^{-1} at room temperature −140°C; 100% RH	Honma et al. 2002
ORMOSIL	HPA	10^{-3} S cm^{-1} at 25°C	Lavrencic et al. 2001
PEO	Tungsten acid	10^{-2} S cm^{-1} at 120°C; 1.4×10^{-2} at 80°C, 100% RH	Nakajima and Honma 2002
PEO, PPO, PTMO	PWA	10^{-2} S cm^{-1} at 140°C, 100% RH	Honma et al. 2001; 2003
		10^{-3}–10^{-2}S cm^{-1} at 60°C, 95% RH	
PVA/glycerin	ZrP + AA	10^{-3}–10^{-4} S cm^{-1} at 20–140°C, <1% and 61 RH	Vaivars et al. 1999
PTFE	ZrP	$>10^{-3}$ S cm^{-1} at 23°C, immersed in distilled water	Park et al. 2000.

Table A.4-a3. PEM fuel cell data of acid-doped PBI membranes (Zhang *et al.* 2006) (Courtesy of Elsevier)

Membranes	Testing conditions ($T_{anode}/T_{cell}/T_{cathode}$/pressure)	Performance (V (V)/I (mA cm^{-2}))	Reference
H$_3$PO$_4$/PBI	190, 170, 150, and 100°C, H$_2$/O$_2$, 1 bar 150°C, H$_2$/O$_2$, 1 bar 150°C, H$_2$/air, 1 bar	0.6/630, 430, 300, and 160 0.54/250 0.41/250	Li *et al.* 2001 Wang *et al.* 1996
H$_3$PO$_4$-SPSF/PB	170°C, H$_2$/O$_2$, 1 bar 130°C, H$_2$/O$_2$, 1 bar	0.6/350 0.6/180	Deimede *et al.* 2000
H$_3$PO$_4$-SPSF/PBI	200°C, H$_2$/O$_2$, 1 bar 200°C, 3% CO– H$_2$/O$_2$, 1 bar 190°C, H$_2$/O$_2$, 1 bar	0.6/700 0.6/570 0.6/430	Li *et al.* 2002 Hasiotis *et al.* 2001

*SOURCES

Adjemian, K.T., Lee, S.J., Srinivasan, S., Benzinger, J., Bocarly, A. B., *J. Electrochem. Soc. 149* (2002): A256–61.

Bonnet, B., Jones, D.J., Roziere, J., Tchicaya, L., Alberti, G., Casciola, M., Massinelli, L., Baner, B., Peraio, A., Ramunni, E., *J. New Mater. Electrochem. Syst.* **3** (2000): 87.

Boysen, D.A., Chrisholm, C.R.I., Haile, S.M., Sekharipuram, V., Narayanan, R. *J. Electrochem. Soc. 147* (2000): 3610.

Cosamagna, P., Yang, C., Bocarsly, A.B., Srinivasan, S. *Electrochim.Acta 47* (2002): 1023–33.

Deimede, V., Voyiatzis, G.A., Kallitsis, J.K., Qingfeng, L., Bjerrum, N.J. *Macromolecules 33* (2000) 7609–17.

Genova-Dimitrova, P., Baradie, B., Foscallo, D., Poinsignon, C., Sanchez, J. Y. *J. Membr.Sci. 185* (2001): 59.

Hasiotis, C., Qingfeng, L., Deimede, V., Kallitsis, J.K., Kontoyannis, C.G., Bjerrum, N. J. *J. Electrochem. Soc. 148* (2001): A513–19.

He, R., Li, Q., Xiao, G., Bjerrum, N.J. *J. Membr.Sci. 226* (2003): 169.

Hickner, M., Kim, Y.S., Wang, F., Zawodzinski, T.A., McGrath, *J. E. Proc. 33rd Int. SAMPE Techn.Conf. 33* (2001): 1519.

Honma, I., Nakajima, H., Nishikawa, O., Sugimoto, T., Nomura, S. *J. Electrochem.Soc.149* (2002): A1389.

Honma, I., Nomura, S., and Nakajima, H. *J. Membr. Sci. 185* (2001): 83.

Honma, I., Nomura, S., Nakajima, H., Sugimoto, T., and Nomura, S. *Solid State Ionics 162–163* (2003): 237–45.

Kim, Y.M., Choi, S.H., Lee, H.C., Hong, M.Z., Kim, K., Lee, H.I. *Electrochim.Acta 49* (2004): 4787–96.

Kwak, S.H., Yang, T.H., Kim, C.S., Yoon, K.H. *Electrochim. Acta 50* (2004): 653–57.

Kwak, S.H., Yang, T.H., Kim, C.S., Yoon, K.H. *Solid State Ionics 160* (2003): 309–15.

Lavrencic S.U., Groselj, N., Orel, B., Schmitz, A., Colomban, P. *Solid State Ionics 145* (2001): 109.

Li, Q.F., He, R.H., Jensen, J.O., and Bjerrum, N.J. *Chem. Mater. 15* (2003): 4896–915.

Li, Q.F., Hjuler, H.A., Bjerrum, N.J. *J. Appl. Electrochem. 31* (2001): 773–79.

Li, Q.F., Hjuler, H.A., Hasiotis, C., Kallitsis, J.K., Kontoyanni, C.G., and Bjerrum, N. J. *Electrochem. Solid State Lett.* **5** (2002): A125–28.

Li, Q., He, R., Jensen, J.O., and Bjerrum, N.J. *Chem. Mater. 15* (2003): 4896–4915.

Mikhailenko, S.D., Zaidi, S.M.J., and Kaliaguine, S. *Catal. Today 67* (2001): 225.

Nakajima, H., and Honma, I. *Solid State Ionics 148* (2002): 607.

Nunes, S.P., Ruffmann, B., Rikowski, E., Vetter, S., Richau, K. *J. Membr. Sci. 203* (2002): 215.

Park, Y., and Nagai, M. *Solid State Ionics 145* (2001): 149.

Park, Y.-I., Dong, K.J., and Nagai, M. *J. Mater. Sci. Lett. 19* (2000): 1735.

Peled, E., Duvdevani, T., Aharon, A., and Melman, A. *Electrochem. Solid State Lett. 3* (2000): 525.

Peled, E., Duvdevani, T., and Melman, A. *Electrochem. Solid StateLett. 1* (1998): 210.

Poinsignon, C., Amodio, I., Foscallo, D., Sanchez, J.Y. *Mater. Res. Soc. Symp. Proc. 548* (2000). 307.

Poinsignon, C., Le Gorrec, B., Vitter, G., Montella, C., and Diard, J.P. *Mater. Res. Soc. Symp. Proc. 575* (2000): 273.

Poltarzewski, Z., Wieczorek, W., Przyluski, J., and Antonucci, V. *Solid State Ionics 119* (1999): 301.

Ramani, V., Kunz, H.R., and Fenton, J.M. *Electrochim. Acta 50* (2004): 1181–87.

Ramani, V., Kunz, H.R., and Fenton, J.M. *J. Membr. Sci. 232* (2004): 31–34.

Roziére, J., Jones, D.J., Tchicaya-Bouckary, L., and Bauer, B., *WO 02/05370*: (2000).

Shao, Z.G., Xu, H., Li, M., and Hsing, I. -M. *Solid State Ionics 177* (2006): 779–785.

Si, Y., Kunz, H.R., and Fenton, J.M. *J. Electrochem. Soc.151* (2004): 623–31.

Song, Y., Fenton, J.M., Kunz, H.R., Bonville, L.J., and Williams, M.V. *J. Electrochem. Soc.152* (2005): A539–44.

Staiti, P., Minutoli, M., and Hocevar, S. *J. Power Sources 90* (2000): 231.

Staiti, P., and Minutoli, M. *J. Power Sources 94* (2001): 9.

Thanganathan, U., Parrondo, J., and Rambabu, B. *J Solid State Electrochem 16* (2012): 2151–58.

Vaivars, G., Azens, A., and Granqvist, C.G. *Solid State Ionics 119* (1999): 269.

Wang, J.T., Savinell, R.F., Wainright, J.S., Litt, M., and Yu, H. *Electrochim. Acta 41* (1996): 193–97.

Yang, C., Srinivasan, S., Bocarsly, A.B., Tulyani, S., and Benzinger, J.B. *J. Membr. Sci. 237* (2004): 145–61.

Zaidi, S.M.J., Mikhailenko, S.D., Robertson, G.P., Guiver, M.D., and Kaliaguine, S. *J. Membr. Sci. 173* (2000): 17.

Zhai, Y.F., Zhang, H.M., Hu, J.W., and Yi, B.L. *J. Membr. Sci.* (2006) in press.

Zhang J. *et al*. High temperature PEM fuel cells. *J. Power Sources, 160* (2006): 872–91.

Zhang, Q.M., Furukawa, T., Bar-Cohen, Y., and Scheinbeim, J. Eds.; *Materials Research Society*: Warrendale, PA, 2000; *vol. 600* (2000) 305.

APPENDIX IV.B

Ion Transport in Electrolytes

1 INTRODUCTION

Electrolytes are the media in which any electric currents are carried by dissolved ions. In PEM fuel cells, solid polymers serve as the electrolyte to conduct protons. Water transport occurs, which interacts with proton migration. In addition, transport in aqueous electrolytes is critical to many electrochemical processes, colloidal systems, various separation and purification methods, and cell/organisms functions. In this appendix, we present and discuss the general theory of ion transport in electrolyte solutions. Detailed discussions on the subject are given by Newman and Thomas-Alyea (2004), and Deen (2012).

2 MACROSCOPIC FORMULATION

2.1 NERNST–PLANCK EQUATION

Differing from neutral species, ions carry certain amount of charges, which are subjected to an electric force upon any electric field $(\vec{E})$. The force on an ion equals to the product of the electric field and the ionic charge. The charge per mole of ion i is $z_i F$, where z_i is the valence (a positive or negative integer). Thus, the electric force is expressed as $z_i FE$. In general, the proportionality constant that relates a force to a velocity is called mobility, and for ion i the molar mobility is denoted as u_i. The mean velocity that results from the electric force is then $z_i u_i F\vec{E}$, and the corresponding flux, generally referred to as migration, is superimposed on those caused by convection and diffusion. Accordingly, the total flux of ion i in a dilute solution is given by

$$\underset{\text{flux}}{\vec{N}_i} \quad = \quad \underset{\text{convection}}{C_i \vec{v}} \quad \underset{\text{difusion}}{- D_i \nabla C_i} \quad \underset{\text{migration}}{+ z_i u_i F C_i \vec{E}} \qquad \text{(Eq. A.4-b1)}$$

where D_i is the pseudobinary diffusivity. Because $\nabla \times \vec{E} = 0$ under electroquasistatic condition, a scalar Φ, usually called the electric potential, can be introduced and defined as $\vec{E} = -\nabla\Phi$ (note $\nabla \times \nabla() = 0$). The above is rewritten as

$$\vec{N}_i = C_i \vec{v} - D_i \nabla C_i - z_i u_i F C_i \nabla \Phi. \qquad \text{(Eq. A.4-b2)}$$

The diffusivity and mobility in an ideal solution are related and stated by the Nernst–Einstein equation:

$$D_i = RTu_i. \qquad \text{(Eq. A.4-b3)}$$

Then, the following Nernst–Planck equation is derived:

$$\vec{N}_i = C_i \vec{v} - D_i \left(\Delta C_i + \frac{z_i F}{RT} C_i \nabla \Phi \right). \qquad \text{(Eq. A.4-b4)}$$

2.2 ELECTRONEUTRALITY ASSUMPTION

The volumetric charge density in an electrolyte solution (C/m^3) is given by

$$\rho_e = F \sum_i z_i C_i \qquad \text{(Eq. A.4-b5)}$$

where the summation in this and the subsequent equations is over all ions in the electrolyte. Local electroneutrality assumption states:

$$\sum_i z_i C_i = 0. \qquad \text{(Eq. A.4-b6)}$$

This assumption is the basis for macroscopic modeling of electrolyte systems. For microscopic modeling, Poison's equation is frequently adopted to replace the electroneutrality assumption:

$$\nabla \cdot (\varepsilon \nabla \Phi) = -\rho_e, \qquad \text{(Eq. A.4-b7)}$$

where ε is the dielectric permittivity, and in vacuum, it is equal to 8.854×10^{-12} C V^{-1} m^{-1}. Inside double layers, the electroneutrality assumption is generally invalid.

2.3 CURRENT DENSITY

The movement of dissolved ions (charged particles) in electrolyte solutions yields current flow, which is similar to that of free electrons. The flux of charge is the electric current density, given by

$$\vec{i} = F \sum_i z_i \vec{N}_i. \qquad \text{(Eq. A.4-b8)}$$

Substituting the Nernst–Planck equation (Eq. A.4-b4) into the above equation yields

$$\vec{i} = F\vec{v}\sum_i z_i C_i - F\sum_i z_i D_i \nabla C_i - \frac{F^2}{RT}\nabla\Phi\sum_i z_i^2 D_i C_i$$

$$\text{or } \vec{i} = F\vec{v}\sum_i z_i C_i - F\sum_i z_i D_i \nabla C_i - F^2\nabla\Phi\sum_i z_i^2 u_i C_i$$

(Eq. A.4-b9)

with the assumption of the electroneutrality (i.e., Eq. A.4-b6), the convection contribution vanishes such that

$$\vec{i} = -F\sum_i z_i D_i \nabla C_i - \frac{F^2}{RT}\nabla\Phi\sum_i z_i^2 D_i C_i$$

$$\text{or } \vec{i} = -F\sum_i z_i D_i \nabla C_i - F^2\nabla\Phi\sum_i z_i^2 u_i C_i$$

(Eq. A.4-b10)

In the absence of any concentration variations or when the diffusion's contribution to current flow is much small as opposed to that due to the electric potential, the above equation simplifies to

$$\vec{i} = -\kappa\nabla\Phi$$

(Eq. A.4-b11)

where κ is the electrical conductivity given by

$$\kappa = \frac{F^2}{RT}\sum_i z_i^2 D_i C_i \quad \text{or} \quad \kappa = F^2\sum_i z_i^2 u_i C_i.$$

(Eq. A.4-b12)

Equation A.4-b11 is analogous to Ohm's law, with κ being inversely proportional to the ohmic resistance. We can further define a tranference number t_j to quantify the contribution of current carried by species j:

$$t_j\vec{i} = -F^2 z_j^2 u_j C_j \nabla\Phi = \frac{F^2 z_j^2 u_j C_j}{F^2\sum_i z_i^2 u_i C_i}\vec{i}.$$

(Eq. A.4-b13)

Thus,

$$t_j = \frac{F^2 z_j^2 u_j C_j}{F^2\sum_i z_i^2 u_i C_i}.$$

(Eq. A.4-b14)

With the tranference number t_j, the mirgration flux of species j can be written by

$$\vec{N}_j^{\text{migr}} = -z_j F u_j C_j \nabla\Phi = \frac{t_j}{z_j F}\vec{i}.$$

(Eq. A.4-b15)

In the presence of any concentration gradients, the current density is not propotional to the electric field due to diffusion, and thus Ohm's law does not hold. In this case, a diffusion term needs to be added to the gradient of potential:

$$\nabla\Psi = -\frac{\vec{i}}{\kappa} - \frac{F}{\kappa}\sum_i z_i D_i \nabla C_i.$$ (Eq. A.4-b16)

It is clear that in the absence of current flow the gradient of potential can arise due to diffusion, known as the diffusion potential. If D_i is the same for all the species:

$$-\frac{F}{\kappa}\sum_i z_i D_i \nabla C_i = -\frac{FD}{\kappa}\sum_i z_i \nabla C_i,$$ (Eq. A.4-b17)

which is equal to zero due to electroneutrality.

2.4 CONSERVATION EQUATIONS

The conservation equation for ionic species is similar to that for chemical species as derived in Chapter 3, following the conservaton law. The major difference is that ion transport is affected by electric potential, the effect of which is included in the Nernst–Planck equation. The equation for ionic species is thus given by

$$\frac{DC_i}{Dt} = D_i\left[\nabla^2 C_i + \frac{z_i F}{RT}\nabla\cdot(C_i\nabla\Phi)\right] + R_{Vi}.$$ (Eq. A.4-b18)

Multiplying the above equation by z_i and summing over all the species yields

$$\frac{D}{Dt}\sum_i z_i C_i = \nabla\cdot\left(\sum_i z_i D_i \nabla C_i + \frac{F}{RT}\nabla\Phi\sum_i z_i^2 D_i C_i\right) + \sum_i z_i R_{Vi}.$$ (Eq. A.4-b19)

The left side equals to zero, as stated by the electroneutrality assumption. The last term on right-hand side is also zero due to charge conservation in a reaction. The term in the parentheses on the right-hand side is actually equal to $\vec{i}/F$, that is,

$$\nabla\cdot\vec{i} = 0.$$ (Eq. A.4-b20)

The above equation states that there is no net charge accumulation even in transient because of local electroneutrality.

In the electrolyte membrane of a PEM fuel cell, no reaction occurs. In the absence of concentration gradients or when diffusion is negligible as opposed to the migration force, the following equation will yield:

$$\nabla\cdot\vec{i} = \nabla\cdot(\kappa\nabla\Phi) = 0.$$ (Eq. A.4-b21)

In the catalyst layer of a PEM fuel cell, ions and electrons are produced or consumed by the electrochemical reactions, and the divergence of $\vec{i}$ is equal to the production or consumption rate of charges j; thus,

$$\nabla \cdot (\kappa \nabla \Phi) = j. \qquad \text{(Eq. A.4-b22)}$$

*SOURCES

1. Newman, J., and Thomas-Alyea, K.E. *Electrochemical Systems*, 3rd edn. Wiley-Inter-science, (2004).
2. Deen, W.M. *Analysis of Transport Phenomena*. Oxford University Press, 2012.

TRANSPORT PROPERTIES OF TYPICAL GASES AT ATMOSPHERIC PRESSURE*

T (K)	ρ (kg/m^3)	$\mu \times 10^7$ (N s/m^2)	$\nu \times 10^6$ (m^2/s)	$k \times 10^3$ (W/m K)	$\alpha \times 10^6$ (m^2/s)	Pr
Air						
250	1.3947	159.6	11.44	22.3	15.9	0.720
275‡	1.284	172.5	13.43	24.28	18.8	0.713
300	1.1614	184.6	15.89	26.3	22.5	0.707
325‡	1.086	196.2	18.07	28.16	25.8	0.701
350	0.9950	208.2	20.92	30.0	29.9	0.700
375‡	0.9413	218.1	23.17	31.86	33.5	0.692
400	0.8711	230.1	26.41	33.8	38.3	0.690
Carbon dioxide (CO$_2$)						
240†	2.2581	120.66	5.34	12.243	6.80	0.785
243†	2.2293	122.16	5.48	12.453	6.99	0.784
273†	1.9779	137.03	6.93	14.663	9.55	0.726
280	1.9022	140	7.36	15.20	9.63	0.765
300	1.7730	149	8.40	16.55	11.0	0.766
320	1.6609	156	9.39	18.05	12.5	0.754
340	1.5618	165	10.6	19.70	14.2	0.746
353†	1.5237	175.42	11.5	21.177	15.4	0.747
360	1.4743	173	11.7	21.2	15.8	0.741
380	1.3961	181	13.0	22.75	17.6	0.737
393†	1.3674	193.80	14.2	24.552	19.2	0.739
400	1.3257	190	14.3	24.3	19.5	0.737

(Continued)

Carbon monoxide (CO)

240	1.4055	147	10.5	20.6	14.1	0.744
243[†]	1.4064	151	10.7	22.5	15.3	0.699
260	1.2967	157	12.1	22.1	16.3	0.741
273[†]	1.2512	165	13.2	24.7	18.9	0.696
280	1.2038	166	13.8	23.6	18.8	0.733
300	1.1233	175	15.6	25.0	21.3	0.730
320	1.0529	184	17.5	26.3	23.9	0.730
340	0.9909	193	19.5	27.8	26.9	0.725
353[†]	0.9669	200	20.7	30.1	29.8	0.694
360	0.9357	202	21.6	29.1	29.8	0.725
380	0.8864	210	23.7	30.5	32.9	0.729
393[†]	0.8684	216	24.9	32.7	35.9	0.692
400	0.8421	218	25.9	31.8	36.0	0.719

T (K)	ρ (kg/m^3)	$\mu \times 10^7$ (N s/m^2)	$\nu \times 10^6$ (m^2/s)	$k \times 10^3$ (W/m K)	$\alpha \times 10^6$ (m^2/s)	Pr
Hydrogen (H$_2$)						
240[†]	0.10230	76.8	75.1	155	108.4	0.693
243[†]	0.10103	77.5	76.7	156	110.2	0.696
250	0.09693	78.9	81.4	157	115	0.707
273[†]	0.08993	83.9	93.3	173	135.6	0.688
300	0.08078	89.6	111	183	158	0.701
350	0.06924	98.8	143	204	204	0.700
353[†]	0.06956	100.0	143.8	212	211.2	0.681
393[†]	0.06248	107.6	172.2	231	255.5	0.674
400	0.06059	108.2	179	226	258	0.695
Nitrogen (N$_2$)						
240[†]	1.4239	150.1	10.54	21.7	14.62	0.721
243[†]	1.4062	151.6	10.78	21.9	14.95	0.721
250	1.3488	154.9	11.48	22.2	15.8	0.727
273[†]	1.2511	166.3	13.29	24.0	18.41	0.722
300	1.1233	178.2	15.86	25.9	22.1	0.716
350	0.9625	200.0	20.78	29.3	29.2	0.711

(Continued)

353[†]	0.9669	202.5	20.94	29.3	29.1	0.720
393[†]	0.8684	219.3	25.25	31.8	35.07	0.720
400	0.8425	220.4	26.16	32.7	37.1	0.704

Oxygen (O_2)

240[†]	1.6275	171.2	10.52	21.8	14.65	0.718
243[†]	1.6073	172.9	10.76	22.0	14.97	0.719
250	1.542	178.6	11.58	22.6	16.0	0.723
273[†]	1.4298	190.5	13.32	24.5	18.68	0.713
300	1.284	207.2	16.14	26.8	22.7	0.711
350	1.100	233.5	21.23	29.6	29.0	0.733
353[†]	1.1049	233.9	21.17	30.9	30.07	0.704
393[†]	0.9923	254.1	25.61	34.1	36.59	0.700
400	0.9620	258.2	26.84	33.0	36.4	0.737

Water vapor (steam)

373[†]	0.5976	122.7	20.53	25.1	20.13	1.02
380	0.5863	127.1	21.68	24.6	20.4	1.06
393[†]	0.5654	130.2	23.03	26.5	23.26	0.99
400	0.5542	134.4	24.25	26.1	23.4	1.04

*SOURCE

F.P. Incropera, D.P. DeWitt, T.L. Bergman, and A.S. Lavine. *Fundamentals of Heat and Mass Transfer*, 6th edn, John Wiley & Sons, 2006.

[†] From NIST Chemistry WebBook.

[‡] From Engineering ToolBox.

APPENDIX VI. A

Governing Equations of Multiphase Flow and Heat Transfer In Porous Media*

This appendix provides a quick summary of the governing equations for multiphase flow and heat transfer in porous media.

1. Mass conservation in phase k:

$$\varepsilon \frac{\partial(\rho_k s_k)}{\partial t} + \nabla \cdot (\rho_k \vec{u}_k) = \bar{m}_k, \qquad \text{(Eq. A.6-a1)}$$

where ε is the porosity of the porous medium, s_k is the phase saturation or the volumetric fraction of the void space occupied by phase k, and $\vec{u}_k$ is the superficial (or Darcian) velocity based on the total cross-sectional area of the porous medium, including the void space and solid matrix. The term $\bar{m}_k$ represents an interfacial mass transfer rate from all other phases to phase k. In the absence of any mass source or sink, the summation of $\bar{m}_k$ over all the phases is equal to 0:

$$\sum_k \bar{m}_k = 0. \qquad \text{(Eq. A.6-a2)}$$

For two-phase flows of liquid water–air system in PEM fuel cells, the two-fluid model of the mass conservation reads

$$\varepsilon \frac{\partial(\rho_g s_g)}{\partial t} + \nabla \cdot (\rho_g \vec{u}_g) = \bar{m}_{lg},$$
$$\varepsilon \frac{\partial(\rho_l s_l)}{\partial t} + \nabla \cdot (\rho_l \vec{u}_l) = -\bar{m}_{lg}. \qquad \text{(Eq. A.6-a3)}$$

2.   Momentum conservation in phase k:

$$\vec{u}_k = -K\frac{k_{rk}}{\mu_k}\left(\nabla p_k - \rho_k \vec{g}\right),$$

(Eq. A.6-a4)

where the presence of the gravitational force is taken into account. This generalized Darcy's equation is valid if inertia as well as (macroscopic) viscous effects can be neglected.

For two-phase flows of liquid water–air system in PEM fuel cells, the two-fluid model of the momentum conservation reads

$$\vec{u}_g = -K\frac{k_{rg}}{\mu_g}\left(\nabla p_g - \rho_g \vec{g}\right),$$

$$\vec{u}_l = -K\frac{k_{rl}}{\mu_l}\left(\nabla p_l - \rho_l \vec{g}\right).$$

(Eq. A.6-a5)

3.   Mass conservation of species α in phase k:

$$\frac{\partial}{\partial t}\left(\varepsilon \rho_k s_k C_k^\alpha\right) + \nabla\cdot\left(\rho_k \vec{u}_k C_k^\alpha\right) = -\nabla\cdot \vec{j}_k^\alpha + \bar{J}_k^\alpha,$$

(Eq. A.6-a6)

where C_k^α is the mass concentration of species α in phase k, and $\vec{j}_k^\alpha$ is a diffusive flux of species α in phase k due to molecular diffusion, hydrodynamic dispersion, or both. The latter is usually expressed in Fickian form:

$$\vec{j}_k^\alpha = -\varepsilon \rho s_k D_k^\alpha \nabla C_k^\alpha,$$

(Eq. A.6-a7)

where D_k^α is a macroscopic second-order tensor and depends on the molecular diffusion coefficient and the definition of fluid velocity.

The last term in Eq. A.6-a6, $\bar{J}_k^\alpha$ denotes the species source arising from interphase species transfer, chemical reaction, and phase change. Without extrinsic species generation by chemical or biological reactions, it follows:

$$\sum_k \bar{J}_k^\alpha = 0.$$

(Eq. A.6-a8)

For two-phase flows of liquid water–air system in PEM fuel cells, the water transport equations in the gas and liquid read

$$\frac{\partial}{\partial t}\left(\varepsilon \rho_g s_g C_g^w\right) + \nabla\cdot\left(\rho_g \vec{u}_g C_g^w\right) = -\nabla\cdot \vec{j}_g^w + \bar{J}_g^w,$$

$$\frac{\partial}{\partial t}\left(\varepsilon \rho_l s_l\right) + \nabla\cdot\left(\rho_l \vec{u}_l\right) = \bar{J}_l^w,$$

(Eq. A.6-a9)

where

$$\bar{J}_g^w + \bar{J}_l^w = 0.$$

(Eq. A.6-a10)

4. Energy conservation in phase k:

$$\frac{\partial}{\partial t}\left(\varepsilon\rho_k s_k h_k\right)+\nabla\cdot\left(\rho_k\vec{u}_k h_k\right)=\nabla\cdot\left(s_k k_k\nabla T\right)+\overline{q}_k, \qquad \text{(Eq. A.6-a11)}$$

where local thermal equilibrium among phases has been assumed ($T_k=T$), and k_k and $\overline{q}_k$ represent the effective thermal conductivity of phase k and the interphase heat transfer rate associated with phase k, respectively. Hence,

$$\sum_k \overline{q}_k = \dot{q}, \qquad \text{(Eq. A.6-a12)}$$

where $\dot{q}$ stands for an external volumetric heat source or sink. The phase enthalpy h_k is related to the common temperature T via

$$h_k = \int_0^T c_k dT + h_k^0, \qquad \text{(Eq. A.6-a13)}$$

where c_k and h_k^0 represent the effective specific heat and the reference enthalpy of phase k, respectively.

These basic conservation equations provide a full system of governing equations for the unknown vector velocities $\vec{u}_k$, scalar pressures P_k, scalar liquid saturation s_k, mass concentration C_k^α, and temperature T.

*SOURCE

C.Y. Wang, and P. Cheng. Multiphase flow and heat transfer in porous media. *Adv. Heat Transf. 30* (1997): 93–196.

MULTIPHASE MIXTURE MODEL IN POROUS MEDIA*

1 BASIC DEFINITIONS

In classic multicomponent mixture models, the mixture's physical properties are based on the properties of its constituents; however, their functional forms are not assumed as *a priori*, instead derived strictly from a rigorous multiphase flow model. The mixture density, velocity, species concentration, enthalpy, and pressure are defined, respectively, as

$$\rho = \sum_k \rho_k s_k \qquad \text{(Eq. A.6-b1)}$$

$$\rho \vec{u} = \sum_k \rho_k \vec{u}_k \qquad \text{(Eq. A.6-b2)}$$

$$\rho C^\alpha = \sum_k \rho_k s_k C_k^\alpha \qquad \text{(Eq. A.6-b3)}$$

$$\rho h = \sum_k \rho_k s_k h_k \qquad \text{(Eq. A.6-b4)}$$

$$p = p_k + \sum_i \int_0^{s_i} Ca_{ik} ds_i + \sum_\alpha \int_0^{C^\alpha} Cs_{\alpha k} dC^\alpha + \int_0^T Ct_k dT, \qquad \text{(Eq. A.6-b5)}$$

where a quantity without a subscript is defined for the multiphase mixture, and the subscript k denotes phase k. The superficial velocity of phase k, $\vec{u}_k$, denotes its intrinsic velocity through a mass-weighted average.

The definition of the mixture pressure in Eq. A.6-b5 is somewhat nonconventional but is indeed consistent with a mixture theory. The definition given in Eq. A.6-b5 greatly simplified the momentum conservation equation for a bulk mixture. The terms Ca_{ik}, $Cs_{\alpha k}$, and Ct_k in Eq. A.6-b5 are called the capillary, solutal-capillary, and thermocapillary factors, respectively, and can be expressed as functions of capillary pressure gradients with respect to saturation, species concentration, and temperature.

In addition, the mixture kinematic viscosity and the phase mobility are introduced, respectively, as follows:

$$\nu = \left(\sum_k \frac{k_{rk}}{\nu_k} \right)^{-1} \qquad \text{(Eq. A.6-b6)}$$

$$\lambda_k = \frac{k_{rk}}{\nu_k} \nu, \qquad \sum_k \lambda_k = 1. \qquad \text{(Eq. A.6-b7)}$$

2 GOVERNING EQUATIONS

The multiphase mixture (M^2) model, consisting of the governing equations, is set up based on the above mixture variables, which are summarized as follows.

2.1 MASS CONSERVATION OF THE MULTIPHASE MIXTURE:

$$\varepsilon \frac{\partial \rho}{\partial t} + \nabla \cdot \left(\rho \vec{u} \right) = 0. \qquad \text{(Eq. A.6-b8)}$$

It is clear that this continuity equation is just a duplicate of the corresponding equation for a single-phase fluid flow. Thus, the above two-phase mass conservation equation is applicable to the single-phase flow.

2.2 MOMENTUM CONSERVATION OF THE MULTIPHASE MIXTURE:

$$\rho \vec{u} = -\frac{K}{\nu} \left(\nabla p - \gamma_\rho \rho \vec{g} \right), \qquad \text{(Eq. A.6-b9)}$$

where γ_ρ is called the density correlation factor, defined as

$$\gamma_\rho = \frac{\sum\limits_k \rho_k \lambda_k}{\sum\limits_k \rho_k s_k} \qquad \text{(Eq. A.6-b10)}$$

It should be noted that this correction factor is a sole function of phase saturation and thus can be regarded as a property of the multiphase mixture. The fundamental reason that a correction factor appears in the body force for a multiphase mixture is the relative motion between phases so that the effective mixture density for the gravitational force contains certain dynamic properties of individual phase (i.e., λ_κ and hence ν_κ).

2.3 DIFFUSIVE FLUX

As in a traditional mixture theory, we can define a diffusive mass flux of phase k relative to the whole multiphase mixture, such that

$$\vec{j}_k = \rho_k \vec{u}_k - \lambda_k \rho \vec{u}, \quad \sum_k \vec{j}_k = 0, \tag{Eq. A.6-b11}$$

or alternatively,

$$\rho_k \vec{u}_k = \vec{j}_k + \lambda_k \rho \vec{u}. \tag{Eq. A.6-b12}$$

The diffusive flux can be expressed as

$$\vec{j}_k = \sum_i \left[-\rho_k D_{cik} \nabla s_i + \frac{\lambda_k \lambda_i K (\rho_k - \rho_i)}{\nu} \vec{g} \right] + \sum_\alpha \left[-\rho_k D_{s\alpha k} \nabla C^\alpha \right] + \left(-\rho_k D_{tk} \right) \nabla T , \tag{Eq. A.6-b13}$$

where the capillary diffusion coefficient D_{cik}, the solutal-capillary diffusion coefficient $D_{s\alpha k}$, and the thermocapillary diffusion coefficient D_{tk}, are given by

$$D_{cik} = \frac{K}{\rho_k \nu} \lambda_k \sum_j \lambda_j \left[-\frac{\partial p_{cjk}}{\partial s_i} \right], \tag{Eq. A.6-b14}$$

$$D_{s\alpha k} = \frac{K}{\rho_k \nu} \lambda_k \sum_j \lambda_j \left[-\frac{\partial p_{cjk}}{\partial \sigma_{jk}} \frac{\partial \sigma_{jk}}{\partial C^\alpha} \right], \tag{Eq. A.6-b15}$$

$$D_{tk} = \frac{K}{\rho_k \nu} \lambda_k \sum_j \lambda_j \left[-\frac{\partial p_{cjk}}{\partial \sigma_{jk}} \frac{\partial \sigma_{jk}}{\partial T} \right]. \tag{Eq. A.6-b16}$$

Physically, Eq. A.6-b13 implies that the diffusive flux of phase k within the multiphase mixture results from the capillary forces due to saturation gradients, as well as concentration and temperature gradient through their effects on the interfacial tensions. In addition, the second term on the right-hand side of Eq. A.6-b13 reflects the gravitational separation due to the difference between phase densities. Note also that the diffusion coefficients defined by Eqs. A.6-b14 to A.6-b16 depend only on phase saturation and, hence, are properties of the multiphase mixture.

2.4 CONSERVATION OF SPECIES α IN THE MULTIPHASE MIXTURE:

$$\varepsilon \frac{\partial}{\partial t}\left(\rho C^{\alpha}\right)+\nabla \cdot\left(\gamma_{\alpha}\, \rho \vec{u} C^{\alpha}\right)$$

$$= \nabla \cdot\left[\varepsilon \rho D \nabla C^{\alpha}\right]+\nabla \cdot\left\{\varepsilon \sum_{k}\left[\rho_{k} s_{k} D_{k}^{\alpha}\left(\nabla C_{k}^{\alpha}-\nabla C^{\alpha}\right)\right]\right\} \quad \text{(Eq. A.6-b17)}$$

$$-\nabla \cdot\left(\sum_{k} C_{k}^{\alpha}\, \vec{j}_{k}\right)+S^{\alpha}\ ,$$

where the correction factor for species advection γ_{α} is defined as

$$\gamma_{\alpha}=\frac{\rho \sum_{k}\lambda_{k}C_{k}^{\alpha}}{\sum_{k}\rho_{k}s_{k}C_{k}^{\alpha}}, \qquad \text{(Eq. A.6-b18)}$$

and the effective diffusion coefficient for the multiphase mixture is defined as

$$\rho D=\sum_{k}\rho_{k}s_{k}D_{k}^{\alpha}. \qquad \text{(Eq. A.6-b19)}$$

S^{α} denotes the water source term. The second term on the left-hand side of Eq. A.6-b17 indicates that species α is advected, on the mixture level, by a modified velocity field $\gamma_{\alpha}\vec{u}$ rather than by the original mixture velocity field. This peculiar feature resembles that related to the gravitational term in the momentum equation, Eq. A.6-b9. The first two terms on the right-hand side of Eq. A.6-b17 combine to represent the net Fickian diffusion fluxes within various phases, whereas the last term on the right-hand side represents the diffusive flux across phases.

In PEM fuel cells, water is present in gas phase as vapor, and the liquid phase is pure water (i.e. $C_{l}^{\alpha}=1$). Assuming the capillary action is a major driving force for liquid flow, the water species equation can be written as

$$\varepsilon \frac{\partial}{\partial t}\left(\rho C^{w}\right)+\nabla \cdot\left(\gamma_{w}\, \rho \vec{u} C^{w}\right),$$

$$= \nabla \cdot\left(\varepsilon \rho D^{eff}\nabla C_{g}^{w}\right)-\nabla \cdot\left(C_{g}^{w}\vec{j}_{g}+\vec{j}_{l}\right)+S^{w}\ , \qquad \text{(Eq. A.6-b20)}$$

Both hydrogen gas and oxygenhave extremely low solubility in liquid water; thus their transport in the liquid phase can be neglected, and only the gaseous transport needs to be accounted for:

$$\varepsilon \frac{\partial}{\partial t}\left(\rho C^{\alpha}\right)+\nabla \cdot\left(\gamma_{\alpha}\, \rho \vec{u} C^{\alpha}\right)=\nabla \cdot\left[\varepsilon \rho D^{eff}\nabla C_{g}^{\alpha}\right]+S^{\alpha}\ , \qquad \text{(Eq. A.6-b21)}$$

where the correction factor for species advection γ_{α} is expressed by for water: $\gamma_w = \dfrac{\lambda_g C_g^w + \lambda_l}{C^w}$,

(Eq. A.6-b22)

for oxygen and hydrogen gas: $\gamma_\alpha = \dfrac{\rho \lambda_g}{\rho_g s_g}$,

2.5 ENERGY CONSERVATION FOR THE MULTIPHASE MIXTURE:

$$\frac{\partial}{\partial t}\left[(1-\varepsilon)\rho_s h_s + \varepsilon \rho h\right] + \nabla \cdot \left(\gamma_h \, \rho \vec{u} \, h\right)$$

$$= \nabla \cdot \left(k_{\text{eff}} \nabla T\right) + \nabla \cdot \left[\sum_k \left(h_k \vec{j}_k\right)\right] + \dot{q} \, ,$$

(Eq. A.6-b23)

where local thermal equilibrium among contiguous phases is invoked, and the correction factor for energy advection γ_h is defined as

$$\gamma_h = \frac{\rho \sum_k \gamma_k h_k}{\sum_k \rho_k s_k h_k}.$$

(Eq. A.6-b24)

The second term on the right-hand side of Eq. A.6-b23 describes the energy flux due to relative phase motions, including both sensible and latent heat transport.

*SOURCES

1. C.Y. Wang, and P. Cheng. Multiphase flow and heat transfer in porous media. *Adv. Heat Transf. 30* (1997): 93–196.
2. Y. Wang. Porous-media flow fields for polymer electrolyte fuel cells: II. Analysis of channel two-phase flow. *J. Electrochem. Soc.* 156: B1134–B1141 (2009).

APPENDIX VIII. A

THERMAL PROPERTIES OF SELECTED MATERIALS*

Table A.8-a1. Thermal properties of selected metallic solids

Composition	Melting point (K)	Properties at 300 K/353 K†				k(W/m K)/c_p(J/kg K)						
		ρ (kg/m³)	c_p (J/kg K)	k (W/m K)	α·10⁶ (m²/s)	100	200	400	500	600	700	800
Aluminum												
Pure	933	2702	906 / 901†	237 / 240†	97.1	302/485	237/802	240/935	237/996	232/1042	226/1091	220/1149
Alloy 2024-T6 (4.5% Cu, 1.5% Mg, 0.6% Mn)	775	2770	875	177	73.0	65/473	163/787	186/925		186/1042		
Alloy 195, Case (4.5% Cu)		2790	883	168	68.2			174/—		185/—		
Copper												
Pure	1358	8933	386 / 398†	398 / 394†	117	482/252	413/356	392/400	388/404	383/414	377/423	371/438
Commercial bronze (90% Cu, 10% Al)	1293	8800	420	52	14		42/785	52/460		59/545		
Phosphor gear bronze (89% Cu, 11% Sn)	1104	8780	355	54	17		41/—	65/—		74/—		
Catridge brass (70% Cu, 30% Zn)	1188	8530	380	110	33.9	75/—	95/360	137/395		149/425		
Constantan (55% Cu, 45% Ni)	1493	8920	384	23	6.71	17/237	19/362					
Iron												
Pure	1810	7870	443 / 441†	80.3 / 74.1†	23.1	132/216	94.0/385	69.4/486	61.3/495	54.7/566	48.7/619	43.3/686

Material	Melting Point (K)	ρ (kg/m³)	c_p (J/kg·K)	k (W/m·K)	α·10⁶	100 K (k/c_p)	200 K (k/c_p)	400 K (k/c_p)	600 K (k/c_p)	800 K (k/c_p)
Armco (99.75% pure)		7870	447	72.7	20.7	95.6 / 215	80.6 / 384	65.7 / 490	53.1 / 574	42.2 / 680
Carbon steels										
Plain carbon (Mn ≤ 1%, Si ≤ 0.1%)		7854	434	60.5	17.7			56.7 / 487	48.0 / 559	39.2 / 685
AISI 1010		7832	434	63.9	18.8			58.7 / 487	48.8 / 559	39.2 / 685
Carbon-silicon (Mn ≤ 1%, 0.1% < Si ≤ 0.6%)		7817	446	51.9	14.9			49.8 / 501	44.0 / 582	37.4 / 699
Carbon-manganese-silicon (1% < Mn ≤ 1.65%, 0.1% < Si ≤ 0.6%)		8131	434	41.0	11.6			42.2 / 487	39.7 / 559	35.0 / 685
Stainless steels										
AISI 302		8055	480	15.1	3.91			17.3 / 512	20.0 / 559	22.8 / 585
AISI 304	1670	7900	477	14.9	3.95	9.2 / 272	12.6 / 402	16.6 / 515	19.8 / 557	22.6 / 582
AISI 316		8238	468	13.4	3.48			15.2 / 504	18.3 / 550	21.3 / 576
AISI 347		7978	480	14.2	3.71			15.8 / 513	18.9 / 559	21.9 / 585
Nickel										

(Continued)

Table A.8-a1.(Contd.) Thermal properties of selected metallic solids

Properties at various temperatures (K)

| Composition | Melting point (K) | Properties at 300 K/353 K† | | | | k(W/m K) / c_p(J/kg K) | | | | | | |
		ρ (kg/m³) · c_p (J/kg K) · k (W/m K) · $\alpha \cdot 10^6$ (m²/s)				100	200	400	500	600	700	800
Pure	1728	8900	444	90.5	23.0	158	106	80.1	72.1	65.5	65.3	67.4
			461†	84.7†		232	383	477	527	590	524	524
Nichrome (80% Ni, 20% Cr)	1672	8400	420	12	3.4			14		16		21
								480		525		545
Inconel X-750 (73% Ni, 15% Cr, 6.7% Fe)	1665	8510	439	11.7	3.1	8.7	10.3	13.5		17.0		20.5
						–	372	473		510		546
Platinum Pure	2045	21450	132	71.4	25.1	77.5	72.4	71.6	72.2	73.0	74.1	75.5
			134†	71.5†		100	125	136	139	141	144	146
Alloy 60Pt–40Rh (60% Pt, 40% Rh)	1800	16630	162	47	17.4			52		59		65
												—
Silicon	1685	2330	712	148	89.2	884	264	98.9	76.2	61.9	50.8	42.2
			754†	118†		259	557	785	831	852	869	886
Silver	1235	10500	237	427	174	450	430	420	413	405	397	389
			237†	423.8†		187	225	239	243	248	253	258
Tungsten	3660	19300	133	178	68.3	235	197	162	149	139	133	128
			136†	169.5†		87	122	136	138	140	143	145

†: Properties at 353K

Table A.8-a2. Thermal properties of selected nonmetallic solids

| | | Properties at various temperatures (K) | | | | | | | | |
| | | Properties at 300 K/353K | | | | k(W/m K)/c_p (J/kg K) | | | | |
Composition	Melting point (K)	ρ (kg/m³)	c_p (J/kg K)	k (W/m K)	$\alpha \cdot 10^6$ (m²/s)	100	200	400	600	800
Carbon Amorphous	1500	1950	—	1.60	—	0.67	1.18	1.89	2.19	2.37
Diamond, type IIa insulator	—	3500	516	2000	—	9,800	4,300	1540	—	—
				1831†		36	197	853	1344	1626
Graphite, pyrolytic	2273	2210								
k, ∥ to layers				2000 / 1606†		4980	3250	1460	930	680
k, ⊥ to layers				9.5		39	15	7.0	4.4	3.2
c_p			709	4.73†		136	411	992	1406	1650
Graphite fiber epoxy (25% vol) composite	450	1400								
k, heat flow ∥ to fibers				11.1		5.7	8.7	13.0		
k, heat flow ⊥ to fibers				0.87		0.46	0.68	1.1		
c_p			935			337	642	1216		

(Continued)

Table A.8-a2. Thermal properties of selected nonmetallic solids

			Properties at various temperatures (K)							
			Properties at 300 K/353 K			k(W/m K)/c_p (J/kg K)				
Composition	Melting point (K)	ρ (kg/m³)	c_p (J/kg K)	k (W/m K)	$\alpha \cdot 10^6$ (m²/s)	100	200	400	600	800
Nafion‡		2000								
Water level 3				0.185						
Water level 22				0.25						
Sulfur	392	2070	704	0.206	0.141	0.165	0.185			
			753†	0.216†		403	606			

†: Properties at 353 K

‡: Water level = H_2O/HSO_3

Table A.8-a3. Thermal properties of common solids

Description/composition	Temperature (K)	Density, ρ (kg/m^3)	Thermal conductivity, k (W/m K)	Specific heat, c_p (J/ kg K)
Ice	273	920	1.88	2040
	263		2.48	
	253	–	2.57	1945
	243		2.68	
Paper	300	930	0.180	1340
	353	–	0.186	–
Teflon	300	2200	0.35	1050
	353		0.41	
	400		0.45	–
Polycarbonate	296	1200	0.29	1250
Soil	300	2050	0.52	1840
Sand	300	1515	0.27	800
	353		0.27	
Snow	273	110	0.049	–
	500		0.190	–
Glass Pyrex (borosilicate)	60-100	2210	1.3	753
Rock				
Granite, Barre	300	2630	2.79	775
Limestone, Salem	300	2320	2.15	810
Limestone, Indiana	373	2300	1.1	900
Marble, Halston	300	2680	2.80	830
Quartzite, Sioux	300	2640	5.38	1105
Sandstone, Berea	300	2150	2.90	745

* SOURCE

1. F.P. Incropera, D.P. DeWitt, T.L. Bergman, and A.S. Lavine. *Fundamentals of Heat and Mass Transfer*, 6th edn. John Wiley & Sons (2006).
2. Touloukian, Y.S., and C.Y. Ho, Eds. *Thermophysical properties of Matter*, Vols. 1, 2, 4, 5.
3. *CRC Handbook of Chemistry and Physics*. Cleveland, OH: CRC Press (1977).
4. Lienhard IV, J.H., and Lienhard V, J.H. *A Heat Transfer Textbook*, 3rd Ed. Phlogiston Press (2008).
5. Petrenko, V.F., and Whitworth, R.W. *Physics of Ice*. Oxford University Press (2003).

INDEX

Check Out The Other Mechanical Engineering Titles We Have!

Automotive Sensors, *John Turner*
Centrifugal and Axial Compressor Control, *Gregory McMillan*
Virtual Engineering, *Joe Cecil*
Chemical Sensors: Fundamentals and Comprehensive Sensor Technologies, Volumes 1 through 6, as well as **Chemical Sensors: Simulation and Modeling, Volumes 1 through 5**, *Ghenadii Korotcenkov*, Editor
Biomedical Sensors, *Deric P. Jones*
Acoustic High-Frequency Diffraction Theory, *Federic Molinet*
Bio-Inspired Engineering, *Chris Jenkins*
The Essentials of Finite Element Modeling and Adaptive Refinement: For Beginning Analysts to Advanced Researchers in Solid Mechanics, *John O. Dow*
Aerospace Sensors, *Alexander Nebylov*

Announcing Digital Content Crafted by Librarians

Momentum Press offers digital content as authoritative treatments of advanced engineering topics, by leaders in their fields. Hosted on ebrary, MP provides practitioners, researchers, faculty and students in engineering, science and industry with innovative electronic content in sensors and controls engineering, advanced energy engineering, manufacturing, and materials science. **Momentum Press offers library-friendly terms:**

- perpetual access for a one-time fee
- no subscriptions or access fees required
- unlimited concurrent usage permitted
- downloadable PDFs provided
- free MARC records included
- free trials

The **Momentum Press** digital library is very affordable, with no obligation to buy in future years.

For more information, please visit **www.momentumpress.net/library** or to set up a trial in the US, please contact **Adam Chesler**, *adam.chesler@momentumpress.net*.